The Impact of Chemistry on Biotechnology

ACS SYMPOSIUM SERIES **362**

The Impact of Chemistry on Biotechnology

Multidisciplinary Discussions

Marshall Phillips, EDITOR
Agricultural Research Service
U.S. Department of Agriculture

Sharon P. Shoemaker, EDITOR
Genencor, Inc.

Roger D. Middlekauff, EDITOR
McKenna, Conner & Cuneo

Raphael M. Ottenbrite, EDITOR
Virginia Commonwealth University

Developed from a symposium sponsored
by the Biotechnology Secretariat
at the 192nd Meeting
of the American Chemical Society,
Anaheim, California,
September 7-12, 1986

American Chemical Society, Washington, DC 1988

Library of Congress Cataloging-in-Publication Data

The Impact of chemistry on biotechnology: multidisciplinary discussions
Marshall Phillips, editor...[et al.];

"Developed from a symposium sponsored by the Biotechnology secretariat at the 192nd Meeting of the American Chemical Society, Anaheim, California, September 7-12, 1986."

p. cm.—(ACS symposium series; 362)
Includes bibliographies and indexes.

ISBN 0-8412-1446-8
1. Biotechnology—Congresses. 2. Chemistry, Technical—Congresses.

I. Phillips, Marshall. II. American Chemical Society. Biotechnology Secretariat. III. American Chemical Society. Meeting (192nd: 1986: Anaheim, Calif.) IV. Series.

TP248.14.I46 1988
660'.6—dc19 87-30751
CIP

PRINTED IN THE UNITED STATES OF AMERICA

ACS Symposium Series

M. Joan Comstock, *Series Editor*

Foreword

The ACS SYMPOSIUM SERIES was founded in 1974 to provide a medium for publishing symposia quickly in book form. The format of the Series parallels that of the continuing ADVANCES IN CHEMISTRY SERIES except that, in order to save time, the papers are not typeset but are reproduced as they are submitted by the authors in camera-ready form. Papers are reviewed under the supervision of the Editors with the assistance of the Series Advisory Board and are selected to maintain the integrity of the symposia; however, verbatim reproductions of previously published papers are not accepted. Both reviews and reports of research are acceptable, because symposia may embrace both types of presentation.

Contents

Preface

BIOTECHNOLOGY, BROADLY DEFINED, includes any technique that uses living organisms or parts of organisms to make or modify products, to improve plants or animals, or to develop microorganisms for specific uses. Historically, techniques applying biotechnology have followed time-consuming methods of locating mutations with the desired characteristics, which may be naturally occurring or enhanced. These have involved relatively slow processes that have allowed the scientific community and the regulators of new products to keep pace with the progress. Progress changed abruptly in the 1970s when the first gene was cloned, and the first expression of a gene cloned from a different species in bacteria was accomplished. A precise alteration of the DNA nucleotide process bypasses the slow process of seeking improvements through mutation.

Chemists and chemical engineers since then have played an important role in realizing the potential of biotechnology through advancement in the molecular-level understanding of biomolecular structure, function, and mechanism, and their application to chemical problems. Increasing numbers of chemists and chemical engineers from a broad range of disciplines have identified with this mission.

History of the Biotechnology Secretariat

The American Chemical Society is also playing a major role in the use of chemistry in biotechnology by providing a forum for the exchange of research results, application information, analytical uses, and other benefits of biotechnology. The role of biotechnology is spread across many disciplines within the Society. Consequently, a full discussion of the subject required cooperation by many of the constituent divisions of the ACS. After two years of discussion and meetings, the Biotechnology Secretariat was formed. It evolved from the attractive feature of coordinated programming of biotechnology presentations. Through this organization, the programming would be presented to minimize conflicts in schedules, to permit members of the Society to attend many presentations on the topic, and to allow for a planned progression of the overall programming on a regular basis over the next several years. Through the Secretariat, divisions would develop symposia on common topics to ensure that all relevant issues are covered.

A large number of divisions are participating in the activities of the Biotechnology Secretariat: Agricultural and Food Chemistry; Agrochemicals; Analytical Chemistry; Biological Chemistry; Carbohydrate Chemistry; Cellulose, Paper, and Textile; Chemical Education, Inc.; Chemical Health and Safety; Chemical Information; Chemical Marketing and Economics; Chemistry and the Law; Environmental Chemistry; the History of Chemistry; Industrial and Engineering Chemistry; Medicinal Chemistry; Microbial and Biochemical Technology; Petroleum Chemistry, Inc.; Physical Chemistry; Polymer Chemistry, Inc.; and Polymeric Materials: Science and Engineering.

The initial coordinated activity of the Secretariat occurred during the ACS national meeting in Anaheim, California, in September 1986. The Secretariat invited participation from the divisions for the symposium cluster titled "The Impact of Chemistry on Biotechnology." From the symposia that were presented, an outline was developed for this book.

The book is divided into several sections, each of which has a focus related to the interests of the contributing division. The contribution of the book is to illustrate by means of examples, which are not all-inclusive, the impact of chemistry on biotechnology. Our purpose was to demonstrate that biotechnology encompasses a broad spectrum of chemistry with tangible quantitative examples. We hope that the overview and the divisional sections collectively address the multidisciplinary topic of biotechnology.

Acknowledgments

The program chairs of the participating divisions deserve a special note of recognition. Their consideration and cooperation were vital parts in enabling the Program Chair of the Secretariat to organize and present the symposium cluster. John Crum, Executive Director of the ACS, and Randall Wedin, Special Assistant to the Executive Director, provided important assistance in the organization of the Biotechnology Secretariat.

The Committee on Science chaired by Paul Gassman provided a financial grant to the Biotechnology Secretariat to initiate our organization and to launch our first symposium. Numerous people in the divisions, in the staff of the Society, and in the Society in general provided time, counsel, ideas, and patience. The editors acknowledge the following:

Donald W. DeJong
USDA-ARS
Tobacco Research Lab
Oxford, NC 27565

Ronald A. Rader
OMEC International, Inc.
Washington, DC 20005

Peter C. Hall
SRI International
Menlo Park, CA 94025

John Landis
Upjohn Company
Kalamazoo, MI 49009

James N. Seiber
Department of Environmental
Toxicology
University of California at Davis
Davis, CA 95616

Randall W. Swartz
Swartz Associates
Biotechnology Consultants
Winchester, MA 01890

These six divisional organizers of the subsymposia for the cluster symposium in Anaheim provided the introductory chapters for this book. Their efforts and time made the symposium and this book possible. To those many individuals we say thanks.

MARSHALL PHILLIPS
USDA-ARS-NADC
P.O. Box 70
Ames, IA 50010

SHARON P. SHOEMAKER
Genencor, Inc.
180 Kimball Way
South San Francisco, CA 94080

ROGER D. MIDDLEKAUFF
McKenna, Conner & Cuneo
1575 Eye Street, N.W.
Washington, DC 20005

RAPHAEL M. OTTENBRITE
Department of Chemistry
Virginia Commonwealth University
Richmond, VA 23284

OVERVIEW

Chapter 1

The Role of Chemistry

Leroy Hood

Division of Biology, California Institute of Technology, Pasadena, CA 91125

There has been a revolution in the techniques of biology over the past ten years that has dramatically changed the face of modern biology. Three biotechnologies, recombinant DNA techniques, monoclonal antibody techniques and microchemical instrumentation, have made especially important contributions to this revolution that has come to be called the new biology. These biotechnologies are synergistic and together they have greatly shortened the time between many fundamental biological observations and their clinical applications in the world of medicine. Most have some perception of the recombinant DNA and monoclonal antibody techniques. I will focus on the third, microchemical instrumentation, machines that synthesize and sequence genes and proteins, because it demonstrates the power of chemical techniques in biology.

DNA and proteins

The double-stranded DNA molecule makes up the central core of the 23 pairs of human chromosomes and is the blueprint repository for information on how to construct human organisms. DNA has four subunits, the G, C, A, and T nucleotides, and one strand of a DNA molecule always exhibits molecular complementarity for its partner because an A subunit always pairs with a T and the G subunit always pairs with a C. The fundamental units of information in the DNA molecules, the genes, are translated into intermediate messenger RNA molecules, essentially working computer tapes capable of enormous amplification, and these in turn are fed into specialized organelles that synthesize proteins. The four-letter language of DNA is read three units at a time into the 20-letter language (amino acids) of proteins. The primary sequence of amino acids in a protein dictates how the individual protein folds into a particular three-dimensional configuration. Thus, the linear information of the genes is changed into the three-dimensional information of proteins. Proteins are the molecular genes that give the body size, shape, and form, and catalyze the chemical reactions of life.

0097–6156/88/0362–0002$06.00/0

The genetic code dictionary relates the DNA and protein languages so that knowledge of the DNA sequence (order of nucleotides in a gene) allows one to predict precisely the amino acid sequence of its corresponding protein, and conversely, knowledge of the amino acid sequence of a protein permits one, with some ambiguity, to predict the gene sequence. The ability to employ the genetic code dictionary to determine protein sequences from gene sequences and conversely gene sequences from proteins is important in later methodologies we will discuss.

Each human cell has sufficient DNA in its 23 pairs of chromosomes, its genome, to encode 3 million average-sized genes 1,000 nucleotides in length. This calculation is somewhat misleading in that most genes are divided into two or more coding regions (exons) and intervening sequences (introns). Moreover, it has been estimated that only 3% of the genome is actually involved in coding regions for the estimated 100,000 genes necessary to construct a human organism. Nevertheless, recombinant DNA techniques permit individual genes to be isolated and characterized.

Recombinant DNA techniques

The discovery of DNA cutting enzymes, the restriction endonucleases which cut precisely at particular double-stranded DNA sequences, and the DNA joining enzymes, the ligases which join together DNA fragments, paved the way for the new recombinant DNA techniques. These enzymes enable one to take a particular human gene, γ-interferon, and splice it into the DNA sequence of an appropriate DNA vector, such as a circular bacterial plasmid, which is approximately 5,000 nucleotides in length. Then this hybrid DNA molecule may be placed in bacteria where it replicates and thus amplifies the number of interferon genes which can accordingly be removed from the vector sequences by restriction endonucleases. The challenge in this procedure is to isolate from among the potential 3 million genes worth of DNA in the human genome a particular gene of interest. This goal is relatively straightforward for genes that transcribe ample quantities of messenger RNA. It is far more difficult for the so-called rare-message genes that make little messenger RNA. It was a consideration of this problem that led in part to the development at Caltech of the collection of instruments that have come to be known as the microchemical facility.

Microchemical Facility

Over the past ten years my laboratory has developed a series of instruments (Table I) that permit us to sequence and analyze genes more effectively than had been possible heretofore. Let me discuss several of these instruments and consider the impact they have had and will continue to have on problems of central importance to modern biology.

Table I. Michrochemical Facility at Caltech

FINISHED	
GAS PHASE PROTEIN SEQUENATOR	- SEQUENCES PROTEINS
PEPTIDE SYNTHESIZER	- SYNTHESIZES PROTEINS
DNA SEQUENATOR	- SEQUENCES GENES (DNA)
DNA SYNTHESIZER	- SYNTHESIZES GENES
COMPUTER	- UNIQUE PROGRAMS FOR THE ANALYSIS OF PROTEINS AND GENES
HEXAGONAL ARRAY PULSE FIELD GEL ELECTROPHORESIS	- SEPARATES LARGE FRAGMENTS OF DNA
UNDER DEVELOPMENT	
GENE ANALYZER	- INSTRUMENT FOR FINGER-PRINTING HUMAN GENES
SECOND GENERATION PROTEIN SEQUENTATOR	- 10^3 TIMES MORE SENSITIVE THAN CURRENT MACHINES
COMPUTER SUPERCHIP	- APPLICATION OF TRW SUPERCHIP TO NEW APPROACHES FOR DNA AND PROTEIN ANALYSIS

Automated Protein Sequence Analysis. Automated protein sequence analysis employs Edman chemistry to sequentially cleave individual amino acid residues from the N-terminus of the polypeptide chain. Since the introduction of the automated spinning cup sequenator by Per Edman in 1967 (1) revolutionary improvements have led to striking increases in the sensitivity of protein sequence analysis (Table II). Our objective at Caltech is to further increase the level of protein sequencing sensitivity by 2-3 orders of magnitude. This level of sensitivity will permit us to routinely use two-dimensional polyacrylamide gel electrophoresis (2), currently the most sensitive and highly resolving analytic technique for the separation of complex mixtures of proteins, as a preparative device for isolating polypeptides for sequence analysis.

Table II. Improvements in Protein Sequenator Sensitivity

nmol	μg*		Instrument
100	5,000	1967	Edman Spinning Cup
10	500	1971	Commercial Spinning Cup
0.1	5	1978	CIT-modified Commercial Spinning Cup
0.02	1	1979	CIT Microsequencing Spinning Cup
0.005	0.25	1980	CIT Gas Phase Sequenator

*Weight of 50,000 dalton protein, assuming 100% sequenceable material.

To improve the overall sensitivity level of protein sequencing, simultaneous improvements in all aspects of the complex protein sequencing process are required. Improvements must be made in the methods for the isolation of submicrogram quantities of protein, in the Edman chemistry, in the automated sequenator, and in the analysis of the resulting amino acid derivatives (Kent, S.; Hood, L.; Aebersold, R.; Teplow, D.; Smith, L.; Farnsworth, V.; Cartier, P.; Hines, W.; Hughes, P.; Dodd, C. In Modern Methods in Protein Chemistry; L'Italien, J. J., Ed. Plenum Press: New York, in press.

We have addressed these aspects of the protein sequencing problems as follows. First, we have developed a second generation gas phase sequencer that has several unique features including programmable temperature control for optimized couplings/cleavage temperatures, an extremely flexible program for optimization of the existing Edman chemistry and the developing of fluorescent sequencing chemistries, and highly improved valving instrumentation. Second, we have evaluated several different approaches for the development of fluorescent or fluorogenic sequencing chemistry. The introduction of fluorescent tags on amino acid residue derivatives has resulted in a 10^4-fold increase in inherent sensitivity in the analysis of these deriyatives. Third, we have developed an electroblotting procedure to isolate proteins in a suitable form for direct sequence analysis (3). This technique permits us to transfer proteins with high yield from one- or two-dimensional polyacrylamide gels to activated glass fiber filter paper. The electroblotting procedure has several advantages over previous methods of protein isolation. First, the proteins are transferred directly to a matrix in which gas phase sequencing is carried out. Second, because of the minimization of protein handling, the purification of picomole quantities of protein can be achieved with relatively high yields. Third, the procedure avoids the artifactual blocking of N-terminal residues if contamination-free reagents are used. Fourth, this procedure generates very low background in the analysis of the resulting PTH amino acid residues by high performance liquid chromatography. Fifth, as a parallel isolation method, this technique in principle permits the purification of thousands of polypeptides from analytical two-dimensional gels in a state suitable for sequencing.

For conventional gas phase sequencing using phenylisothiocyanate as the coupling reagent, proteins can be immobilized on the glass fiber matrix by noncovalent interactions. In fluorescent sequencing techniques, however, proteins must be covalently attached to the sequencing matrix because the reactions must be carried out under conditions which would normally result in the loss of noncovalently attached protein samples through the use of strong washing solvents. Fluorescent subpicomole techniques will allow the use of conventional two-dimensional gels as preparative tools for the routine isolation of proteins directly for sequencing. This in turn will allow cloning of the corresponding genes by techniques that I will subsequently discuss.

Protein Synthesis. Our laboratory has optimized the chemistry of solid phase peptide synthesis so that in manual synthesis a repetitive yield of 99.6-99.8% has routinely been achieved. In addition, appropriate conditions have been developed such that peptide synthetic procedures are very general, permitting the synthesis of virtually any peptide sequence. Success in these two areas has enabled the automation by Dr. Stephen Kent of this efficient synthetic chemistry in collaboration with Applied Biosystems, Inc. Together, we have developed a peptide synthesizer that can carry out the addition of 15 amino acid residues per day in a fully automated fashion with a typical repetitive yield of 99.4%.

This instrument has provided us with a highly efficient tool for the synthesis of long peptides and proteins. We have applied this to the study of structure-function relationships and mitogenic growth factors. For example, we have achieved the total chemical synthesis of human transforming growth factor α (TGF-α)(Woo, D. D.; Clark-Lewis, I.; Kent, S. B. H. , California Institute of Technology, unpublished data.). This is a 50 amino acid residue protein mitogen which shows homology, particularly in the disposition of its 3 disulfide bridges with epidermal growth factor. The automated assembly of the protected peptide chain took 3.3 days. After cleavage and protection and oxidative refolding, the protein was purified to virtual homogeneity by high performance liquid chromatography procedures. The purity of the resulting product was vividly demonstrated by isoelectric focusing procedures which demonstrated it was more than 95% homogeneous. The biological activity of this material in three different assays was identical with an experimental error to that of native epidermal growth factor. Thus, the optimized synthetic technology is suitable for the generation of small proteins in large quantities (hundreds of milligrams) for structure-function studies.

In order to explore the application of these synthetic procedures to structure-function studies for larger peptides, we have undertaken the complete synthesis of IL3, a protein mitogen that promotes differentiation in a variety of different hematopoetic cells (4). Each synthesis of this 140-residue polypeptide chain took approximately 12 days. The assembly of the protected chain showed an average repetitive yield efficiency of 99.4% per residue, yielding material with 42% of the target 140-residue amino acid sequence. After cleavage and deprotection, followed by oxidative refolding, the group protein was analyzed by SDS gel electrophoresis. Two compact bands of protein were visible, corresponding to the unfolded material and the more compact refolded material. This folded material had all of the activities of native IL3 in a variety of in vitro and in vivo assays. It had full maximal activity in a thymidine incorporation assay, but only approximately 1% of the specific activity of the native IL3. More recently, we have used higher resolution purification procedures to achieve a specific activity of 10% of the native protein. It is possible to rapidly synthesize a derivative of this hormone to study the structure-function relationships of individual residues. We have carried out this procedure to demonstrate that only 1 of the 2 disulfide bridges is essential for biological activity.

DNA Synthesis. In 1983 we cooperated with Dr. Marvin Caruthers, who had developed the phosphoramidite chemistry for the synthesis of DNA, and Applied Biosystems, to develop a DNA synthesizer (5). This instrument has helped to revolutionize molecular biology (Table III). It gives us the capacity to synthesize DNA fragments that arrange from a few nucleotides up to more than 100 or so nucleotides; these fragments can be synthesized so that they constitute a staggered or overlapping set of fragments which readily self anneal to form larger DNA fragments that eventually can be ligated and thus assembled into genes (6). In addition, oligonucleotides can be constructed for in vitro mutagenesis (7), primer sequences can be synthesized to permit primer directed DNA sequencing (8), oligonucleotides can be generated from reverse translated amino acid sequences to facilitate the cloning of genes which encode low levels of messenger RNA (rare-message genes), as will be discussed subsequently, and oligonucleotide probes can be generated for the analysis of the expression of messenger RNA molecules during development.

Table III. Uses of the DNA Synthesizer

SYNTHESIS OF COMPLETE GENES
CLONING OF RARE-MESSAGE GENES
SPECIFIC PRIMER-DIRECTED DNA SEQUENCING
IN VITRO MUTAGENESIS
GENE STRUCTURE MAPPING

Another important application of phosphoramidite chemistry has been to modify the chemistry to allow the synthesis of tailor made oligonucleotide derivatives. We have used this approach to prepare a number of fluorescent oligonucleotide primers for the use in automated DNA sequence analysis (see below). The basic strategy we have employed is to synthesize protected amino derivatives of nucleoside phosphylamidites, which when used in DNA synthesis yield corresponding amino nucleotide products (9). This amino oligonucleotide may then be reacted with a wide variety of compounds particularly amino reactive fluorescent dyes. This strategy is general in nature, allowing many different oligonucleotide derivatives to readily be prepared.

DNA Sequence Analysis. The sequence analysis of DNA is one of the critical procedures in modern molecular biology. There are two general strategies for sequencing DNA, the chemical degradation method developed by Maxam and Gilbert (10) and the enzymatic method developed by Sanger and his colleagues (11). We have focused on the partial automation of the enzymatic method. In this method the DNA fragment to be analyzed is cloned into an M13 bacteria phage vector. A primer DNA oligonucleotide complementary to the vector adjacent to the insert fragment is used to initiate enzymatic replication of the unknown DNA. One of the chain terminating dideoxy analogues of the four nucleotides A,C,G, and T, is added to each of four reaction mixtures to give four distinct sets of DNA fragments, each ending with a specific base in the sequence. In the conventional method of DNA sequence analysis a radioisotopic label on the oligonucleotide primer and the four reactions are separated by electrophoresis in adjacent lanes of a high resolution polyacrylamide gel. Autoradiography of the gel yields a thin image of the DNA fragments produced in the four reactions and the radioautogram may then be visually interpreted to yield the DNA sequence.

Our approach to the automation of this process was to devise a system in which the DNA fragments could be detected in real time during their electrophoretic separation within the gel (12). We used four fluorescent dyes, each of a different color, to tag the four products of the four sequencing reactions. The four reactions are combined and coelectrophoresed in a single lane of a polyacrylamide gel and the order in which the different colored DNA fragments proceeds down the gel is determined using a high sensitivity laser based detection apparatus coupled to a minicomputer. The DNA fragments are labeled with a fluorescent tag by chemical synthesis of the appropriate dye oligonucleotide primers such as described above and by use of the fluorescent primers in the enzymatic sequencing reactions. In typical runs on our prototype apparatus, we have obtained sequence information up to 250-350 bases in a single run. The commercial version of this instrument, developed at Applied Biosystems, Inc. in collaboration with our group, permits data to be simultaneously obtained from 16 lanes on a slab gel, providing an instrument with enormous capacity for automated analysis of DNA.

Although this new DNA sequencing approach is very powerful, we believe this technology is still in its infancy. In years to come we expect dramatic improvements in many aspects of the automated DNA sequencing process. Increases in sensitivity of detection of the four fluorofors, involving advances both in the optics and instrumentation as well as in the dye chemistry itself, will permit ever higher resolution data to be obtained, allowing sequences to be determined from longer and longer stretches of DNA in a single operation. Development of a chemistry for automated fluorescent sequencing by the chemical degradation method, in addition to the enzymatic method, will allow higher accuracy data to be obtained. The new degradation chemistry will also facilitate the automation of the sequencing reactions themselves. The development of solid phase chemistry for the rapid synthesis of a set of the four dye oligonucleotide conjugates will permit the rapid specific primer directed sequencing strategies to be used in conjunction with fluorescent detection.

Summary

The microchemical facility, with its allied set of instruments for the synthesis and analysis of DNA and proteins, operates in a synergistic manner with two other biotechnologies that have been developed in the last ten years: recombinant DNA technology and monoclonal antibody technology. For example, we can now sequence picomole quantities of a polypeptide. This protein sequence can be reverse translated into DNA sequence via the genetic code dictionary and the corresponding gene cloned by conventional recombinant DNA techniques. If the oligonucleotide fragment leads to the cloning of a multigene family, each of the individual genes can be sequenced on the DNA sequencer to determine whether they are pseudogenes or possible functional genes. Alternatively, short peptide fragments of the unknown protein can be synthesized, coupled to larger carrier proteins and used to raise monoclonal antibodies that recognize the original peptide fragment and sometimes the intact protein from which the fragment was derived. Also given the sequence of a protein, the genetic code dictionary can be used to determine an appropriate DNA sequence and the corresponding gene can be synthesized by the DNA synthesizer. It is our belief the development of new instrumentation and approaches for biological research will change science over the next ten years much as it has during the past decade. These new instruments, in conjunction with the existing ones, will permit us to approach each of the four fundamental problems in biology with enormous power and effectiveness. These problems include the complete characterization of the human genome, understanding the molecular mechanisms of development, the prediction of three-dimensional structure from primary amino acid sequence information, and its corresponding corollary of how three-dimensional structures can perform their respective functions, and understanding how macromolecules can self-integrate into higher order assemblages that can then intercommunicate one with another to create cells and organisms. Truly, biology has moved into the realm of big science, but in a very interesting fashion. The big science of microchemical facilities is certainly far more modest in its dimension than its counterpart in physics, yet it permits an enormous amplification and expansion of the research possibilities for individual biologists. Chemistry is playing a leading role in the biological revolution.

Acknowledgments

This paper was in part adapted from L. Hood et al. Sixth MPSA 1986, Walsh, K., Ed.

Literature Cited

1. Edman, P.; Begg, G. Eur. J. Biochem. 1967, 1, 80-91.
2. O'Farrell, P. H. J. Biol. Chem. 1975, 250, 4007.
3. Aebersold, R. H.; Teplow, D. B.; Hood, L. E.; Kent, S. B. H. J. Biol. Chem. 1986, 261, 4229-4238.
4. Clark-Lewis, I.; Aebesold, R.; Ziltener, H.; Schrader, J. W.; Hood, L. E.; Kent, S. B. H. Science 1986, 231, 134-139.
5. Horvath, S. J.; Firca, J. R.; Hunkapiller, T.; Hunkapiller, M. W.; Hood, L. Meth. Enzymol. 1987, 154, 314-326.
6. Normanly, J.; Ogden, R.; Horvath, S.; Abelson, J. Nature 1986, 321, 213-219.
7. Schultz, S. C.; Richards, J. H. Proc. Natl. Acad. Sci. USA 1986, 83, 1588-1592.
8. Strauss, E. C.; Kobori, J. A.; Siu, G.; Hood, L. Anal. Biochem. 1986, 154, 353-360.
9. Smith, L. M.; Fung, S.; Hunkapiller, M. W.; Hunkapiller, T. J.; Hood, L. E. Nucl. Acids. Res. 1986, 13, 2399-2412.
10. Maxam, A. M.; Gilbert, W. Meth. Enzmol. 1980, 65, 499-560.
11. Smith, A. J. H. Meth. Enzymol. 1980, 65, 560-580.

12. Smith, L. M.; Sanders, J. Z.; Kaiser, R. J.; Hughes, P.; Dodd, C.; Connell, C.; Heiner, C.; Kent, S. B. H.; Hood, L. E. _Nature_ 1986, _321_, 674-679.

RECEIVED August 6, 1987

Chapter 2

Impact of Biotechnology on the Chemical Industry

Jonathan J. MacQuitty

Genencor, Inc., 180 Kimball Way, South San Francisco, CA 94080

Recent rapid developments in some of the newer technologies such as protein engineering, high level protein secretion systems, and microbial pathway engineering, have reduced or eliminated many of the technical obstacles to the application of biotechnology to the production of specialty chemicals. Indeed, some of these developments are already being translated into commercial applications. It is predicted that the impact of biotechnology on the chemical industry will increase and extend well into the next century.

There has been a tendency in the popular press and in the biotech industry itself to classify many different technologies as "biotechnology". Clearly, this is an oversimplification. From a technical standpoint there are a wide variety of techniques involved and moreover, from a commercial point of view, there are distinct differences in the way these technologies should be applied. This paper attempts to classify the technologies involved with particular reference to the production of specialty chemicals and to review the technical and commercial limitations that might exist in applying these technologies.

In the first section, the production of chemical products by a variety of biological approaches is examined. These approaches are characterized in terms of chronological phases or "waves" of development. The waves have rather different commercial characteristics and require different business strategies if the technologies involved are to be applied successfully.

In the second section, the technical barriers or limitations that exist in applying these biological approaches to the chemical industry are reviewed particularly as applied to the production of specialty chemicals. The thesis put forward is that most, if not all, of these barriers can now be overcome because of a number of recent and very impressive technical advances.

0097-6156/88/0362-0011$06.00/0

In the third section, some commercial barriers that exist today and limit the full impact of biotechnology on the chemical industry are reviewed. Several approaches are suggested for minimizing these barriers and some examples are given where such limitations have been removed resulting in improvements in the cost or performance of specific products.

The Four "Waves" of Biotechnology

There has been an evolution in biological approaches to chemical problems, and these can be separated somewhat crudely into different chronological phases or "waves" (Table I). These waves have different technical bases and, more important, require different commercial strategies in order to be utilized successfully.

Table I. The Production of Chemicals Using Biological Approaches

1st Wave	1940s	Classical Fermentation	e.g. penicillin, citric acid
2nd Wave	1970s	rDNA Technology	e.g. insulin, hGH
3rd Wave	1980s	Protein Engineering	e.g. modified subtilisins
4th Wave	1980s	Pathway Engineering	e.g. new pathway to ascorbic acid

The First Wave: Classical Fermentation. The first wave started in the mid-forties with the scaleup and commercial production of primary and secondary metabolites, such as citric acid and penicillin and has now evolved into a well developed fermentation industry (1-2). Production is achieved using fermentation, both surface and submerged, with submerged fermentation now the more important.

The biological techniques involved in this first wave include:

- -- isolating a microorganism which produces the chemical of interest by means of screening/selection procedures.
- -- improving production yields by means of random mutagenesis of the microorganism.
- -- increasing yields still further by optimization of culture media and fermentation conditions.

From a commercial standpoint, the first wave is limited to those chemicals produced in nature. It is also limited by its trial and error approach: it can take years or even decades to achieve a significant improvement in product yields. For example, it took 20 years applying these techniques to improve the fermentation yield of penicillin from an initial level of a few milligrams per liter to 7 grams per liter (1).

The techniques involved here are often referred to as "classical" techniques. However, with modern improvements in the ability to screen and assay for desired properties, this approach is now more powerful and more targeted than when originally applied. Classical mutagenesis and selection still remains a highly viable approach to the development of biological production processes.

The Second Wave: rDNA Technology. The second wave started in the mid-seventies with the pioneering work of Cohen and Boyer which established techniques for recombinant technology in simple procaryotes such as E. coli (3). This was followed by intensive efforts to demonstrate the expression of simple proteins from these procaryotes (4). The techniques involved here include:

- -- isolating the gene coding for the protein of interest from a natural source.
- -- cloning the gene into appropriate vectors for expression including necessary upstream and downstream regulatory elements.
- -- transforming a desired production microorganism with these vectors.
- -- expressing the gene using these regulatory elements so as to produce high yields of the protein of interest.

The implications of this second wave of rDNA technology are to open up a large range of proteins for consideration as specialty chemicals. Previously, these chemicals might have existed only in minute quantities in nature and thus might not have been available for commercial use. Another difference from the classical fermentation techniques described above is that the timeframe for yield improvement can be made much faster.

The application of these techniques began with the cloning and expression of human genes such as insulin (5) but has since been extended to other mammalian genes such as rennin (6), microbial genes such as glucoamylase (7), and even plant genes such as thaumatin (8).

The Third Wave: Protein Engineering. The third wave, protein engineering, started in the early eighties as a spinoff of rDNA technology. However, it is rather different from the second wave since it is concerned with producing new (i.e., man-made) proteins which have been modified or improved in some way over their natural counterparts. The techniques involved in protein engineering are also more complicated than those described previously.

They involve:

- -- introducing changes into specific locations or regions of a gene in order to produce a new gene.
- -- expressing the new protein from such a gene using the approach as described above.

-- characterizing the structure of this new protein by crystallographic or other techniques.
-- determining the functional parameters for the new protein.
-- selecting new locations or regions to modify as a result of this structure:function comparison.

This is shown schematically in Figure I. As can be seen, such an approach is necessarily an iterative approach, and might appear to be time consuming as a result. Fortunately there are some ingenious short cuts which have been developed recently that can keep these iterations to a minimum. For example, using "cassette" mutagenesis it is possible to produce, in a single experiment, some 19 proteins differing only by a single amino acid at a designated site in the protein (9). Hybrid enzymes can also be produced where parts of two or more "parent" genes are combined to form novel hybrids (10). These and other mutagenesis techniques allow different proteins to be produced in large numbers, with controlled diversity from the original natural protein, and in fairly rapid fashion.

It is also possible to accelerate the characterization of these new proteins both structurally and functionally. For example, by using area detectors, the collection of x-ray crystallographic data can be accelerated significantly. Indeed, if the structure of a parent protein is well resolved, the structures of related proteins can now be analyzed in 2-3 weeks using Fourier difference techniques (Bott, R., Genentech, So. San Francisco, CA, unpublished data). Similarly, new assay techniques have been developed for screening functional parameters, and new computer software can be employed to speed up the process of analyzing these data. For example, new computational methods have been developed for the spectrophotometric determination of enzyme kinetics. This has enabled important kinetic parameters (e.g. k_{cat}, K_m) to be determined in a matter of minutes compared to the several days required by previous techniques (11).

From a commercial point of view it is worth noting that since these novel proteins are man-made, excellent patent protection should be obtainable. The composition-of-matter type claims which can be made for these new proteins, could potentially provide a much stronger proprietary base than the process claims available from rDNA technology on its own.

The Fourth Wave: Pathway Engineering. The fourth wave, pathway engineering, has really only just begun. This wave involves modification of various metabolic pathways using recombinant techniques so as to enhance production of a particular metabolite or indeed to include pathways not indigenous to the organism. Major steps in this fourth wave include:

-- identifying the enzymes involved in the metabolic pathway of interest both from within the production host and from other natural sources.
-- obtaining the genes coding for these enzymes and modifying them, if necessary, using the techniques described above.

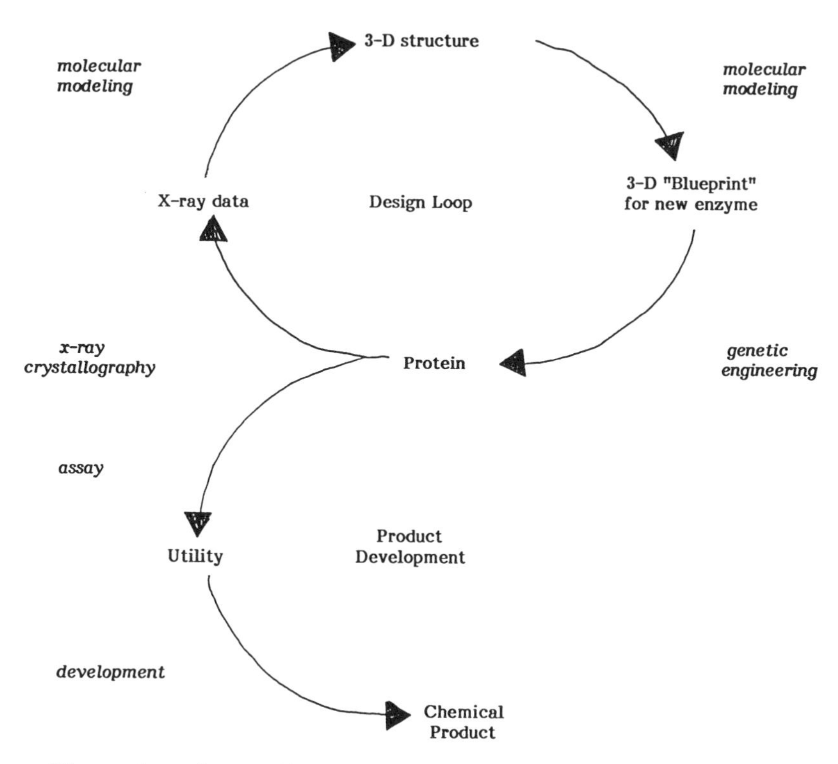

Figure 1. The Application of Protein Engineering to Chemical Products.

-- cloning the genes coding for these new improved enzymes into the production host along with the appropriate regulatory elements.
-- optimizing the effective interaction between these regulatory elements and the culture and fermentation media conditions in order to achieve proper regulation of the key metabolic pathways.

The last step alone is obviously critical and only recently has the feasibility of this been demonstrated. There has, for example, been excellent work on modifying amino acid production using this technique (12). 2-Keto-gulonic acid, a precursor to Vitamin C, has also been produced by using this approach (13).

From a commercial standpoint, there are some clear differences between pathway engineering and classical fermentation. For one thing, since the Supreme Court has allowed engineered organisms to be patented, it is now possible to establish a better proprietary position. Secondly, pathway engineering allows a much more targeted approach to yield improvement. The approach allows research to be conducted on a much sounder technical basis than the classical trial and error methodology.

Increasing Impact of Biotechnology. With the development and growth of the four "waves" described above, the impact of biotechnology on the chemical industry has increased dramatically. A clear example of this increasing impact is in the area of industrial enzymes. This is summarized somewhat figuratively in Table II. As can be seen, each wave of biotechnology has expanded the potential availability of different enzymes. This should allow the development of a number of enzyme product lines and expand the commercial application of enzymes to many new industrial areas.

Table II. The Accelerating Impact of Biotechnology

Example: Availability of Industrial Enzymes

	Classical Fermentation	rDNA Technology	Protein Engineering	Pathway Engineering
Approximate Number of Enzymes Available	15-20	5-10,000	20^{300}	∞

Technical Limitations to the Application of Biotechnology

In the first section, four different waves of biotechnology were identified. In this section some of the technical factors are described which limit the application of these waves of biotechnology to the chemical industry. In particular the technical requirements for the production of specialty chemicals using these biotechnology approaches are reviewed. The thesis being proposed is that many of the technical requirements limiting the application of biotechnology to the chemical production have now been reduced or eliminated.

Lower Cost Biochemical Processes. One of the common barriers cited against biological approaches to chemical production is the high cost of the biochemical processes involved compared to chemical alternatives and the lack of methods available to reduce these costs.

Biochemical processes for the production of chemicals can be divided into two major categories:

1. Direct fermentation from a carbon feedstock. The desired end product is recovered from the fermentation broth after fermentation. The Fermentation has typically been performed in a batch or modified batch mode but continuous fermentation processes are now emerging.

2. Bioconversion using a biocatalyst. Here the biocatalyst can be microbial cells (both living and dead) or enzymes and the biocatalyst can be freely suspended or immobilized.

The fermentation processes are heavily dependent on carbon feedstock costs and will be discussed below. Bioconversion processes, however, need not be expensive processes and indeed several inexpensive bioconversions have been commercialized at very large scale. Some examples are summarized in Table III. The thermolysin dependent production of aspartame is now being scaled up in Holland by a joint venture between Toyo Soda and DSM (14). The penicillin acylase route has now almost totally displaced the more expensive chemical alternatives (15). In the case of high fructose corn syrup (HFCS) production, the cost of the immobilized glucose isomerase is now around 0.2¢/lb. of product (Crabb, D. Genencor, Inc., So. San Francisco, CA, unpublished data). One of the reasons for the low cost of these processes is the very long life of the biocatalyst: the immobilized lactase in the conversion of whey, for example, has a half life ($t_{1/2}$) greater than a year (16). Another reason for low cost is the continuing increase in production yields for these industrial enzymes.

The processes shown in Table III can now be run economically at very large scale. The earlier development of effective immobilization techniques substantially reduced the cost barrier to using biochemical routes on a large scale for chemical production. Now with the development of protein engineering, biocatalysts with improved performance can be developed which can lead to still lower costs for bioconversions and the gradual replacement of certain chemical steps by biochemical steps.

Table III. Examples of Commercial Scale Bioconversions

Substrate(s)	Product	Immobilized Enzyme
Phenylalanine Aspartic Acid	Aspartame	Thermolysin
Glucose Syrup	42% High Fructose Corn Syrup (HFCS)	Glucose Isomerase
Benzyl-penicillin	6-Amino Penicillanic Acid (6-APA)	Penicillin Acylase
Lactose (Whey)	Glucose, Galactose	Lactase

Lower Cost Feedstocks. As mentioned above, direct fermentation routes are heavily dependent on the cost of the carbon feedstock. For a fermentation route to be viable, its feedstock costs must compare favorably to those of petroleum based feedstocks.

Clearly with oil at $15/bbl, the feedstock advantage shifts towards those based on petroleum. However, with oil prices rising again the feedstock pendulum has started to swing back towards fermentation feedstocks. In addition, with the continuing increase in agricultural productivity, the real costs of agricultural feedstocks (especially surpluses and waste materials) have been declining year by year. As a result, "crossovers" are increasingly occurring where the raw material of choice switches from a petroleum based source to an agricultural carbon source.

One area Genencor has worked on is the use of lignocellulose as a feedstock. Potentially, this is a very low cost and abundant carbon feedstock (e.g., as wood, corn stover, sugar cane bagasse). However, lignocellulose is not easy to use as feedstock, since the lignin component is intimately associated with carbohydrate polymers. In addition, an inexpensive way of liberating glucose and other monomer sugars from the cellulose and noncellulose polymers is required.

Figure 2 shows a scheme for utilizing lignocellulose as a feedstock. There are several pretreatment techniques being studied with the best choice processes being currently based on steam explosion. The largest cost element in this process is the cost for producing the cellulase enzyme complex. At Genencor we have had considerable success in reducing this cost. This has been accomplished by improving the production yields and by modifying the enzyme complex itself. Another feature of this process is the emphasis on high value uses for the enzyme-generated lignin. The higher value possible here could be used as a credit towards the cost of the overall process. Further work is planned to upgrade the lignin value still higher by additional enzyme mediated modifications.

The increasing availability and declining costs of agricultural feedstocks, coupled with the potential for

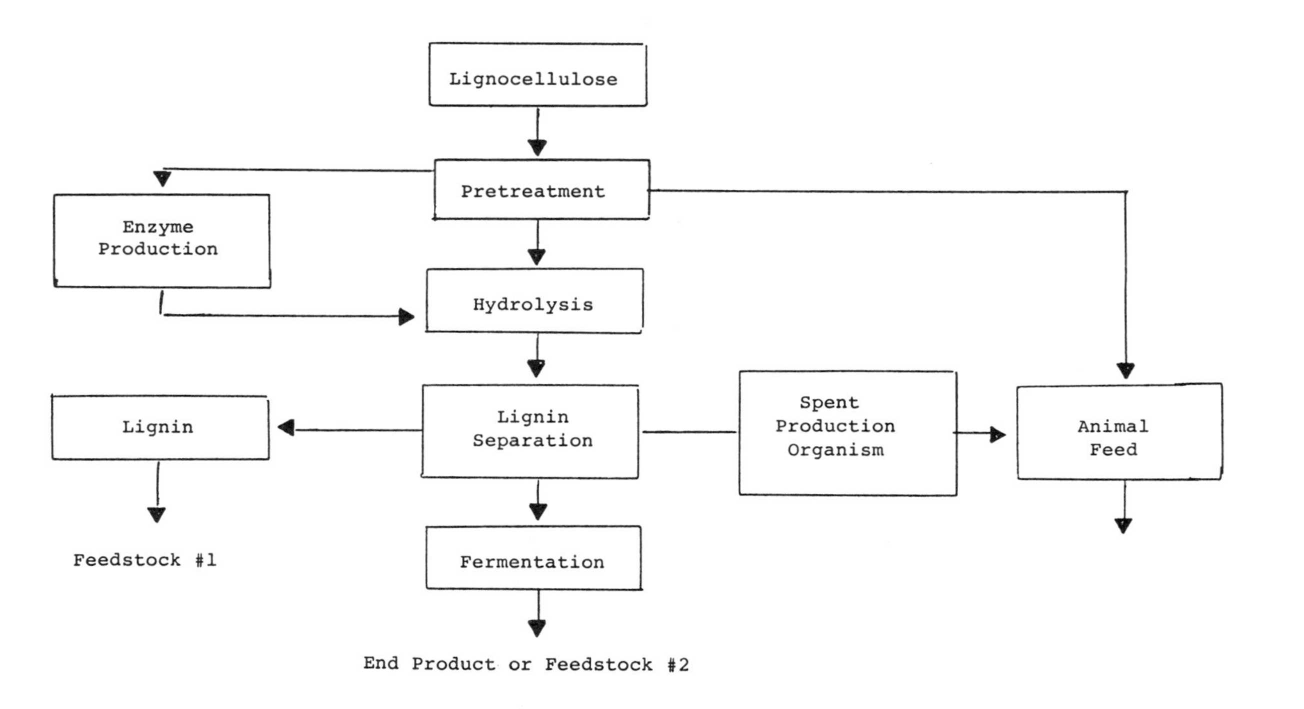

Figure 2. The Use of Lignocellulose as Low Cost Feedstock.

lignocellulose, will one day produce lower feedstock costs for biomass than for those based on petroleum. With oil prices continuing to rise that day may not be too distant.

Lower Cost Recovery Systems. The use of rDNA techniques for producing specialty chemicals has been limited by high costs due in large part to the complexity of recovery and purification. For example, a significant portion of the cost for enzymes produced using rDNA technology comes from the intracellular nature of expression in organisms such as *E.coli*. In addition, the incorrect folding of enzymes or proteins so expressed, especially those containing disulphide bonds can often be a problem. Thus, the protein of interest must be purified away from intracellular materials and then, if necessary, be refolded back into its correct structural conformation.

There are production strains of microorganisms known which might avoid these problems. However, the genetics of these microorganisms have been poorly understood until recently. One family of production hosts that we have studied intensively is the filamentous fungi. These fungi have been used for producing food-grade enzymes for many years. Production strains have been modified classically so that high yields of enzyme product (over 30 grams/liter) can be obtained. The product is generally secreted by the fungi into the culture medium so that purification involves a simple separation of enzyme broth from production microorganisms. In addition, fungi are eucaryotes and might fold proteins in the correct conformation during the process of secretion.

We have worked with research strains of filamentous fungi, notably from *A. nidulans*, but have preferred using industrial production strains of *T. reesei*, *A. awamori*, and *A. niger* because of the inherently better characteristics of these strains for industrial production. One protein we have studied in some detail is the enzyme rennin, which is used in the manufacture of cheese. Rennin contains three disulphide bonds and is incorrectly folded when expressed from standard rDNA production hosts such as *E. coli*, *Bacillus*, and even yeasts. In *Aspergillus*, rennin is expressed and then secreted into the culture medium in the correct conformation (17). This allows recovery of rennin using fairly simple procedures and the process is now being scaled up for commercial production.

One of the advantages of biological approaches is that they tend to operate at ambient temperatures and pressures and use aqueous systems. However, aqueous product streams raise considerable problems in recovery even when, as discussed above, the product is secreted from the production organism. The development of new bioengineering approaches, especially the use of more sophisticated membranes, has offered some promise in minimizing this problem. These membranes allow passage of certain compounds based on type or molecular weight. Under development are more advanced bioreactor systems that use membrane bound enzymes or electrostatic principles to help separate product and substrate streams (18).

The use of more effective production organisms and more sophisticated recovery systems will help minimize recovery costs for biological approaches to chemical production. However, unless the product is to be shipped as an aqueous mixture, there is likely to be a "floor" of approximately 10¢/lb. for these type of costs and this must be considered during the project planning stage (19).

Synthesis Reactions. Biological systems are obviously capable of synthesis but their utility in chemical production to date has largely been in hydrolysis reactions. With synthesis reactions there can be a co-factor requirement and often an equilibrium that needs to be driven to completion. These technical barriers have both proved in the past to be costly to overcome.

Recently, though, progress has been made in improving the *in vitro* recycling of co-factors. For example, the production of phenylalanine and other amino acids by continuous enzymatic conversion of the corresponding α-keto-carboxylic acid has now moved to the pilot scale stage (20). In addition, *in vivo* methods of recycling the co-factor appear achievable as shown in the recycling of NADPH in the production of a Vitamin C precursor (13).

The need to "drive" reactions can also be overcome by using two-phase systems. At the research scale reverse miscelles have recently been used as one way of producing an excess of substrate compared to product thus shifting the reaction towards completion (21). At the commercial scale, high substrate concentrations have been used to drive the enzymatic modification of fats and oils in various "transesterification" processes (22).

$$R_1\overset{O}{\overset{\|}{C}}\text{-O-}R_2 + R_3\overset{O}{\overset{\|}{C}}\text{-O-H} \text{ --- } R_3\overset{O}{\overset{\|}{C}}\text{-O}R_2 + R_1\overset{O}{\overset{\|}{C}}\text{-OH}$$

The two-phase system can be liquid:solid as well as liquid:liquid. Here product precipitation can help "pull" the equilibrium to completion. One example of this at the commercial scale is a process developed by Toyo Soda for producing a derivative of the dipeptide sweetener aspartame. Thermolysin is used as the coupling enzyme and the derivative produced precipitates under the reaction conditions pulling the reaction forward (23).

Again there has been significant progress in overcoming the technical barriers involved. Reactions based on biological approaches can now be controlled and driven towards desired products. In addition, *in vivo* co-factor recycling appears achievable. A low cost *in vitro* recycling methodology is still missing but a number of research groups are now working towards this goal.

Faster Yield Improvement. One of the problems with the classical fermentation industry has been the long lead time necessary for yield improvement and the uncertainty of the process. The penicillin example given before, where yields took 20 years or so to reach 7 gm/liter, is typical. These technical barriers limited commercial fermentations to those fermentations where metabolites were naturally produced in high yield or to those where the products had especially high value uses (e.g., pharmaceuticals). However, modern techniques for screening production organisms, and the improved biochemical understanding and control of fermentations now possible, can shorten this time significantly. Figure 3 for instance shows the rapid increase in cellulase yields that were produced from T. reesei using this approach.

Even more impact can be obtained once recombinant techniques are applied. Figure 4 shows the dramatic yield increase obtained from a recombinantly modified Bacillus production strain. The strain was improved by introducing more powerful regulatory elements and by tying production of the subtilisin to different stages of growth in the Bacillus strain.

The greater control that is now possible in the genetics of production organisms and in the metabolic regulation of fermentations can lead to much faster scale up of production yields. This in turn affects the planning stage for projects by reducing some of the concerns about yield improvement. Nonetheless a hyper-producing production strain is not sufficient for economic production without a large-scale, low cost fermentation facility and the ability to operate it efficiently.

"Tailor-Made" Products. Speciality chemicals are largely performance chemicals, that is to say their value is dependent on how well they solve a specific problem for an end user. The last technical barrier examined here concerns the ability to tailor products for these specific end users. The traditional approach with products made biologically has been to take whatever could be made in large quantities and find markets that would accept it. From a commercial standpoint, it is clearly more desirable to do the reverse: to identify key performance parameters required and to go back and modify the product to solve the problem.

With the advent of protein engineering, the ability to modify products, especially proteins to solve specific problems has markedly increased. Table IV shows some performance chemicals where protein engineering will allow targeting of certain proteins towards a specific market. Indeed all of the performance chemicals listed are being worked on currently using this approach. For example, in the protease area, enzymes have now been developed with improvements in the following characteristics.

- Turnover number (k_{cat})
- Substrate affinity (k_m)
- Substrate specificity (k_{cat}/k_m)
- Stability (to bleaches, high pH, temperature)
- pH optimum (shift to higher or lower; narrow or broaden
- reaction (shift from synthesis to hydrolysis)

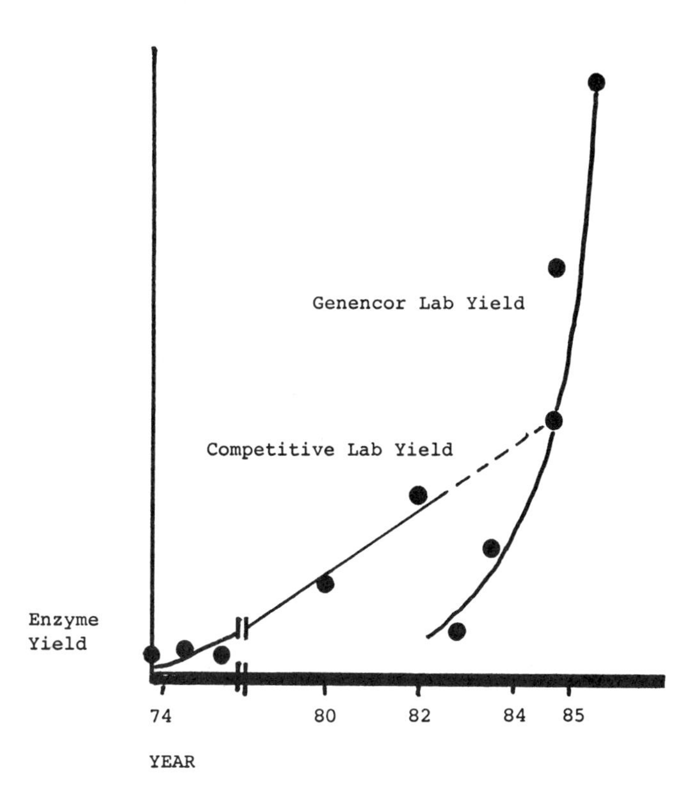

Figure 3. Yield Improvement Using Improved "Classical" Techniques.

EXAMPLE: A novel subtilisin produced from a Bacillus production organism using rDNA techniques

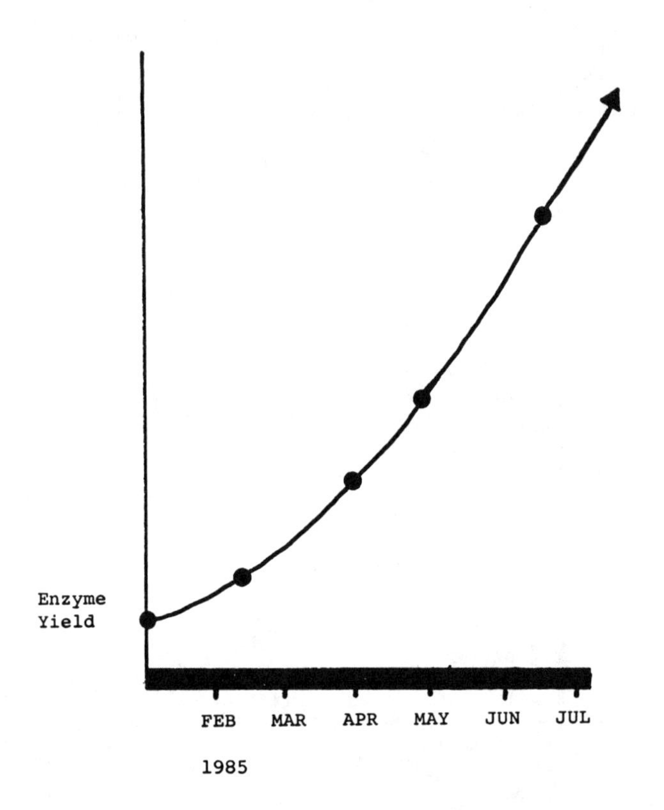

Figure 4. Yield Improvement Using rDNA Techniques.

The ability to make composition of matter patent claims for these improved proteases should provide a strong proprietary base for the enzymes and the products containing them.

Table IV. Examples of Potential "Tailor Made" Products

1. Biological Ingredients
   - Detergent enzymes (e.g. proteases)
   - Protein sweeteners (e.g. thaumatin)
   - Insecticidal toxins (e.g. B.t. toxin)
2. Biological Materials
   - Underwater adhesives
   - Natural teeth filling
3. Biological Catalysts
   - Glucose isomerase for HFCS production
   - Esterases for transesterification reactions

Commercial Limitations to the Application of Biotechnology

In the previous section the technical factors limiting the application of biological approaches to the production of chemicals were reviewed. This review showed that with the recent progress in the biological sciences a number of these technical factors have been overcome or at least minimized. So from a technical standpoint, the tools exist for biotechnology to make a dramatic impact on the chemical industry. And yet commercial progress in this area has been slow to date. This is due, to some extent, to constraints of a more commercial nature and some of these are examined below.

Large Critical Mass for R&D. The research and development resources required to identify, develop and scale up new biologically based production processes for chemical products are considerable. At Genencor, for example, there are over 80 technical people involved in these activities with a wide variety of disciplines represented, extending from molecular biologists and geneticists at one extreme to mechanical engineers at the other (Table V).

Table V. Some of the Disciplines Necessary for the Application of Biotechnology to Chemical Production

- Molecular Biology
- Genetics
- Microbial Physiology
- Microbiology
- Assay Development
- Protein Chemistry
- Enzymology
- Organic Chemistry
- X-ray Crystallography
- Fermentation Science
- Process Engineering
- Mechanical Engineering

In addition the costs to put together such teams are very large. Genencor has spent almost $70M over the last 5 years on facilities, equipment and R&D expenditures, not including the costs Genentech has borne in establishing some of the basic technology. For a chemical company to put in this level of expenditure requires a sizable commitment to biotechnology and only a few companies such as Monsanto and Du Pont have been willing to justify this expenditure.

Integrated Approach. The nature of these technologies is to require a strong multi-disciplinary approach with team members working closely together. In addition, to compare biological approaches to more traditional chemical approaches, it is important to have easy access to comparative cost information of these chemical alternatives. The newer biotechnology companies can rarely do this especially in obtaining up-to-date data that are based on real world experience. Frequently these data can lead to more integrated approaches to the development of new chemical production processes. A combined process using both chemical and biological steps can often give the lowest cost. Without a mechanism in place to facilitate data exchange between the traditional chemical industry and the newer biotechnology industry such integrated approaches are unlikely to be developed.

Difficulties in Exchanging Data and Capturing Value. With chemical companies holding some of the pieces and biotechnology firms the others, it has been difficult to establish strong proprietary positions for particular processes. Traditional enzyme companies, for example, have generally chosen to develop catalysts on their own and then sell them to a wide range of competing chemical companies. This has proved to be very unsuccessful even when technical successes have been achieved, thereby causing complaints from both chemical and enzyme companies (24).

Part of the reason for this lack of success is the commercial difficulties associated with exchanging data and ultimately capturing value from an integrated approach. Enzyme companies would prefer to sell biocatalysts and processes to the largest number of chemical companies possible. Chemical companies would prefer to have several different biocatalyst suppliers. This produces a traditional supplier:purchaser standoff. Chemical companies are unwilling to provide data helpful to such projects if direct competitors will have access to the improved process. Enzyme companies will not release much information about their biocatalysts for fear chemical companies will then use these data with other enzyme suppliers or go in-house with their own biocatalyst development.

The consequences of this standoff are illustrated in comparing the HFCS and aspartame sweetener businesses. Both are billion dollar businesses with similar markets. In HFCS there has been a commodity strategy in the industry both for the end product and for the biocatalysts involved. The resulting margins both for the corn processor as end user and for the enzyme firm as biocatalyst supplier have been average at best, with margins in the enzyme business being actually below average recently. With aspartame, proprietary positions in the end product but also in the production process have allowed, and will perhaps continue to allow, above average returns to be made.

Difficulties in overcoming this information barrier and in establishing proprietary positions for products and processes will need to be resolved if real value is to be derived from biological production techniques.

Managing Technical and Marketing Risk. The effective management of risk is an important paradigm for success in many technical industries, and the application of biotechnology to the chemical industry is no exception. There are many examples of technical hurdles being overcome to develop a new product or process only to find there was little or no market. If both technical and market risk need to be borne at the same time, the potential returns from the project must rise accordingly to provide a sufficient reward for the risk entailed.

The prevailing attitude with many chemical companies is to concentrate efforts on developing new chemical products, especially in the specialty chemical area. This clearly makes sense when dealing with established technologies. However, some of the newer biotechnologies appear very risky and, if targeted towards new products, will produce a combination of marketing and technical risk which might prevent or slow the application of biotechnology.

Some Solutions to the Above. Clearly there are biotechnology companies that have the critical R&D mass to tackle chemical production problems and, with some, this even translates to an ability to scale up such production processes to industrial scale. There are also chemical companies with the capability to evaluate existing chemical approaches, to identify steps where a biological approach would have the most impact, and to help develop an integrated production process. What is needed is a mechanism for companies from both groups to collaborate effectively.

One mechanism which does seem to meet these needs is a 1:1 exclusive relationships between the biotech firm and the chemical company. This can be simply a series of agreements between the companies for a specific market area ensuring that information and materials are exchanged on a confidential basis, that the value of the improved product or process is shared appropriately and that parties agree to work exclusively with each other for a specific product or market. It can of course go beyond this, right up to forming a joint venture between the companies. Although such commercial arrangements have been less common in the past, several companies have recently formed collaborations. For example, Shell Chemical Co. and Gist-Brocades, an enzyme supplier, joined forces to form a joint venture in the specialty chemical area. Another example is the collaboration between Lubrizol, Genentech, and Pfizer on the production of Vitamin C.

In terms of managing risk, there are several ways that this problem could be solved. One is to manage the marketing risk by initially targeting existing market areas. This is the approach

Genencor has adopted. By improving processes and products which are currently being sold, the marketing risk can be reduced significantly, provided the dynamics of the market are well understood. Of course, in time biological approaches will become better known and the ability to understand and predict what problems should be tackled and what should not, will have improved. At this point the risks for tackling new markets will become less burdensome and new product development could become much more aggressive.

Conclusions

Recent technical advances in the biological sciences have produced an accelerating pace of change as new "waves" of technology have grown. Such techniques as protein engineering, high yield secretion and metabolic pathway engineering have contributed to this acceleration. They have also enabled some of the technical barriers that existed previously in applying biological approaches to the chemistry to be overcome or minimized. Indeed, as was mentioned above, several commercial developments are now underway.

The impact of biotechnology on the chemical industry is now mostly limited by commercial barriers that prevent the development of improved products and processes in a way where the real value of these developments can be fully captured. The formation of collaborations and joint ventures between chemical companies and biotechnology firms should help address this problem. These collaborative mechanisms should increase the number of chemical projects being tackled biologically and lead to an increasing impact of biotechnology on the chemical industry that will continue well into the next century.

Although the health care aspects of biotechnology have attracted the most interest to date from the business community, the production of chemicals using biotechnology may turn out to be an equally valuable aspect of biotechnology. In 1986, the sales from pharmaceuticals products produced using recombinant biological approaches topped $200M. Although there was little or no production of chemicals using such technology in 1986, the Office of Science and Technology Policy has forecasted some $14 billion of chemicals a year will be produced by biological approaches by the year 2000 (25).

If the promise of biotechnology is to unfold in the ways described, much more extensive interaction will be needed between the growing biotechnology industry and the chemical industry, and between the fields of biology and chemistry. The American Chemical Society certainly has a role to play in providing a forum for these interactions and for facilitating the growth of research and commerce in this area for the public good.

Acknowledgments

I should like to thank Dr. Sharon Shoemaker for constructive comments and Leslie Flint for typing and editing the manuscript.

Literature Cited

1. Aharonowitz, Y.; Cohen, G. Scientific American 1981, 245, 141.
2. Eveleigh, D. E. Scientific American 1981, 245, 155.
3. Cohen, S. N.; Boyer, H.W. U.S. Patent 4 468 464, 1984.
4. Wetzel, R. American Scientist 1980, 68, 664.
5. Goeddel, D. V.; Kleid, D. G.; Bolivar, F.; Heyneker, H.L.; Yansura, D. G.; Crea, R; Hirose, T.; Kraszewski, A.; Itakura, K; Riggs, A. D. Proc. Natl. Acad. Sci. USA 1979, 76, 106.
6. Emtage, J. S.; Angal, S.; Doel, M. T.; Harris, T. J. R.; Jenkins, B.; Lilley, G.; Lowe, P. A. Proc. Natl. Acad. Sci. USA 1983, 80, 3671.
7. White, T. J.; Meade, J. H.; Shoemaker, S. P.; Koths, K. E.; Innis, M. A. Food Technology 1984, 38, 90.
8. Edens, L.; Lederboer, A.; Verrips, C.; Van den Berg, J.A. U.S. Patent 4 657 857, 1987.
9. Wells, J. A.; Vasser, M.; Powers, D. B. Gene 1985, 34, 315.
10. Gray, G. L.; Mainzer, S. E.; Rey, M. W.; Lamsa, M. H.; Kindle, K. L.; Carmona, C.; Requadt, C. J. Bacteriol. 1986, 166, 635.
11. Estell, D. A.; Graycar, T. P.; Wells, J. A. J. Biol. Chem. 1985, 260, 6518.
12. Edwards, M. R.; Taylor, P. P.; Hunter, M. G.; Fotheringham, I. G. PCT Patent Publication WO87/00202, 1987.
13. Anderson, S.; Marks, C. B.; Lazarus, R.; Miller, J.; Stafford, K.; Seymour, J.; Light, D.; Rastetter, W.; Estell, D. Science 1985, 230, 144.
14. Chem. Brit. 1985, June, 520.
15. Industrial Enzymology; Godfrey, T.; Reichelt, J., Ed; The Nature Press: New York, 1983; Chapter 4, p 446.
16. Food Technology 1984, 38, 26.
17. Cullen, D.; Gray, G. L.; Wilson, L. J.; Hayenga, K. J.; Lamsa, M. H.; Rey, M. W.; Norton, S.; Berka, R. M. Bio/Technology 1987, 5, 369.
18. Bratzler, R. L. Biotech '87 1987, Vol. 1, Part 5, 23.
19. Busche, R. M. Biotechnology Progress 1985, Vol. 1, No. 3, 165.
20. Wandrey, C.; Wichmann, R.; Leuchtenberger, W.; Kula, M. R.; Buckman, A. U.S. Patent 4 304 858, 1981.
21. Hilhorst, R.; Spruijt, R.; Laane, C.; Veeger, C. Eur. J. Biochem. 1984, 144, 459.
22. Macrae, A. R.; How, P. PCT Patent WO 8303844, 1983.
23. Toyo Soda. U.S. Patent 4 212 945, 1977.
24. Royer, G. P. Biotech '85 1985, Vol. 2, 201.
25. Impacts of Applied Genetics, Office of Technology Assessment, 1981, 261.

RECEIVED August 3, 1987

Chapter 3

Evolution and Future of Biotechnology

Paul Schimmel

Department of Biology, Massachusetts Institute of Technology, Cambridge, MA 02139

Technical innovation, in the form of new methods and procedures, gave birth to biotechnology. The field continues to be technology driven. Because of their central role in most biological systems, proteins are the focus of much of the research. While protein engineering is now established as a discipline, its most important contributions lie in the future.

Technological innovation drives the field of biotechnology. The development of new methods and techniques is the single greatest force behind the creation and continued progress of the field. I illustrate this below with a few examples. The examples show that were we to subtract one critical innovation, the character of biotechnology would be entirely different or the field would not even exist.

Research and development, in one way or another, usually focuses on proteins as products or uses proteins as tools. The first pharmaceutical products--human insulin and human growth hormone--and many of the new pharmaceuticals under development, the array of diagnostic products based on monoclonal antibodies, enzymes that are used for commercial applications, biorational pesticides, and many other products and product candidates, are proteins. Even when they are not products, proteins are often key components of a procedure or process. For example, restriction enzymes are used in medical research to map DNA molecules in search of restriction fragment length polymorphisms that can be used as genetic markers; and some production processes use large amounts of specific enzymes *in vivo* or *in vitro*. Because of the role of proteins, much effort in biotechnology is directed at protein engineering. Examples of the direction and status of protein engineering are given below.

ROLE OF TECHNOLOGICAL INNOVATION

DNA SEQUENCING. There would be no biotechnology, as we know it today, without the powerful and elegantly simple methods for determination of the nucleotide sequences of large pieces of DNA (1,2). This invention, like others, was rapidly implemented into a format that makes it routine for the average bench scientist. This single development, which can be viewed as a discontinuous leap ahead in methodology, accelerated by decades the pace of research.

We now know the entire nucleotide sequence of human mitochondrial DNA--about 17,000 nucleotides (3). We also probably know about 20% of the nucleotide sequence of the entire genome of a common bacterium--*E. coli*. This genome is comprised of about four million base pairs and no one doubts that, should we wish, the rest could be determined in short order. The human genome is much larger--about two billion base pairs. We are a long way from knowing even one percent of this genome, but that is not the point. The point is that we are in a position with the present technology, and with straightforward improvements in the technology, to debate seriously whether we should allocate resources so that the sequence of the entire human genome can be determined.

DNA SYNTHESIS. The gradual, incremental improvements in DNA synthesis have led to highly efficient procedures for synthesis of genes and gene fragments(4). While DNA sequencing is typically an end in itself, DNA synthesis is used as a means to one of many ends. DNA synthesis is used routinely to make genes or to alter small segments of cloned genes. DNA synthesis also drives DNA sequencing, at least with the popular "dideoxy" method of sequence determination. This is because synthetic oligonucleotide probes are used as primers for enzymatic synthesis of DNA fragments from templates of unknown sequence(2,4). Synthetic oligonucleotides are also the tools by which new genes are isolated, by hybridization procedures in which the synthetic probe binds selectively to a complementary sequence in a large mixture or bank of DNA molecules(5).

SEQUENCING AND SYNTHESIS OF PROTEINS AND PEPTIDES. The sequencing of protein chains can now be done with picomoles of material that are electro-blotted directly onto membranes(6). This, in turn, gives access to information about scarce, but vital, proteins and serves as the basis upon which they can be cloned, expressed, and produced in sufficient quantity to be studied in depth and potentially to be used. The chemical synthesis of peptides and small proteins, by automated and semi-automated procedures, has given further impetus to deeper investigations of protein design and to the creation of peptides with biological activity(7). Peptide synthesis is also now a core technology for the field of immunology.

SEPARATION OF DNA MOLECULES. Methods for separation of molecules by gel electrophoresis have experienced a renaissance that has played a major role in accelerating the pace of biotechnology. We now can separate by electrophoresis a set of DNA molecules, up to a few hundred nucleotides in size, that differ in length by increments of only one nucleotide. This has important applications and ramifications, including the method by which nucleotide sequences are visualized on gels(1,2). An entirely different problem is how to separate giant DNA molecules of hundreds of thousands or millions of base pairs. Recent inventions have made it possible to separate virtually all seventeen of the chromosomes of the yeast *S. cerevisiae*, by a single gel electrophoresis protocol(8-10). Among other applications, these procedures rapidly accelerate the rate at which specific genes are mapped to specific chromosomes.

OTHER ADVANCES. While biotechnology is now established and well recognized, the innovation that characterized the early stages of the field has continued. These include new ways to express genes and to isolate and stabilize the gene products, to make mutations, to transfer DNA into foreign hosts, and to map genetic markers on large(for example, human) chromosomes. With respect to the latter, the positions in the human genome of many of the loci responsible for specific heritable diseases are now within reach(11).

PROTEIN ENGINEERING

The overwhelming majority of biological transactions are executed by proteins. For that reason, protein engineering is a central part of biotechnology and there are many different facets as to how mutatgenesis and other forms of protein engineering are applied.

RELATIONSHIP OF MUTAGENESIS TO CHEMICAL MODIFICATION EXPERIMENTS. The ability to manipulate and change amino acid sequences, by virtue of nucleotide substitutions in the gene of a protein, is a tool that supplants some of the traditional approaches of analytical protein chemistry. Chemical modification of reactive side chains, for example, is a longstanding method to probe protein structure. This approach requires isolation and identification of the modified side chain, a procedure which involves protein digestion, multi-step chromatography, and sequencing of the modified peptide. An alternative approach is to change, separately, by mutation those side chains that are likely to be the targets of modification(12). These are changed to structurally similar, but non-reactive residues. An example is the change of a cysteine(with a reactive SH) to an alanine, serine, or threonine. The removal of the correct reactive side chain renders the protein refractory to the specific modification which occurs with the wild-type protein. If the wild-type amino acid is also critical for catalytic activity, then its substitution will lead to a clear change in activity. The ability to make many such changes by site-directed mutatgenesis facilitates a rapid scan of the reactive groups.

STRUCTURE-ACTIVITY RELATIONSHIPS. Much recent research has concentrated on structure-activity relationships, with emphasis on proteins of known three dimensional structure. These investigations typically take advantage of oligonucleotide-directed mutagenesis, so that single amino acid substitutions can be introduced into defined sites. Based on a three dimensional structure, these are sites which are postulated to be critical for the activity or stability of the protein.

These investigations have tested the general concept that enzyme catalysis is a binding phenomenon. That is, the differences in binding energies of the transition state and of the substrate and product determine the rate of catalysis. They also have tested the roles of specific amino acid side chains in the detailed catalytic mechanisms of specific enzymes. These important investigations have not produced big surprises, however. For the most part, they have validated longstanding ideas about the role of the "lock and key" fit between protein and ligand as the prime determinant of specificity and catalysis.

THE POWER OF SELECTION SYSTEMS. The studies of structure-activity relationships have generally been guided by the particular structure that is of interest. A limited number of mutations are made and tested. These mutations are typically at obvious places where an effect of the mutation on the activity would be expected. It is likely that more profound information will come when proteins of known structure are subject to an unbiased random mutagenesis, particular phenotypes are selected, and the structures of the protein variants are determined.

There are thousands of proteins that are of interest for one reason or another and it is not practical to think in terms of determining the three dimensional structure of each. Applications generally do not require detailed knowledge of the structure. Moreover, new properties can be engineered into proteins without knowledge of the three dimensional structure. In these cases, as with a protein of known structure, the challenge is to design selection systems, or highly efficient screens, to obtain mutants that have an improved thermal stability, higher catalytic efficiency, better properties in a particular solvent, and so on. There are recent examples of the use of special selections and mutagenesis to obtain a protein with higher catalytic activity than the starting enzyme(13), and of the selection of an enzyme with higher thermal stability(14,15). In these instances, the three dimensional structure of the respective enzyme is unknown. Because our understanding of structure and activity is still primitive, I suspect that, in any case, selection systems applied to mutagenized genes will uncover better variants of a particular protein than those made by rationale design based on a particular structure. The challenge in this case comes back to designing creative selection schemes.

GLOBAL MUTATIONS. It has long been known that a point mutation in a protein can give rise to a major loss of stability. Single point mutations can also give major improvements in stability. In some cases, these mutations may act as global mutations, that is, mutations that add an incremental amount of free energy of stabilization, whether alone or combined with one or more additional mutations in other parts of the molecule(16). This characteristic suggests that it may be possible to build improved proteins by increments, through the cumulative effects of successive amino acid replacements that each enhance the molecule with respect to a specific property.

PROTEINS IN PIECES. For some, perhaps most, proteins, the biological activity resides in a sub-piece that may be small relative to the total size of the protein. Alternatively, a portion of the activity--such as the binding of a ligand--can be isolated as a small fragment. These fragments constitute structural and, additionally or alternatively, functional domains. They can be identified and isolated without any knowledge whatsoever of the three dimensional structure(17-19). The manipulation of structure and activity can be concentrated, therefore, on a small protein fragment(13) or directed to a small part of the whole protein. Attempts at crystallization and three dimensional structural analysis can be directed at small pieces, which typically are easier to handle and analyze.

Domains that are isolated as fragments are the building blocks for the construction of chimeric proteins. Chimeric antibodies that are mouse-human hybrids have important medical applications(20). Chimeric enzymes that have novel activities and properties present a somewhat greater challenge. The continued delineation of domains within proteins will open a large number of opportunities for making chimeric constructions, but at present there are only limited data on which to predict the success of this particular endeavor for making new enzymes or active polypeptide fusions(21).

Dissections of proteins into pieces plays a crucial role in the development of new vaccines. This is especially evident in the recent research on an AIDS vaccine, where a fragment of the viral envelope protein has been shown to elicit neutralizing antibodies, in an animal host, against the live virus(22). While vaccines(such as the polio vaccine) can be based on an attenuated or inactivated form of the whole virus, there is a small risk of viral reactivation that will lead to the infection against which the vaccine is intended. Use of a protein fragment dissected from a part of the virus eliminates this possibility.

CONCLUDING REMARKS

Technological innovation has made possible the projects that are pursued today. In some respects, the existing technology has created more opportunities than we are in a position to pursue at this time. Were we to freeze technology at its present state of development, it would take years for the field to exploit, with the extant technology, the obvious projects that are at hand. When this circumstance is coupled with the continued development of new technology, there are powerful forces in place for the rapid future expansion of biotechnology.

The major rewards of protein engineering, which impacts on almost all areas of biotechnology, are for the the future. The ability to manipulate structure is leading gradually to a deeper understanding of how proteins are assembled into functional units. This advance in basic understanding is one of the major contributions of protein engineering. There is little doubt that protein engineering will continue to lead to protein variants with altered properties, and new functional species of antibodies, enzymes, and other proteins. We are still exploring the possibilities.

ACKNOWLEDGMENTS

This work was supported by Grant Numbers GM15539 and GM23562 from the National Institutes of Health, and by W. R. Grace and Company.

LITERATURE CITED

1. Maxam, A. M.: Gilbert, W. *Proc. Natl. Acad. Sci. USA* 1977, **74**, 560-564.
2. Sanger, F.: Nicklen, S.; Coulson, A. R. *Proc. Natl. Acad. Sci. USA* 1977, **74**, 5463-5467.
3. Anderson, S.; Bankier, A. T.; Barrell, B. G.; de Bruijn, M. H. L.; Coulson, A. R.; Drouin, J.; Eperon, I. C.; Nierlich, D. P.; Roe, B. A.; Sanger, F.; Schreier, P. H.; Smith, A. J. H.; Staden, R.; Young, I. G. *Nature* 1981, **290** 457-465.
4. Smith, M. *Ann. Rev. Genet*. 1985, **19**, 423-462.
5. Putney, S.; Herlihy, W.; Royal, N.; Pang, H.; Aposhian, H. V.; Pickering, L; Belagaje, R.; Biemann, K.; Page, D.; Kuby, S.; Schimmel, P. *J. Biol. Chem*. 1984 **259**, 14317-14320.
6. Matsudaira, P. *J. Biol. Chem*. 1987, **262**, in press.
7. Merrifield, R. B.; Barany, G. In *The Peptides, Analysis, Synthesis, Biology*; Gross, E.; Meinenhofer, J., Ed.; Academic: New York, 1980; Vol. 2, Chapter 1.
8. Schwartz, D. C.; Cantor, C. R. *Cell* 1984, **37**, 67-75.
9. Carle, G. F.; Frank, M.; Olson, M. V.; Science 1986, **232**, 65-68.
10. Chu, C.; Vollrath, D.; Davis, R. W. Science 1986, 234, 1582-1585.
11. Francomano, C. A.; Kazazian, Jr., H. H. *Ann Rev. Med*. 1986, **37**, 377-395.
12. Profy, A. T.; Schimmel, P. *J. Biol. Chem*. 1986, **261**, 15474-15479.
13. Ho, C.; Jasin, M.; Schimmel, P. *Science* 1985, **229**, 389-393.
14. Masumura, M.; Aiba, S. *J. Biol. Chem*. 1985, **260**, 15298-15303.
15. Liao, H.; McKenzie, T.; Hageman, R. *Proc. Natl. Acad. Sci. USA* 1986, **83**, 576-580.
16. Shortle, D.; Lin, B.; *Genetics*, 1985, **110**, 539-555.
17. Jasin, M.; Regan, L.; Schimmel, P. *Nature*, 1983, **306**, 441-447.
18 Jasin, M.; Regan, L.; Schimmel, P.; *Cell*, 1984, **36**, 1089-1095.
19 Regan, L.; Bowie, J.; Schimmel, P.; *Science*, 1987, in press.
20. Oi, V. T.; Morrison, S. L. *Biotechniques*, 1986, **4**, 214-221.
21. Toth, M.; Schimmel, P. *J. Biol. Chem*. 1986, **261**, 6643-6646.
22. Putney, S. D.; Matthews, T. J.; Robey, W. G.; Lynn, D. L.; Robert-Guroff, M.; Mueller, W. T.; Langlois, A. J.; Ghrayeb, J.; Petteway, Jr., S. R.; Weinhold, K. J.; Fischinger, P. J.; Wong-Staal, F.; Gallo, R. C.; Bolognesi, D. P. *Science* 1986, **234**, 1392-1395.

RECEIVED September 2, 1987

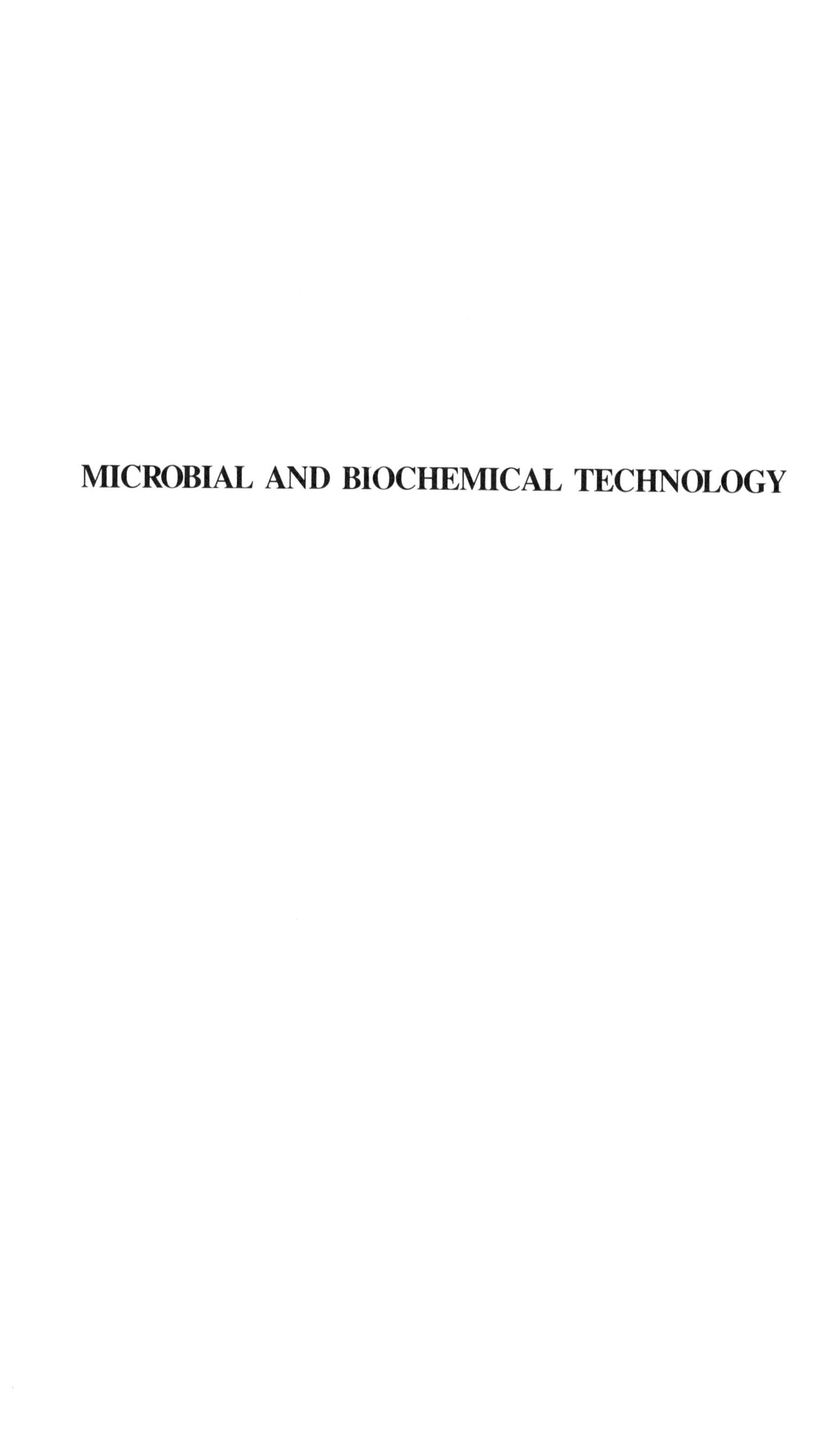

MICROBIAL AND BIOCHEMICAL TECHNOLOGY

Chapter 4

Introduction to Microbial and Biochemical Technology

Sharon P. Shoemaker [1] and Randall W. Swartz [2,3]

[1]**Genencor, Inc., 180 Kimball Way, South San Francisco, CA 94080**

[2]**Tufts University Biotechnology Engineering Center, Medford, MA 02155**

[3]**Swartz Associates, Biotechnology Consultants, 15 Manchester Road, Winchester, MA 01890**

Biotechnology processing has been the focus of the Microbial and Biochemical Technology Division of the ACS since its formation more than 25 years ago. In the tradition of the first biochemical engineers (Babylonian beer makers, circa 6000 B.C.) the division emerged from the Division of Agriculture and Food Chemistry. The founders were fermentation people committed to optimizing the technology for the preparation and purification of antibiotics. This required a multidisciplinary approach and this division brought together microbiologists, biochemists and chemical engineers and defined their relationship along with the new field of BIOCHEMICAL ENGINEERING.

The technology arising from the antibiotic programs was then applied to fermentation pathway engineering and the purification of enzymes, single cell protein, and biopolymers. Most recently this has been extended to the preparation of therapeutically important enzymes and the preparation of monoclonal antibodies. In the intervening period many new refinements have been made but there is a clear and well defined trail leading from the projects begun in various university and industrial labs in the 40's and the process designs used today to prepare recombinant proteins in mammalian and microbial cells. The development of this technology in part is exemplified in the contributions to this section.

RECEIVED September 14, 1987

Chapter 5

Computer-Aided Design of a Biochemical Process

C. L. Cooney, D. Petrides, M. Barrera, and L. Evans

Massachusetts Institute of Technology, Cambridge, MA 02193

There are three types of problems encountered in the development of chemical and biochemical process. First, process synthesis is the determination of the process topology, i.e. the process flowsheet. The screening of various reaction schemes, separation sequences, and general processing routes falls into the area of process synthesis. An excellent review of process synthesis is given by Nishida, Stephanopoulos, and Westerberg (1). Second, process design is the determination of the unit sizes, the system flow rates, and the various operating parameters of the units for a given flowsheet. For example, the design of the system would include the selection of the final cell concentration in the fermentor. Third, proccess optimization is the determination of the best overall process. Before the "best" system can be developed, some measure of the quality of the system must first be established. This measure becomes the objective function for the optimization. Process optimization can include the best selection of design parameters as well as the selection of the best flowsheet and even the process leading to the best or highest quality product. Thus, process optimization in the most general sense would include the activities of process synthesis and process design. The applications of optimization to computer-aided design are discussed in more detail by Westerberg (2).

Flowsheet Simulation

Flowsheet simulation is a computer-based tool that can be used to aide in the solution of the three problem types discussed above. The simulation of a chemical process consists of modeling the appropriate physical and chemical phenomena that occur in a unit. An entire process can be simulated by combining the unit models in some fashion. There are a number of potential benefits to applying computer-aided methodology. The use of flowsheet simulation

0097–6156/88/0362–0039$06.50/0

can speed the design of a process and improve the quality of the final design. For example, the effects of varying process parameters can often be determined more quickly using simulation and at less cost than a series of laboratory or pilot plant experiments. In addition, flowsheet simulation can be used to determine the most sensitive operating parameters and thus help focus process development and pilot plant research work. Finally, flowsheet simulation can be used to optimize existing processes. Performing a number of computer experiments is quicker, cheaper, and safer than experimenting with new operating conditions on-line (3).

Flowsheet simulation has become an accepted tool for the chemical engineer designing or operating processes in the chemical process industries (CPI). There are even standard texts on the subject (4). As the CPI matured, the economic pressures of competition have stimulated the desire to design and operate the most efficient processes. Flowsheet simulation in the CPI has developed over the past thirty years in response to these desires to minimize capital and operating costs for complicated chemical processes. Developments in the areas of process and physical property modeling, mathematical and numerical algorithms, and computer technology have made the development of flowsheet simulation possible. [For reviews on flowsheet simulation, see Motard et al. (5), Hlavacek (6), Rosen (7), and Evans (8)].

The heart of any process simulator is the ability to model the performance of the unit operations that comprise the process. This modeling must include complicated equilibrium phenomena often encountered in chemical systems. A great deal of work has been done over the past few decades to develop good models for unit operations typically found in the more traditional segments of the CPI. Because these chemcial processes are generally operated continuously and at steady state, the equations representing the material and energy balances and the physical property relationships are non-linear and algebraic in nature.

Despite development of good process models, the development of flowsheet simulators would have been impossible without suitable algorithms to solve the systems of equations or without the growth of inexpensive and powerful computers. The 1960's and 1970's saw a great amount of progress in the areas of algorithms and computer power. Sargent (9) presented a review of algorithms for systems for non-linear algebraic equations. The increases in computer speeds and memory sizes have allowed more sophisticated algorithms to be applied to more complicated problems.

The development of commerical flowsheet simulators applicable to industrial problems in the CPI began in the 1970's as a result of these advances in modeling, algorithms, and computer technology. The potential monetary benefits to be gained by applying this type of computer tool has been an important driving force in the development of commerical packages which include: ASPEN (10-12), CONCEPT (13,14), FLOWPACK (15), PROCESS (16), and SPEED-UP (17).

Potential Benefits of Simulation to BPI

Biotechnology is presently an area of high growth, and expectations for further advances in the biochemical process industries (BPI) are high. As the field moves from the laboratory into industrial scale, the role of engineering and process design and development become more important. There are potential benefits to be accrued by applying flowsheet simulation to the design of processes in the BPI. These benefits are analagous to those for the CPI. In addition, the pressure to develop, design, and start-up a working process as soon as possible for a new product is very high where early market penetration is often critical for economic success (18). Flowsheet simulation offers the opportunity to shorten the time required for process development. In addition, there is a high degree of interaction between the fermentation and the downstream recovery sections of typical bioprocesses. Figure 1 shows the many possible interactions between the tow sections of the process and among the units of a recovery section itself. These interactions significantly complicate the design of the process and provide an opportunity for the use of flowsheet simulation to allow a more systematic analysis (19).

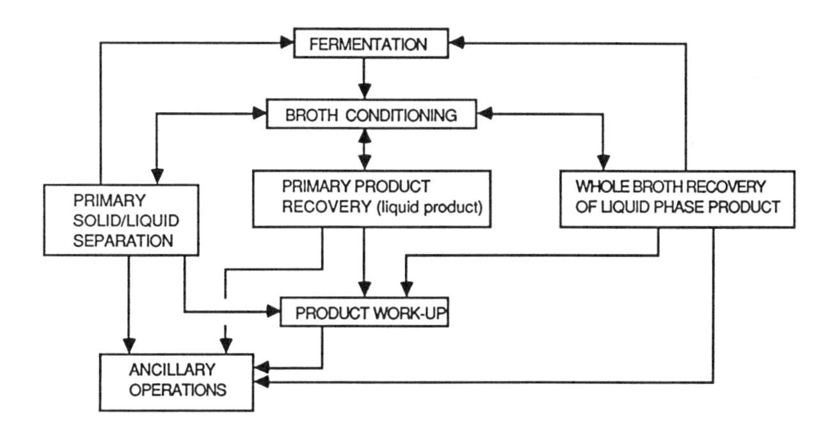

Figure 1. Interactions among upstream and downstream unit operations. (Reproduced with permission from Ref. 19. Copyright 1985. Turret Wheatland).

There are currently a number of difficulties in applying flowsheet simulation to biochemical processes. First, bioprocess contain many unit operations not typically found in the traditional CPI. Some of these processes are poorly understood and predictive quantitative models do not exist for these operations. Second, physical property information is lacking for many typical materials in a bioprocess. Complicated solution chemistry and the presence of solids in the process cause additional difficulties. Third the prevalance of batch and semi-continuous processing units in biochemical flowsheets causes problems for flowsheeting. Batch operations are inherently time-dependent, which means that differential equations must be integrated. Although techniques are available to solve these systems of algebraic and differential equations, interfacing time-dependent units with a program structure assumes steady state operation and the addition of discrete state variables make the simulation problem considerably more difficult.

A general purpose flowsheet simulator that can handle the problems encountered in the BPI does not exist at present. However, a fair amount of work has been done that provides a foundation upon which a general purpose simulation tool can be constructed. Naturally, the advances that led to the development of commercial simulators for the CPI and the experience gained while using these design aides in the CPI will make the task of developing a flowsheet simulator for bioprocess easier. In this review, we concentrate on what has been done in the area of biochemical process modeling and simulation, particularly in regard to potential contributions to the development of a general purpose bioprocess flowsheet simulator. In addition, we review some recent attempts to apply conventional process simulators to problems involving bioprocess operations. These examples serve to highlight the potential benefits that a tool of this type could provide to the BPI.

Process Modeling in the BPI

Most of the work in the literature on process modeling can be placed into one of two categories: modeling a single unit and modeling multiple unit operations. A significant amount of work has been done in the first category. In particular, many investigators have developed models for the growth kinetics of various microorganisms. For example, Constantinides et al. (20,21) developed a model to calculate the optimal temperature profile for a batch penicillin fermentation. Imanaka et al. (22) (a-galactosidase), Zertucke and Zall (23) (ethanol from whey), and Okita and Kirwan (24) (penicillin) represent the type of work that has been done in modeling fermentation growth kinetics. The availability of good models for growth kinetics is essential for a general purpose bioprocess simulator.

A large number of papers have been published on modeling of single unit operations. Most of the common unit operations found in biochemical processes have been covered in the literature. For example, Shimizu and co-workers (25-27) used a general bioreactor model to determine the optimal operating procedures for batch, fed-batch, and repeated batch reactors. Rotary filtration (28), centrifugal separation (29), chromatographic separation (30), semi-continuous counter-current adsorption (31), and ultrafiltration (32) are some of the downstream processing operations that have been modeled.

These models are generally based on principles of transport phenomena and thermodynamics. However, many of these models require physical property parameters that cannot be accurately predicted. For example, the viscosity of a mixture of biomaterials cannot be predicted precisely. Viscosity is an important input parameter for centrifugation and microfiltration models. Thus, the scarcity of physical property data for biomaterials and the accompanying lack of good prediction methods for physical properties has limited the utility of many unit operation models. Experimental studies are required to fit the appropriate model parameters for these cases.

For some unit operations the performance of the unit is not yet well understood in fundamental terms. The parameters for the empirical process models are often dependent on the particular operating conditions and the materials being processed. Cell disruption by high pressure flow devices (33) is just one example of a unit operation that demonstrates this problem. Further work is required to develop predictive models based on fundamental principles for operations of this type.

In the papers of the second category, multiple unit operations are considered. Since many investigators are concerned with the optimization of the overall process, they take a total process approach to the modeling. This allows them to investigate interactions that occur among the various units. The lack of a general flowsheet simulator for the BPI has limited the application of this approach and the literature is rather sparse in modeling of multiple unit operations in bioprocesses.

Okabe and Aiba presented a series of papers (34-37) on flowsheet optimization for antibiotic production. They used simple performance models for each unit operation rather than more rigorous, detailed models. For example, they simulated the batch operation of the bioreacator by curve-fitting experimental results of product and by-product concentrations as a function of time rather than integrating the differential equations that described the system. Their computer program was written specifically for this system and was not intended to be a general purpose tool. The flowsheet optimization was done using a slightly modified complex algorithm.

There are also several papers in the literature on the modeling of waste treatment systems that use biochemical methods. Paterson and Denn's work (38) is characteristic of efforts in this area. In this paper, the authors developed computer models for the various unit operations of an activated sludge process. They considered various processing schemes to determine the best system layout. The program, again written specifically for this application, also included a costing routine which allowed the various alternatives to be compared on an economic basis.

Finally, a few papers have looked at using existing flowsheet simulators (without modifications) to model some bioprocessing systems. The applications are usually limited to simulation of plants which produce bulk biochemical products, such as organic acids and solvents. The downstream processing steps for these biochemicals consist of conventional unit operations found in the CPI. For example, the fermentation and recovery of acetone-butanol (39) and acetone-butanol-ethanol (40) were simulated using the ASPEN simulator. The capabilities of ASPEN were mainly exploited for the simulation and the economic evaluation of the recovery stages of the above processes. Bhattacharya et al. (40) simulated the fermentation with a set of user-supplied subroutines.

Flowsheet Simulation in the BPI

Process flowsheet simulation for the BPI is currently in its infancy. Attempts to apply general simulation packages to bioprocess have just begun to appear in the literature in the past few years (e.g. Bhattacharya et al. [40]). The increased interest in the development of a bioprocess simulator is evidenced by two recent articles describing the requirements for such a tool (18,19). Both pointed out the benefits of such a tool, emphasizing the opportunity to take an overall systems approach to the design and optimization of an integrated process. Many of the additional benefits they cited have been mentioned previously in this chapter.

The unique requirements of a bioprocess simulator include: the need for additional unit operation models, the need for physical property data and property prediction models, and the need to incorporate batch and semi-batch operations. Jackson and DeSilva (19) included a detailed list of the types of unit operations modules required for a flowsheeting package for downstream recovery systems (Table I). The ability to handle solids was an important feature mentioned by both papers. Evans and Field (18) also included a list of the key physical properties for bioprocesses (Table II).

Table I. Unit Modules for a Process Flowsheeting Package Aimed at Downstream Process Operations

Solid-Liquid Separation Modules:
- Batch Filter (operating at constant rate or constant pressure, with or without precoating, and capable of handling compressible filter cakes)
- Rotary Drum Filter (operating either under vacuum or under pressure, without precoating)
- Centrifuge (horizational axis scroll type and veritcal disc type)
- Hydrocyclone (four different standard designs)
- Cross Flow Filter (hollow fibert tangential flow microfiltration unit)
- Gravity Settler
- Froth Flotation Unit

Thermal Concentration/Solids Drying Modules:
- Single Effect/Flash Evaporator
- Multiple Effect Evaporator (operating in forward feed, backward feed or parallel feed mode)
- Spray Dryer
- Flash or Pneumatic Dryer
- Freeze Dryer
- Concurrent Rotary Dryer
- Drum Dryer

Membrane Separation Modules:
- Batch Reverse Osmosis Unit
- Multistage Reverse Osmosis Unit (with both concentration and purification section if necesary)
- Batch Ultrafiltration/Diafiltration Unit
- Continuous Ultrafiltration Unit

Other Separation Modules:
- Adsorption/Ion Exchange Unit(s)
- Crystallization/Precipitation Unit
- Leaching Unit
- Liquid-Liquid Extraction Unit
- Distillation Unit(s)

Support and Service Modules:
- Steam Splitter
- Two Stream Mixer (operating either adiabatically or isothermally)
- Temperature Setter
- Pressure Setter
- Sterilizer
- Freezing/Refrigeration Unit
- General Component Splitter

Exotic Separation Modules:
- Bipolar Membrane Separation
- Electrokinetic Transport Processes
- High Gradient Magnetic Separation

Table II. Physical Properties of Major Importance for Bioprocesses

Thermodynamic Properties
Steam Propterties
Fugacities of Volative Components
Solvent Vapor Pressure
Enthalpy
Heat Capacity
Osmotic Pressure
pH
Ionic Strentgh
Gas Solubility in Aqueous Phase
Solid Solubility in Aqueous Phase
Liquid-Liquid Extraction Distribution Coefficients
Transport Properties:
Density
Surface Tension
Viscosity
Thermal Conductivity
Diffusivity

The difficulty caused by the prevalance of batch unit operations in the BPI has been one of the stumbling blocks in applying conventional flowsheet simulation packages to bioprocesses. A new simulator geared to batch oriented processes has recently been described; Joglekar and Reklaitis (41) described the capabilities of the BOSS system. Although the simulator was not designed specifically for bioprocesses, its ability to handle batch operations and multiproduct plants could provide a good starting point for a bioprocess simulator.

Evans and Field (18) concluded that the best way to develop a general bioprocess simulator would be to modify an existing commercial simulator. They felt that many of the necessary components for a bioprocess simulator already existed in existing packages, such as data base structures, report writing capabilities, numerical methods for equation solving, and some of the unit operation models. The time and expense required to duplicate these previous efforts would be large, ranging from 20 to 60 man-years at a cost in the millions of dollars (8).

We have chosen to adopt ASPEN PLUS to biochemical process simulation; it has the ability to handle solids as well as non-conventional components that cannot be characterized by equations-of-state. Biomass is an example of a non-conventional material. The current version of ASPEN PLUS has some batch processing capabilities, with unit operation models for batch reaction and batch distillation. Finally, the system has a modular structure which allows the user to incorporate his own unit operation models. This modularity facilitates the modifications that are required to handle bioprocesses. The benefits of applying a total systems approach can be demonstrated by some recent work in the simulation of penicillin recovery (42,43).

Simulation of Penicillin Recovery

In a desire to test the ability of ASPEN PLUS to handle a biochemical flowsheet. We examined its applicability to the recovery of penicillin G from a fermentation broth. The flowsheet used for this simulation illustrated in Figure 2. The fermentation broth is clarified rotary vacuum filtration, the filtrate is cooled, acidified and extracted with butylacetate. Penicillin G is then crystallized and dried. We carried out a simulation through all of the unit operations shown in Figure 2 with the exception of the final drying step. The validity of the resulting simulation can be examined by comparing the predicted pH for extraction recommended by the simulator with actual values. Results are shown in Figure 3; here, one can see the sensitivity of the adjusted separation cost to solvent flow rate and process pH during extraction. For instance, as an

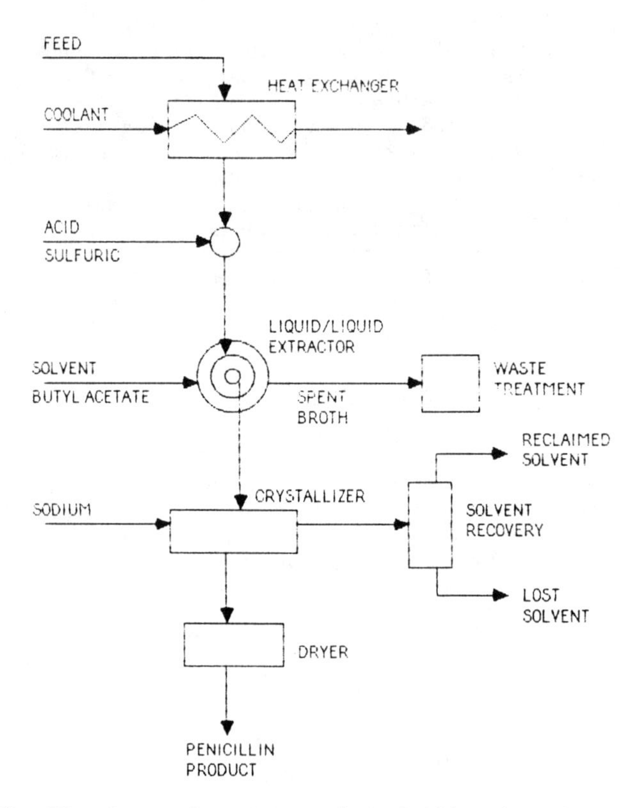

Figure 2. Flowsheet of recovery of penicillin from fermentation broth. (Reproduced with permission from Ref. 43. John Wiley & Sons.)

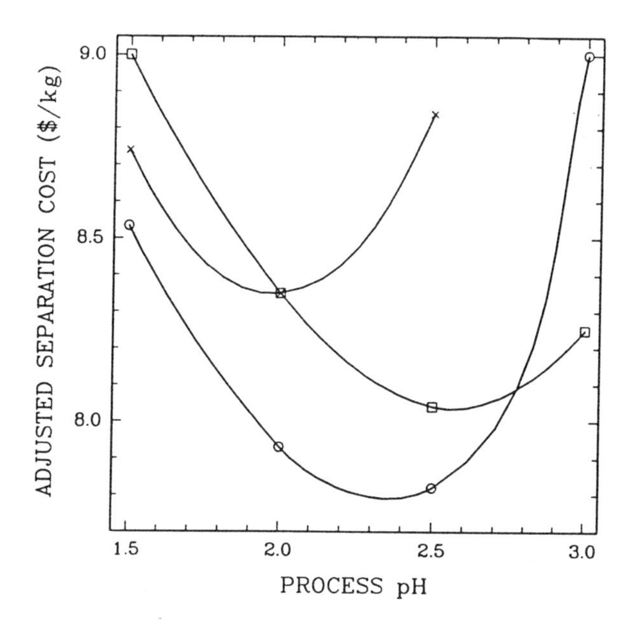

Figure 3. Adjusted separation cost for the one extractor model versus the process pH for several solvent flow rates with the filter recovery constant at 0.985.

increased amount of wash water is used to recovery the maximum amount of penicillin from the rotary vacuum filter; there is a corresponding need to increase the amount of solvent in the centrifugal extractor. As a result, different pH values are suggested for the extraction and there is a direct impact on the adjusted separation cost results. The adjusted separation costs includes the actual operating expense as a well as a penalty taken for penicillin lost in the recovery process. The in process pH value suggested by the simulation corresponds to values used in practice.

The results from such a simulation are summarized in Table III. Summarized here are the process operating variables and results including economics for recovery and production. A summary of the yearly operating costs and required capital investment, if a new facility were to be built, are presented in Tables IV and V.

Table III. Base Case Feed Condition, Optimum Operating Conditions, and Process and Economic Results for the One Extractor Model

Variable Name		Amount
	Broth Flow	4,000 kg/hr
Feed	Penicillin Conc.	25 gm/liter
Conditions	Broth Temp.	25 deg.C
	Broth pH	6.5
	Filter Recovery	0.985
Process	pH	2.31
Variables	Temperature	0.0 deg.C
	Solvent Flow Rate	1,000 kg/hr
	Wash Water Flow	2,992 kg/hr
	H2SO4 Acid Flow	991 kg/hr
Process	Cooler Duty	-44,540 cal/sec
Results	Solvent Recovery:	
	Boiler Duty -	36,071 cal/sec
	Cond. Duty -	-36,071 cal/sec
	Product Yield	90%
Economic	Total Production	259,000 kg/yr
Results	Net Recovery	$ 6.19 per kg
	Total Production Cost	$ 17.94 per kg

Table IV. Yearly Operating Costs for the Recovery Section at Optimum Operating Conditions for the One Extractor Model

Yearly Operating Costs		Percent of Total
Raw Materials	$ 447,000	28%
Operating Labor	102,000	7
Waste Treatment	95,000	6
Utilities	55,000	3
Overhead (incl deprec)	904,000	56
	$1,603,000	

Table V. Intitial Capital Investment for a One Extractor Penicillin Recovery Facility Running at Optimum Conditions

Item	Capital Cost
Process Units	$ 1,389,000
Setting Labor	857,000
Other Directs	1,494,000
Working Capital	1,612,000
Start-up Cost	1,065,000
Total:	$ 6,417,000

One of the benefits of such a process simulator is the ability to provide the investigator with an opportunity to do sensitivity testing. For instance, one can ask what is the impact of the concentration of penicillin in the fermentation on the recovery process. Results are seen in Table VI; one can also ask what would be the effect of choosing an alternative solvent for extraction. The effect of different solvents is shown in Table VII.

Table VI. Adjusted and Actual Separation Costs for Various Feed Penicillin Concentrations and Resulting Penicillin Product Yield

Feed Conc (gm/liter)	Adj Sep Cost ($/kg)	Actual Sep Cost ($/kg)	Separation Cost Per 100 Liters Feed ($/100 L)	Penicillin Yield
20	9.33	7.74	13.94	0.89
25	7.77	6.19	14.00	0.90
30	6.75	5.22	14.12	0.90
35	6.01	4.48	14.22	0.90

Table VII. Adjusted and Actual Separation Costs for Various Solvents Used in the Extractor and the Resulting Penicillin Product Yield

Solvent	Adjusted Sep. Cost ($/kg)	Actual Sep. Cost ($/kg)	Penicillin Yield
Isopropyl Acetate	7.33	6.03	0.91
Butyl Acetate	7.77	6.19	0.90
Amyl Acetate	7.94	6.35	0.89
Chloroform	7.74	6.03	0.90
Ethyl Ether	8.11	6.48	0.89
Ethylene Dichloride	8.19	6.57	0.89

One question often asked in process design relates to the economic impact of alternative processing strategies. In the recovery of penicillin a second counter current extraction is often used to back extract impurities would contaminate penicillin during crystallization. These impurities include the precursor to penicillin G, phenylacetic acid and other penicillin analogs that may be formed as minor contaminants. Using ASPEN PLUS for simulation of the penicillin recovery stream, Strong et al. (43) was able to show that a two extractor model designed to remove 90% plus of phenylacetic and acetic acids increased the net recovery cost by about $1.75/kg. Much of this increased cost was shown to be associated with a decreased product yield, from 90%-85%, between the one and two extractor models, respectively.

Another question often asked in the recovery of pencillin relates to the merits of whole broth extraction vs. filtered broth extraction. While whole broth extraction offers the possibility of eliminating penicillin loss during rotary vacuum filtration and decreasing operating costs by eliminating the need for filter aid and operation of a rotary vacuum filter. There are some negative impacts. In particular, the extraction efficiency in the centrifugal extractor is likely to decrease and solvent entrainment in the extracted broth increases, thus, leading to a greater need for fresh solvent. These trade-offs can be predicted through process simulation as shown in Figure 4. Shown here is the trade-off between plate efficiency in the centrifugal extraction vs. solvent entrainment. Above the curve, whole broth extraction is more economic and below the curve extraction of the filtered broth is prefered.

The power of flowsheet simulation for biochemical systems is clearly demonstrated in these examples.

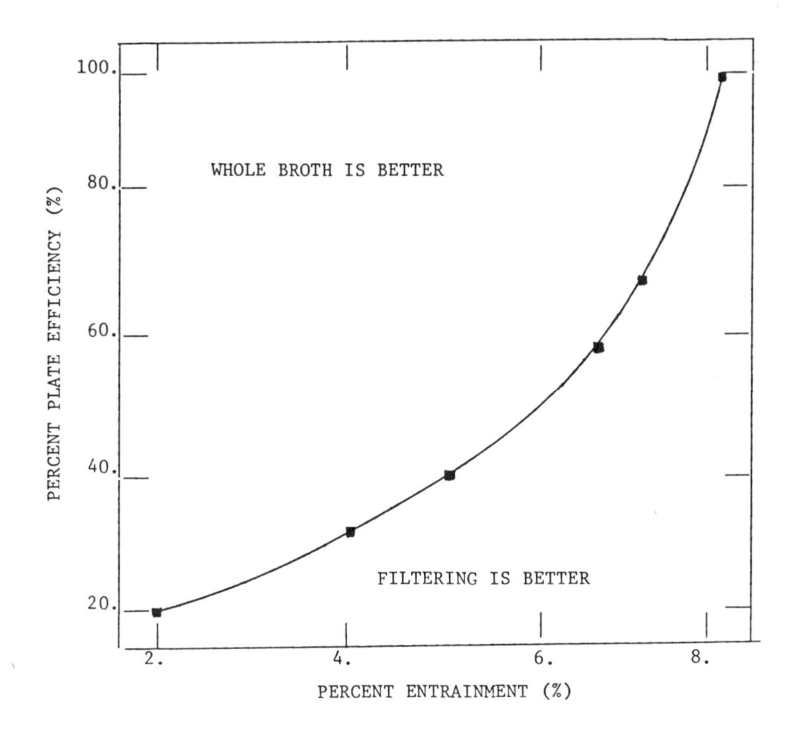

Figure 4. Curve showing all combinations of percent entrainment and percent plate efficiency for which whole broth and filtered broth extractive have the same adjusted separation cost.

Simulation of the Downstream Processing of a Proteolytic Enzyme

This case study is a second example of biochemical flowsheet simulation using ASPEN PLUS. Proteolytic enzymes, which are used in detergents. The downstream processing of these products consists of solids removal and broth concentration. This concentrated enzyme solution can be sold as a liquid product or processed further to a dehydrated form. The flowsheet for this case study is shown in Figure 5.

A drum vacuum filter is used to remove bacterial cells and other insoluble solids in the fermentation broth. Filter-aid is added to enhance the separation, and wash water is used to remove entrapped enzyme from the filter. Although evaporation has been the traditional technique for product concentration, ultrafiltration is an increasingly preferred alternative because of its lower cost, ability to remove residual low molecular weight solutes, and low operating temperature (5C), which reduces product degradation. A spray dryer is used to dehydrate the final product.

Because ASPEN PLUS does not contain unit operation models for ultrafiltration and spray drying, user-defined blocks were written for these units. The results of the base case simulation are shown in Table VIII. Enzyme concentration by the ultrafilter requires a membrane area of 175 square meters. The costing capabilities available in ASPEN PLUS were utilized to perform an economic evaluation of the process, and the results are summarized in Table IX.

One benefit of a simulation tool is the ability to examine the effects of various operating conditions on the overall process economics. In this example, water removal by the ultrafilter was studied. Higher levels of water removal not only decrease the costs incurred during spray drying but also increase product degradation. Enzyme denaturation takes place in the ultrafilter because of high shear stress. A first order loss rate was used to model product degradation.

This example demonstrates the potential of flowsheet simulation to direct experimental studies. If the final cost was insensitive to the value of k, there would be no need to spend the time and the money required to perform the experiments.

The effects of the wash water flow rate on the product purity and the process economics also were examined Figures 6 and 7. Increasing the wash water flow rate decreased product loss in the vacuum filter. In addition, the purification in the ultrafilter was improved because more low molecular weight solutes were removed. The increase in enzyme purity with increasing wash water flow rate is shown in Figure 6. However, there is a corresponding increase in

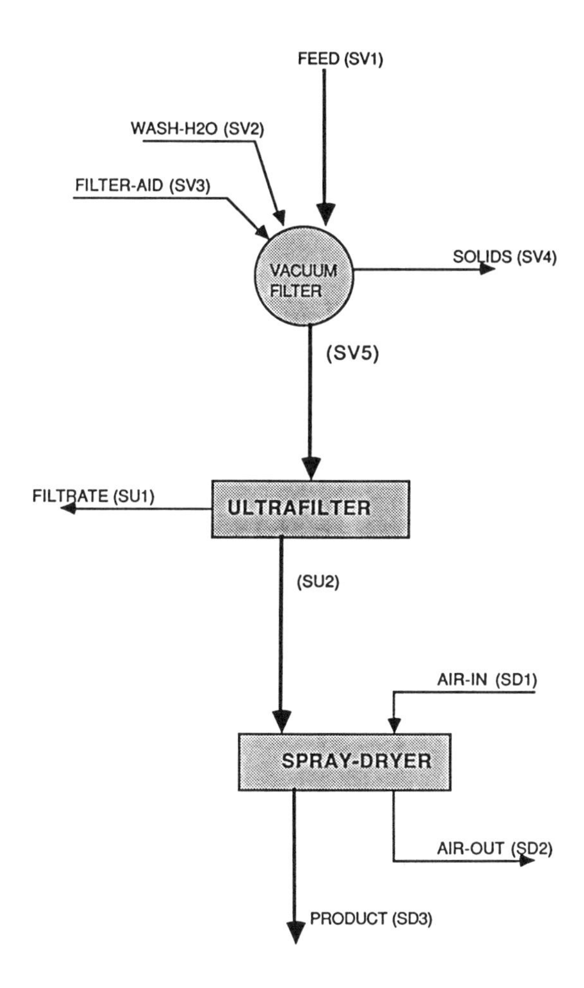

Figure 5. Block diagram of the flowsheet for protease recovery.

Table VIII. Flowsheet simulation results of the base case

	STREAM FLOWRATES AND COMPOSITIONS									
	SV1	SV2	SV3	SV4	SV5	SU1	SU2	SD1	SD2	SD3
Flowrate (kg/hr)	5000	1000	100	810	5290	4200	1090	16600	17550	125
	%	%	%	%	%	%	%	%	%	%
H2O	78.90	100.0	-	15.96	90.80	95.10	88.80	-	5.5	3.84
Cells w.b.	7.00	-	-	47.27	-	-	-	-	-	-
Enzyme	0.60	-	-	0.01	0.55	-	2.48	-	-	21.35
Solutes	13.50	-	-	19.45	8.70	4.90	8.68	-	-	74.81
Filter-Aid	-	-	-	17.31	-	-	-	-	-	-
Air	-	-	-	-	-	-	-	100.0	94.5	-

Table IX. Economic analysis of the base case

Item	Cost
Capital Investment	
Total Fixed Capital	$ 8,530,000
Working Capital	7,775,000
Startup Cost	4,978,000
TOTAL INVESTMENT	$ 21,283,000
Manufacturing costs (annual cost)	
Raw material	$ 23,266,000
Utilities	1,146,000
Labor	1,025,000
Supplies(filter aid, membrane repl.)	714,000
General Works	673,000
Depreciation (15 years, straight line)	558,000
TOTAL ANNUAL COST	$ 27,382,000
Production Volume (pure enzyme)	**211,250 kg/year**
Unit Cost (pure enzyme)	**$ 129.6/kg**
Unit Cost (crude product)	**$ 27.7/kg**

Assumptions

Item	Unit Price
Broth	$/kg 0.5
Filter-aid	0.50
Steam	0.015
Process Water	2.4e-4
Electricity	$/kwh 0.05
Gas	$/cal 5.0e-5
UF Membrane	$/sqm 500.

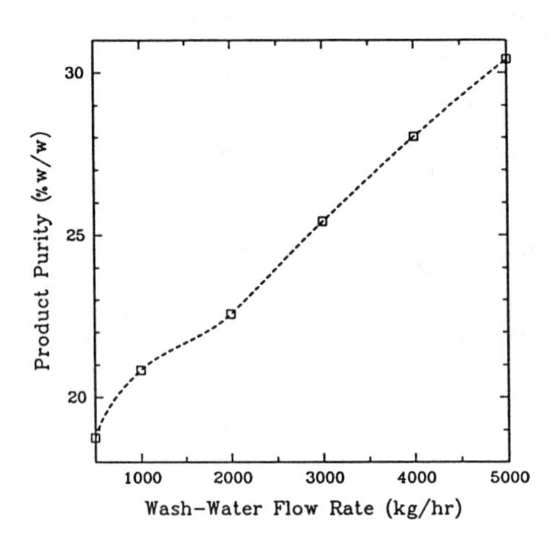

Figure 6. The effect of the wash water flow-rate on the purity of the product.

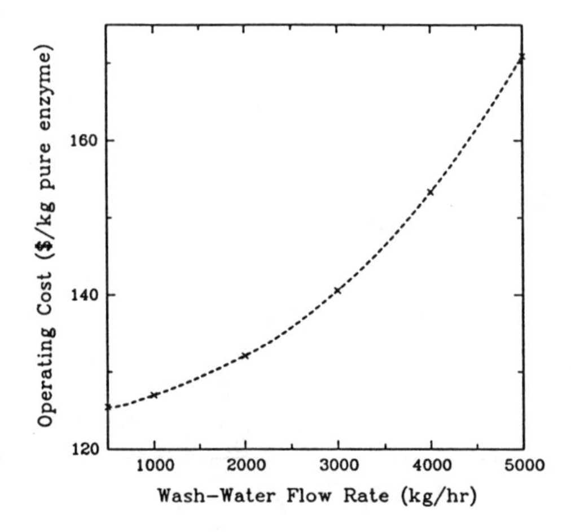

Figure 7. The effect of the wash water flow-rate on the operating cost.

the cost because of higher capital and operating cost of the ultrafilter and the increased cost of wash water. Figure 6 shows the effect of wash water flow rate on the operating cost.

Summary

In conclusion, the results described in this paper illustrate the use of computer aided design and flowsheet synthesis in optimizing biochemical process. We believe that this "tool" is important to the process development engineer and process design engineer. While there remains a need to improve the models describing the individual unit operations and methods to predict and extrapolate the physical properties of biochemical materials, there is sufficient information available to begin using flowsheet synthesis. It is anticipated that it will be possible to substantially decrease the development time and cost associated with a biochemical product. In addition, confidence in a flowsheet will increase as more work is done to experimentally validate the predictions of optimizations such as the ones illustrated here.

Literature Cited

1. Nishida, N.; Stephanopoulos, G.; Westerberg, A.W. AIChE J. 1981, 27, 321-351.
2. Westerberg, A.W. In Foundations of Computer Aided Chemical Process Design (Volume I); Mah, R.S.H.; Seider, W.D., Eds.; Engineering Foundation: New York, 1981; p. 149-184.
3. Wright, A.R. Chem. Eng. Res. Des. 1984, 62, 391-397.
4. Westerberg, A.W.; Hutchinson, H.P.; Motard, R.L. Winter, P. Process Flowsheeting; Cambridge University Press: Cambridge, England 1979.
5. Motard, R.L.; Shacham, M.; Rosen, E.M. AIChE J. 1975, 21, 417-436.
6. Hlavacek, V. Computers and Chemical Engineering 1977, 1, 75-100.
7. Rosen, E.M. In Computer Applications to Chemical Engineering; Squires, R.G.; Reklaitis, G.V., Eds.; ACS Symposium Series No. 124; American Chemical Society: Washington, D.C., 1980 pp 3-36.
8. Evans, L.B. In Foundations of Computer-Aided Chemical Process Design (Volume I); Mah, R.S.H.; Seider, W.D., Eds.; Engineering Foundation: New York, 1981; pp. 425-470.
9. Sargent, R.W.H. In Foundations of Computer-Aided Chemical Process Desgin (Volume I); Mah, R.S.H.; Seider, W.D., Eds.; Engineering Foundation: New York, 1981; pp. 27-76.

10. Evans, L.B.; Boston, J.F.; Britt, H.I.; Gallier, P.W.; Gupta, P.K.; Joseph, B.; Mahalec, V.; Ng, E.; Seider, W.D.; Yagi, H. Computers and Chemical Engineering 1979, 3, 319.
11. Gallier, P.W.; Evans, L.B.; Britt, H.I.; Boston, J.F.; Gupta, P.K. In Computer Applications to Chemical Engineering; Squires, R.G.; Reklaitis, G.V., Eds., ACS Symposium Series No. 124; American Chemical Society: Washington, D.C., 1980; pp. 293-308.
12. Britt, H.I. In Foundations of Computer-Aided Chemical Process Design (Volume I), Mah, R.S.H.; Seider, W.D., Ed.; Engineering Foundation: New York, 1981; p 471-510.
13. Winter, P. "Towards Integration of Process Plant Design," Computer-Aided Design Center, Cambridge, England, 1980.
14. Thambynayagam, R.K.M.; Branch, S.J.; Winter, P. Proc. 227th EFCE Mtg: CHEMPLANT '80, 1980.
15. Berger, F.; Perris, F.A.; Computers and Chemical Engineering, 1979, 3, 309.
16. Brannock, N.F.; Verneuil, V.S.; Wang, Y.L. Computers and Chemical Engineering 1979 3, 29.
17. Hernandez, R.; Sargent, R.W.H. Computers and Chemical Engineering 1979, 363.
18. Evans, L.B.; Field, R.P. Proc. World Congress of Chemical Engineers, 1986.
19. Jackson, A.T.; Desilva, R.L. Process Biochemistry Dec. 1985, p 185
20. Constantinides, A.; Spencer, J.L.; Gaden, E.L. Biotechnology and Bioengineering 1970, 12, 803.
21. Constantinides, A.; Spencer, J.L. Gaden, E.L. Biotechnology and Bioengineering 1970, 12, 1081.
22. Imanaka, T.; Kaieda, T.; Taguchi, H. J. Ferment Technol. 1973, 51, 431.
23. Zertuche, L.; Zall, R.R. Biotechnology and Bioengineering 1985, 27, 547.
24. Okita, W.B.; Kirwan, D.J. Biotechnology Progress 1986, 2, 83.
25. Shimizu, K.; Kobayashi, T.; Nagara, A.; Matsubara, M. Biotechnology and Bioengineering 1985, 27, 743.
26. Hasegawa, S.; Shimizu, K.; Kobayashi, T.; Matsubara, M. J. Chem. Tech. Biotechnol. 1985, 35B, 33.
27. Matsubara, M.; Hasegawa, S.; Shimizu, K. Biotechnology and Bioengineering 1985, 27, 1214.
28. Okabe, M.; Aiba, S. J. Ferment. Technol. 1974, 52 759.
29. Murkes, J.; Carlsson, C.G. Filtration and Separation Jan./Feb. 1978, p 18.
30. McCoy, B.J. In Chemical Separations, Vol. I: Principles; king, C.J.; Navratil, J.D., Eds.; Litavran Literature: New York 1986, pp 113-129.
31. Hidajat, K.; Ching, C.B.; Ruthven, D.M. The Chemical Engineering Journal 1986, 33, 855.

32. Kovasin, K.K.; Hughes, R.R.; Hill, C.G., Jr. Computers and Chemical Engineering 1986, 10, 107.
33. Engler, C.R.; Robinson, C.W. Biotechnology and Bioengineering 1981, 23, 765.
34. Okabe, M.; Aiba, S. J. Ferment. Technol. 1974, 52, 279.
35. Okabe, M.; Aiba, S. J. Ferment. Technol. 1975, 53, 230.
36. Okabe, M.; Aiba, S. J. Ferment. Technol. 1975, 53, 730.
37. Aiba, S.; Okabe, M. Process Biochemistry Apr 1987, p 25.
38. Paterson, R.B.; Denn, M.M. The Chemical Engineering Journal 1983, 27, B13.
39. Marlatt, J.A.; Datta, R. Biotechnology Progress Mar 1986, Vol. 2.
40. Bhattacharya, A.; Motard, R.L.; Dunlop, E.H.; "Simulation of Acetone-Butanol-Ethanol Process by BIO-ASPEN," ACS Meeting, Los Angeles, Sept. 1986.
41. Joglekar, G.S.; Reklaitis, G.V. Computers and Chemical Engineering 1984, 8, 315.
42. Strong, J.E.,Jr., Cooney, C.L.,. "Simulation of Downstream Pencicillin Recovery I. Modeling and Optimization of a Single Extractor Process," Biotechnology and Bioengineering 1987 (accepted).
43. Strong, J.E.,Jr., Cooney, C.L.,. "Simulation of Downstream Pencicillin Recovery II. Comparison of Alternative Processes," Biotechnology and Bioengineering 1987 (accepted).

RECEIVED October 30, 1987

Chapter 6

Bioconversion of Cellulosic Material to Short-Chain Acids

A. A. Antonopoulos and E. G. Wene

Argonne National Laboratory, Argonne, IL 60439

Studies were conducted to determine the feasibility of using cellulosic waste feedstock to produce short-chain organic acids. This study focused on acid production from anaerobic digestion of a simulated municipal solid waste feedstock using 5-L continuous stirred anaerobic digesters. Methane production was inhibited by short retention time (8-12 days), high volatile acid concentration, low pH, and inoculum from a long-term acid adapted culture. Nearly steady state operation was reached with acid concentrations of 15,000 to 18,000 mg/L. Increased nitrogen additions resulted in total concentrations of volatile acids between 27,000 to 30,000 mg/L. The effects of retention time, high-temperature treatment, and pH were also studied.

Municipal solid waste (MSW) contains a substantial amount of cellulosic material that can be used for liquid and gaseous fuels production. Since MSW exists in high amounts everyday, is concentrated, must be disposed and commands a tipping fee for disposal, bioconversion of MSW cellulosics to fuels appears to be economically and environmentally sound. The purpose of this work was to investigate the feasibility of utilizing MSW feedstock to produce short-chain organic acids.

Past research has determined that interacting groups of bacteria in anaerobic environments break down carbohydrates, proteins and lipids and convert them to methane and carbon-dioxide (1-3). During this conversion organic acids are produced which are then metabolized to acetate (and other simple compounds), and acetate is finally catabolized by methanogenic bacteria to methane and carbon dioxide (2-7).

Methane bacteria are extremely sensitive to oxygen, prefer neutral pH and fail to reproduce under acidic conditions, are inhibited by certain toxic substances and certain trace metals,

0097-6156/88/0362-0062$06.00/0

and their reproduction and survival can be suppressed by extreme microenvironmental conditions (2-4, 6, 8-10). Therefore, controlled oxygenation of the anaerobic digester, heat treatment, shortening of the retention time, addition of methanogenic inhibitory compounds, acidification, and other techniques could inhibit methanogenesis and allow the production and accumulation of organic acids. Work with biomass (plant) feedstock has indicated a feasible bioconversion to organic acids and transformation of these acids to alkanes (11-13). Research on using MSW for bioproduction of organic acids and transformation to alkanes has not been reported.

During this investigation we sought to optimize the operating conditions to produce high concentrations of organic acids from cellulosic a (MSW component) feedstock. Results are presented showing the effects of substrate heat treatment, retention time, nitrogen supplementation, and addition of a methanobacterial inhibitor on volatile acid production and accumulation.

Materials and Methods

Reactors. Anaerobic digestion experiments were carried out in continuously stirred tank reactors with once-a-day feeding, using Virtis 43 Series Fermenters (Virtis Co., Gardiner, N.Y. 12525. These fermenters were 5-L glass jars with a stainless steel head assembly, a magnetically coupled agitator, and a main fermenter cabinet. The head assembly consisted of addition and sampling ports, a fluid circulating system with two-sided control for temperature regulation, wells for temperature control and measurement, and a pH probe. Produced gas was collected by water displacement in calibrated plexiglass columns filled with acidified saturated NaCl solution to reduce the solubility of CO_2. Gas volumes were corrected for normal temperature and pressure.

Feedstock. The feedstock for the anaerobic digesters was a simulated MSW material consisting of 77% paper products (60% newsprint, 15% cardboard, and 25% assorted paper), 20% food and garden wastes, and 3% textiles and rags. This feedstock was mixed in large batches, oven-dried and stored in plastic bags in a freezer until used. It was fed to the digesters in this form or mixed with water depending on the amount of feedstock required. During one period of operation the feedstock was treated by soaking at room temperature for 24 hours in a mixture of 2% $Ca(OH)_2$ and Na_2CO_3.

Inoculum. The inoculum for the anaerobic digesters was obtained from an operating anaerobic digester at the Wheaton Municipal Sanitary District, Wheaton, Illinois. This digester operates on sewage and is designed as a front end treatment for solids reduction and methane generation. This culture has been maintained in this laboratory and been periodically reseeded with fresh inoculum from the Wheaton digester, compost piles, and enriched sewage sludge.

Start-up of Digesters. The two 5-L anaerobic digesters were initially fed with 2.5 L of simulated MSW feedstock with a total solids concentration of 30 g/L, and 2.5 L of inoculum from the Wheaton anaerobic digester. This mixture was allowed to slowly stabilize with weekly feedings. The digester temperatures were maintained at 35C with agitation rates at 150 rpm. Agitation rates were varied to maintain the solids in suspension and increased to completely mix the contents prior to sampling. The pH of the digesters was initially controlled to remain above 5.5, and the solids loading rate was varied from 2.0 to 10.0 kg/m^3- day of volatile solids to maintain the desired solids concentration.

Development of Acidogenic Populations. During operation of the digesters we were interested in developing stable acidogenic populations and determining methods of eliminating methane production. The digesters were heated to 80 C for 15 to 30 minutes to reduce the populations of non-spore forming bacteria. Also, 2-bromoethane sulfonic acid (BESA) was added to the digesters to a final concentration of 2 X 10^{-4}. BESA was added at twice weekly intervals to maintain the concentrations in the digesters at that level.

Methods of Analyses. Analyses were made to determine volatile acid concentration and composition, gas composition, alkalinity, total (TS) and volatile solids (VS). The volatile acid concentration, alkalinity and VS were determined in accordance with Standard Methods (14). Gas composition and individual volatile acids were estimated by gas chromatography. Gas composition was determined using a Varian (Varian Instruments Division, Palo Alto, CA) 90P gas chromatograph with a thermal conductivity detector, and a Porapak Q column (6mm X 160 cm). The column temperature was held at 75 C and the thermal conductivity detector was held at 150 C. Alternatively, methane content was determined using a Gow Mac 750P (Gow Mac, Bridgewater, N.J.) gas chromatograph with a flame ionization detector using the same column type and temperatures. Volatile acids were determined using a Gow Mac 750P gas chromatograph with a flame ionization detector and a 3.18mm X 160 cm column packed with 100/120 mesh Chromsorb WAW with 15% SP 1220 and 1% H_3PO_4 (Supelco, Bellefont, PA). The temperature program started with a column temperature of 100C for 2 minutes then ramped 10 degrees/min to 140 C. The injection port was held at 150 C and the detector at 200 C. Helium was used as the carrier gas at a flow rate of 35 mL/min. Samples of digester fluid were prepared as follows. Digester fluid was centrifuged at 4,000 X G for 5 min. Two-mL samples were acidified with two drops of concentrated H_2SO_4, and two mL of diethyl ether were added. The two phases of the mixture were inverted at least 15 times. The samples were centrifuged if they developed a stable emulsion. The gas chromatograph was calibrated using the external standard method with standard solution containing 10 mM each of acetic, propionic, butyric, isobutyric, valeric, isovaleric, caproic, isocaproic, and heptanoic acid.

Results and Discussion

The initial operation of the anaerobic digesters was a start-up period to assess the development of a stable acidogenic culture and to determine methods to limit methane production. During the first 100 days of operation the digester solids loading rate was gradually increased to where the total solids and volatile solids were 30 g/L and 20 g/L respectively. The volatile acid concentration at the end of this period had gradually increased to near 6,000 mg/L. The increases were not steady, there were fluctuations in acid accumulation. Part of the instability was due to fluctuations in pH which were controlled with $NaHCO_3$ to keep the pH above 5.5. Methane production was also fluctuating during this period of time.

Several methods were tried during this period to reduce methane production. Heat treatment of the digester was effective in eliminating methane production for a short period of time, however acid production was also reduced as evidenced by a decrease in acid accumulation following heat treatment. Acid production began to recover 3 to 5 days following heat treatment while methane generation returned about 7 to 10 days after treatment. Therefore it would require frequent heat treatment to eliminate methane production which would also decrease acid accumulation.

There was an immediate elimination of methane production following the addition of BESA to the digester, however, volatile acid accumulation also decreased following this treatment.

Pretreatment of cellulosic materials in alkali has been shown to have a swelling effect on cellulose which makes the cellulose more susceptible to enzymatic attack(15). Beginning on day 120 the feedstock was pretreated with alkali prior to feeding. The digester was operating on an 8-day solid and liquid retention time. The solids loading was increased during this period to raise the total solids to near 7%. This treatment added alkalinity to the digester and pH control was not required to maintain the pH above 5.5.

The volatile acid concentration at the beginning of this treatment was near 8,000 mg/L and rapidly increased to 17,000 mg/L (Figure 1). As this treatment continued the total concentrations of volatile acids decreased from that peak and remained between 10,000 to 15,000 mg/L with minor fluctuations.

The operation of the digester during this period demonstrated that it is possible to operate a digester at relatively high volatile acid concentrations on a continuous basis. Samson and LeDuy (16) have also shown that it is feasible to operate an anaerobic digester with volatile acid concentrations up to 23,000 mg/L using blue-green algae as feedstock. This blue-green algal material has a protein content up to 60%. The authors attribute the stability of this anaerobic digester operating on blue-green alga feedstock to the high protein content which resulted in a high concentration of dissolved ammonia and high alkalinity.

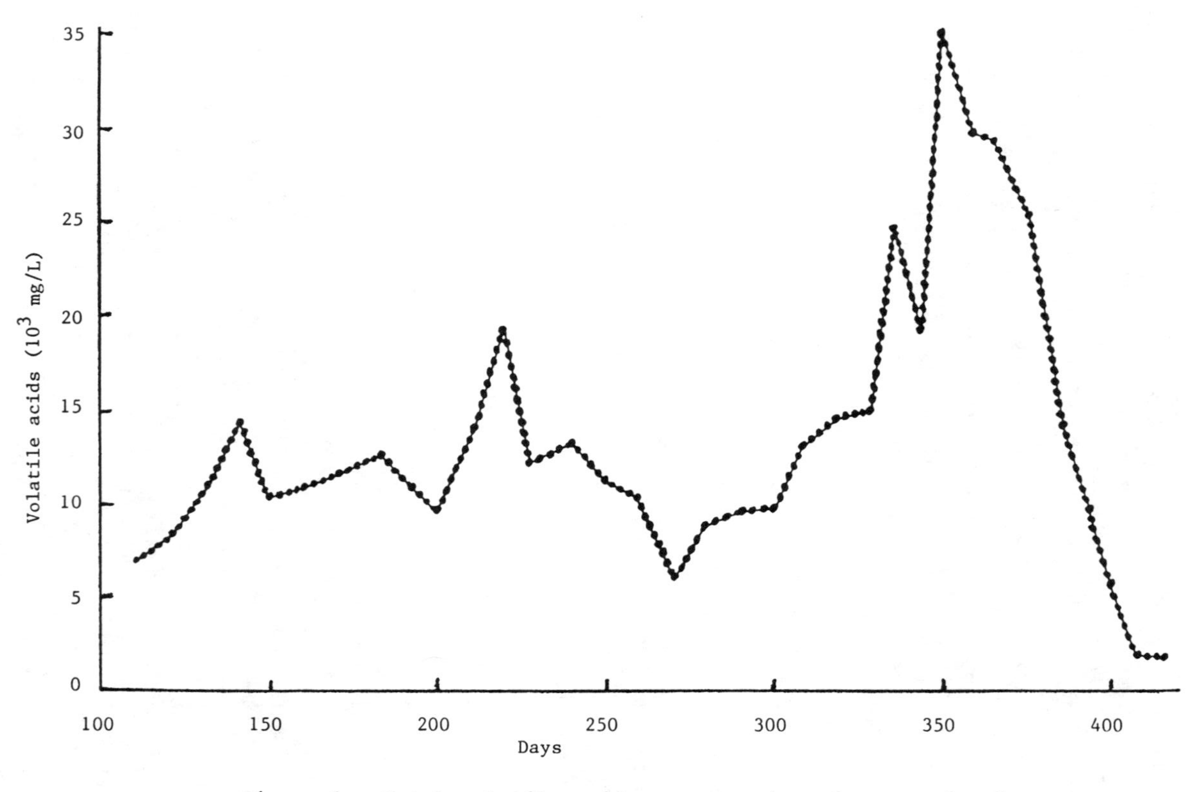

Figure 1. Total volatile acid concentrations in a completely stirred 5-L digester.

On day 200 the alkali pretreatment was discontinued and solids loading was adjusted to maintain the total solids between 6 and 8% (Figures 1, 3). The pH was only adjusted if it fell below 5.5 (Figure 4). Within two retention times after discontinuing the alkali pretreatment the volatile acid concentration reached 20,000 mg/L. This was followed by a rapid increase in pH to near 7.0 . The acid concentration then fell to near 12,000 mg/L and the gas production doubled. This rapid change in digester behavior is largely unexplained but may have been due to a decrease in the alkalinity after discontinuing the alkali pretreatment. After this event the digester reached a fairly steady state operation for the next 70 days during which the volatile acid concentration averaged near 10,000 mg/L. Peak production during this period suggested that higher concentrations and levels of acid production could be achieved under the proper operating conditions. One factor which may have been limiting during this period was nitrogen.

Beginning on day 300 (Figures 1,2) the feedstock was supplemented with proteose peptone. The amount of protein in the supplement was equivalent to 0.1g/L/day and the nitrogen was equivalent to 0.02 g/L/day. The retention time of the digester during this period was increased to 10 days. With the supplement added to the feedstock the volatile acid concentration began to increase from the beginning level of between 10,000 to 12,000 mg/L. After 5 retention times the concentration of volatile acids reached 30,000 mg/L. During this period the total solids in the digester increased from near 70 to 90 g/L (Figure 3).

At the beginning of this period (Figure 2) propionic acid accounted for nearly 10,000 mg/L of a total of 13,000 mg/L of volatile acids. As the supplement was added the proportion of acetic acid to propionic acid began to increase. On day 350 acetic acid accounted for nearly half of the total volatile acid of 30,000 mg/L and the amount of propionic decreased to about 5,000 mg/L. Butyric acid also increased and by day 350 the amount of butyric acid was near 7,000 mg/L.

The pH of the digester increased from an average of 5.8 to over 6.5 as the supplement was added (Figure 4). This may have been due to an increase in ammonia as a result of feeding the supplement.

Total volatile acids began to decline on day 350 and continued to decline until day 370 when feeding was discontinued and digester fluid was replaced with water to continue on a 10-day retention time. When feeding was stopped there was little further acid production.

Conclusions

This study has demonstrated that it is possible to operate an anaerobic digester with high levels of volatile acids, up to 30,000 mg/L. This was done with short retention times of between 8 to 10 days. The results of this work indicate that the rapid production of volatile acids probably results from an easily hydrolyzable portion of the simulated MSW feedstock.

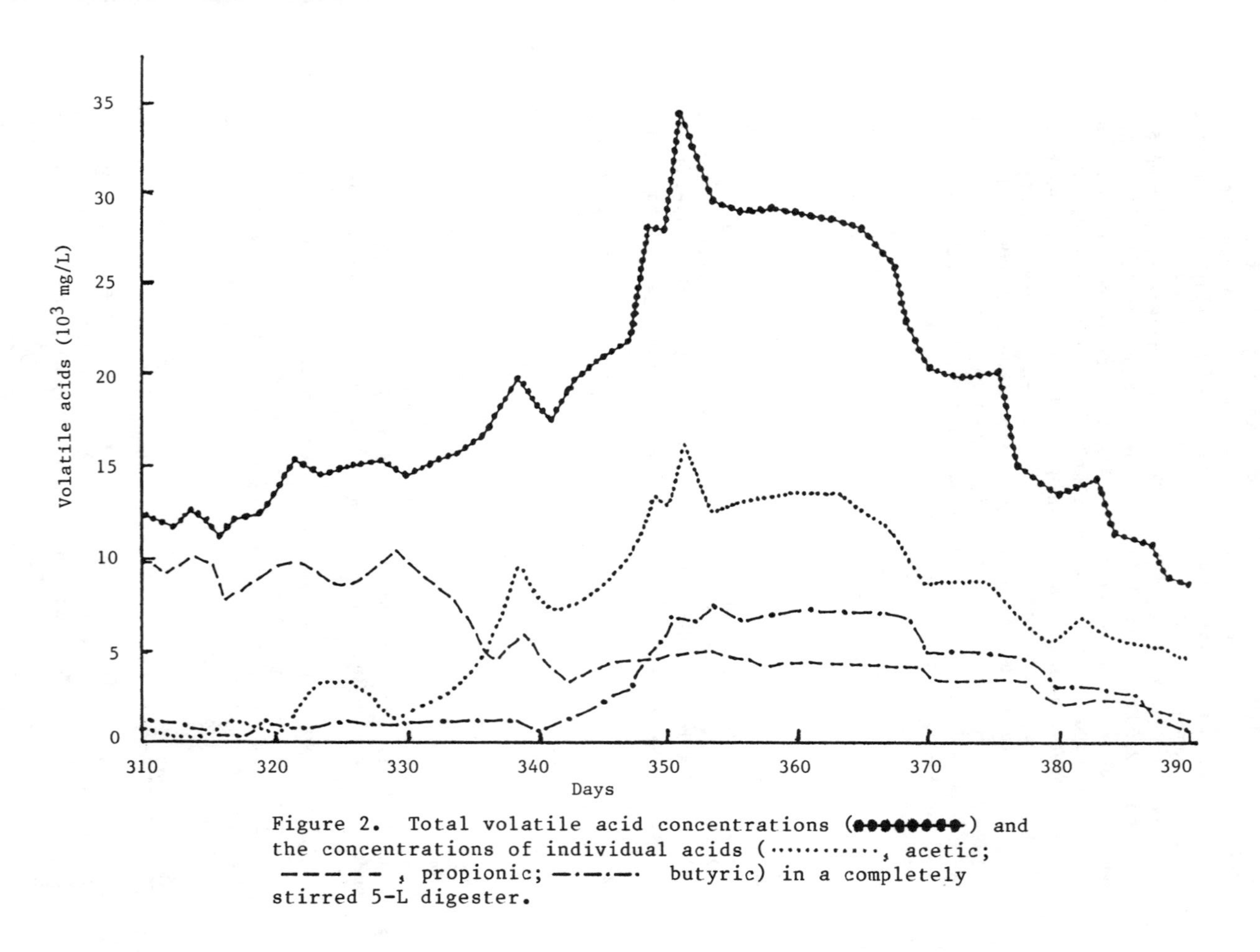

Figure 2. Total volatile acid concentrations (●●●●●●●●) and the concentrations of individual acids (··········, acetic; — — — — —, propionic; —·—·—· butyric) in a completely stirred 5-L digester.

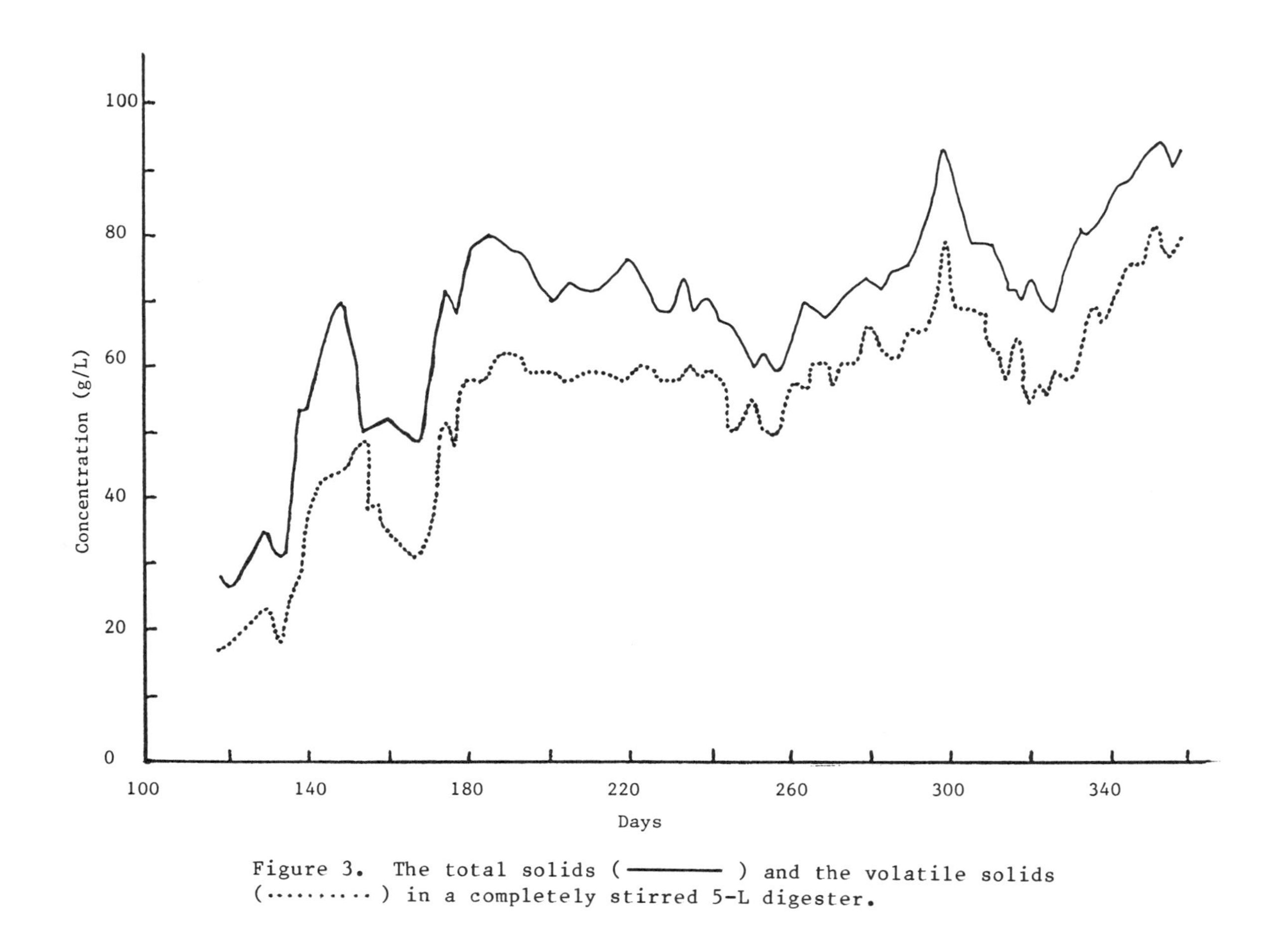

Figure 3. The total solids (——————) and the volatile solids (..........) in a completely stirred 5-L digester.

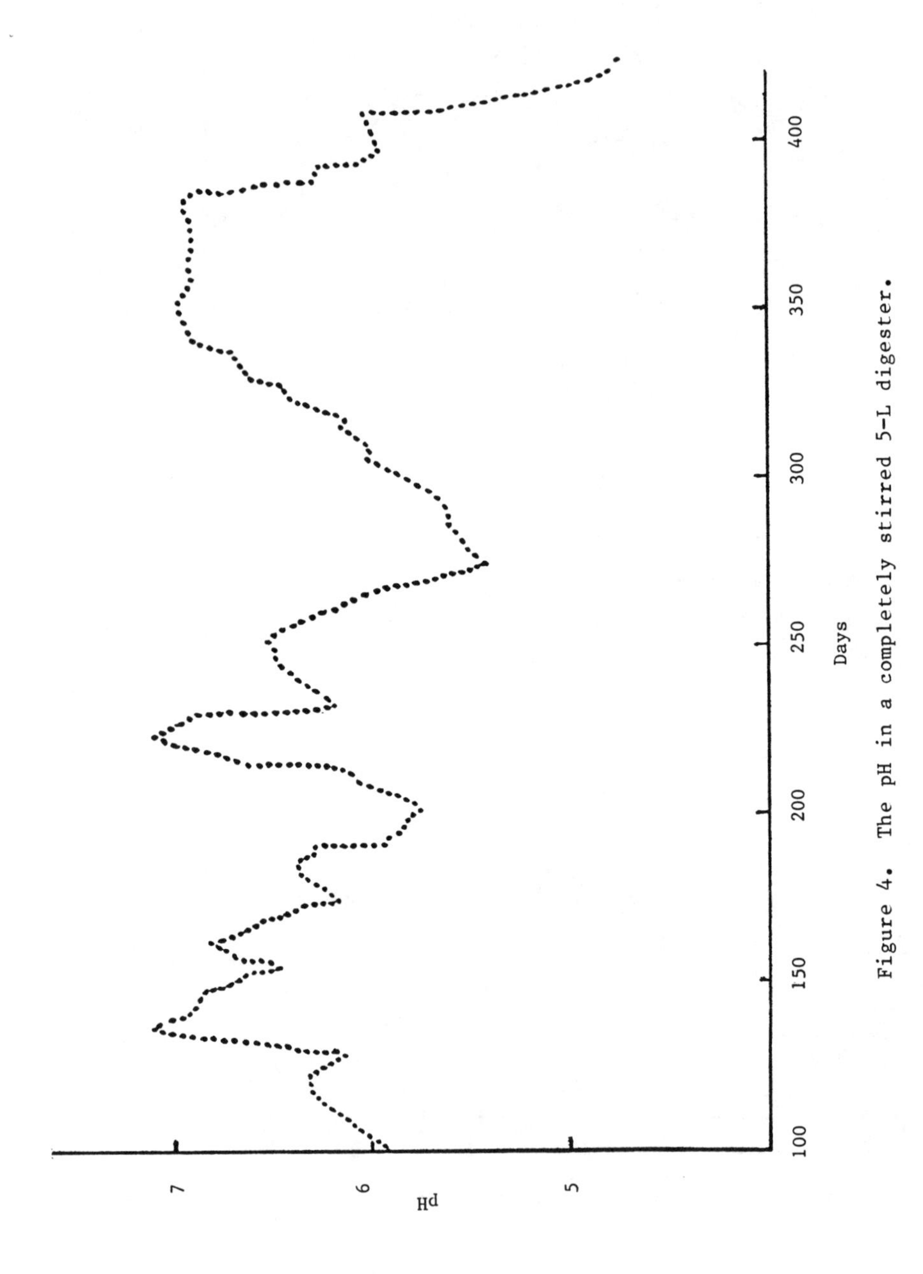

Figure 4. The pH in a completely stirred 5-L digester.

One major problem with operating an anaerobic digester at high loading rates and high volatile acid concentrations is the instability of the system. The anaerobic digester operating on MSW did not develop a stable pH and acid accumulation never did reach steady state operation. However, there were periods of time during operation which demonstrated high potential for rapid production of volatile acids.

Future work needs to be conducted to determine if stable operating conditions can be achieved in a digester operating on MSW feedstock.

Acknowledgments

This research has been supported by the U.S. Department of Energy, Assistant Secretary for Conservation and Renewable Energy, under Contract W-31-109-ENG-38.

Literature Cited

1. Barker, H.A. In Bacterial Fermentations; Barker, H. A., Ed.; John Wiley and Sons: New York, 1956; 1-95.
2. Zeikus, J. G. Bacteriol. Rev. 1977, 41, 514-41.
3. Bryant, M.P. J. Animal Sci. 1979, 48, 193-201.
4. Mah, A. R.; Ward, D. M.; Baresi, L; Glass, T. L. Ann Rev. Microbiol. 1977, 31, 309-41.
5. Wolfe, R. S. Microbiol Biochem. 1979, 21, 270-95.
6. Zeikus, J. G. Ann Rev. Microbiol. 1980,34, 423-64.
7. Wolfe, R. S. Experientia, 1982, 38, 198-201.
8. Hungate, R. E. The Rumen and Its Microbes; Academic Press, New York, 1966: 533.
9. Speece, R. E; McCarty, P. L. In Advances in Water Research; Eckenfelder, W. W., Ed.; Pergamon Press: New York, 1964; Vol. 2, 305-22.
10. Thauer, R. K; Jungermann, K.; Decker, K. Bacteriol. Rev. 1977, 41, 100-80.
11. Sanderson, J. E.; Garcia-Martines, D. V.; Dillon, J. J.; George, G. S.; Wise, D. L. Proc. 3rd Ann Biomass Energy Systems Conf., 1979.
12. Levy, P. F.; Sanderson,J. E.; Ashare, E.; de Riel, S. R.; Wise, D. L. Liquid Fuels Production from Biomass; Dynatech Final Report No. 22147; 1981, 153.
13. Levy, P. F.; Sanderson, J. E.; Kispert, R. G.; Wise, D. L. Enz. Microb. Technol. 1981, 3, 207-15.
14. Standard Methods for the Examination of Water and Wastewater, 14th Ed.; American Public Health Association: Washington, D. C. 1979.
15. Ghosh, S.; Henry, M. P.; Klass, D. L. Biotech. Bioeng. Symp. 10, 1980, 163-87.
16. Samson, R., LeDuy, A. Biotech. Bioeng. 1986, 28, 1014-23.

RECEIVED August 18, 1987

Chapter 7

Scale-Up of Bioseparations for Microbial and Biochemical Technology

Michael R. Ladisch [1] and Phillip C. Wankat [2]

[1]Laboratory of Renewable Resources Engineering and Department of Agricultural Engineering, Purdue University, West Lafayette, IN 47907
[2]School of Chemical Engineering, Purdue University, West Lafayette, IN 47907

Centrifugation, filtration, ultrafiltration, adsorption and chromatography are important unit operations for separations involving biological molecules, i.e., bioseparations. Examples of these unit operations are presented, together with appropriate equations and physical and biochemical properties data which aid development of preliminary scale-up specifications. In cases where the necessary data is not readily available, experimental approaches and appropriate empirical relations are discussed. Various examples are briefly presented and include: throughput of a disk stack centrifuge; broth clarification using a biological processing aid; microfiltration of yeast cells; product concentration using a radial flow cartridge; constant pattern adsorption; linear chromatography; and staging strategies for maximizing chromatography column throughput. This chapter has the objective of providing research scientists and engineers a first introduction to bioseparations scale-up.

The application of bioseparation processes to the recovery and purification of a target molecule (i.e., the product) start in the laboratory where product isolation at a specified purity and activity is the key criterion of success. Yield and cost are often secondary priorities. Once the product is isolated, the decision to scale-up its production may be quickly made with the development of the manufacturing technology to be based on experience gained from laboratory procedures. This chapter addresses the bioseparation techniques of centrifugation, filtration, membranes, adsorption, and chromatography from the point of view of a scientist or engineer trying to use data obtained at a laboratory scale for sizing individual unit operations and developing a preliminary process flowsheet.

0097-6156/88/0362-0072$08.25/0

Physical and biochemical property data needed for "first estimate" calculations are often not available from the literature. Consequently, advance planning with respect to laboratory scale experiments can assist the process designer by assuring that data for preliminary development of process concepts will be available when needed. Examples of key parameters are:

(1) the number of purification steps required to attain a given purity and activity as well as the product yield at the end of each step;

(2) settling and cake-forming characteristics of microbial cell mass:

(3) fouling and concentration polarization properties of clarified product broth with respect to membrane separations;

(4) adsorption behavior of the product (including width of the mass transfer zone) and product stability at the conditions of adsorption/desorption; and

(5) for chromatography, estimates of the Peclet number (which reflects dispersion) and stationary phase capacities for solutes present in the product stream.

The chapter presented here attempts to outline several commonly used separation techniques by providing a brief description of the respective unit operations and the corresponding basic equations. The intent is to introduce the reader to the types of questions which arise during early phases of scale-up, and thus, indicate the type of advance planning of laboratory experiments needed to facilitate measurements of properties which impact bioseparations.

Background: Separation Strategies

Fermentation products, and natural products derived from renewable resources, are usually produced in a dilute aqueous solution or suspension. Consequently, a separation scheme will usually have the steps of (*1,2*):

(i) removal of insolubles;

(ii) primary product isolation;

(iii) purification; and

(iv) final product isolation (polishing).

Processing steps (ii) to (iv) require a clarified feed for efficient purification and reasonable operational stability of the sorbents or membranes used in these steps. Consequently, the removal of insolubles is quite critical to the success of downstream processing steps.

The strategy of how various steps are combined impacts the degree of product recovery, as well as capital and operational costs. If 90% recovery ($\eta = 0.9$) is assumed for each step, the overall maximum recovery which could be anticipated is $(\eta)^m$ ($=(0.9)^m$) for m steps. A minimum number of processing steps is obviously desirable. For example, if purification of a component by liquid chromatography can replace an adsorption step and 2 out of 3 crystallization steps, the net product recovery will be increased by $(\eta)^{m_o-3}/(\eta)^{m_o} = (0.90)^{-3}$ $= 1.37$ or 37%. Although the cost of the chromatography step might be higher than the unit operations it replaces, the additional revenue resulting from increased product recovery could more than offset the added cost. This example illustrates the importance of a systems approach to chromatography scale-up in which changes in the types and number of purification steps are considered in addition to the increased size of individual unit operations during scale-up.

Removal of Insolubles

The treatment of a heterogeneous broth through filtration or centrifugation coincides with step (i). Centrifugation is often the unit operation of choice, although it has fairly high capital and maintenance costs (*2,3*). According to Hacking (*2*), a single centrifugal stage can contribute as much as $80/ton (wet) of bacteria processed, and several stages may be required for processing. Bacteria may have centrifugation costs which are 4X higher than yeasts. In this case, a flotation or flocculation step prior to centrifugation may be needed for bacteria. Some cultures are readily handled using rotary filters although throughputs vary (*2,3*). For example, *Penicillus chrysogenum* has a design filtration rate of 35 to 45 gal $ft^{-2}\ h^{-1}$ as compared to 2.1 for *Streptomyces kanamyceticus.* The throughput of centrifugation and filtration steps are a function of the physical properties of the complex broths being treated and these are often not well defined with respect to physical properties. Hence, scale-up requires tests on large scale equipment before a process can be specified. A basic equation for centrifugal sedimentation is (*4*):

$$q_c = \frac{\pi b w^2 (\rho_p - \rho) D_{pc}^2}{18\mu} \frac{r_2^2 - r_1^2}{\ln (2r_2/(r_1 + r_2))} \quad (1)$$

where q_c is the volumetric flow rate corresponding to the cut diameter; D_{pc} is the particle diameter which will be eliminated by the centrifuge; b is the width (height) of centrifuge basket; ω is the angular velocity; ρ and ρ_p are densities of fluid and particle, respectively; r_1 and r_2 are the inner and outer radii, respectively, of material in the centrifuge; and μ is viscosity. The cut point is defined by the diameter of the particle which just reaches ½ the distance between r_1 and r_2. If the thickness of the liquid layer, relative to the bowl radius, is small (i.e., $r_1 \cong r_2$), the volumetric flowrate may be estimated by:

$$q_c = \frac{2\pi b\,\omega^2(\rho_p - \rho)D_{pc}^2 r_2^2}{9\mu} \tag{2}$$

It may be possible to generate a first estimate of effective centrifuge diameter, r_2, and speed, ω, for a given throughput, q_c, using viscosity data for fermentation broths available in a handbook (cf., Chapter 8 in reference 3), together with general ranges of particle size and density for a given class of microorganism. The result can be compared with the mechanical stress limits of commercial and bench scale machines (*5,6*). Small diameter, tubular bowl, centrifuges such as the Sharples 1P (4.5 cm diameter) and 6P (10.8 cm diameter) attain up to 62,500 x g and 14,000 x g at speeds of 50,000 rpm and 15,500 rpm, respectively (*6*). In comparison, a 76 cm diameter machine has a limit of about 7500 x g. Basket conveyor machines are 30 to 274 cm in diameter with limits of about 2,000 to 260 x g, respectively. Equations 1 and 2 indicate why a large density difference and a large particle size combined with a low viscosity favor efficient separation.

Disk stack centrifuges are able to attain up to 15,000 x g in production scale equipment with throughputs of several thousand liter/hour of cell containing broth (*7*). For example, recovery of *E. coli* cells of 0.5 to 0.8 micron size from a 6% to 7% (by volume) feed is said to be possible at a throughput of 3,000 to 4,000 liter/hour to give a clarified product of 0.02% solids/volume and a 70% to 80% solids/volume solids stream. Processing of *E. coli* cell debris at 4% to 5% (by volume) at 1,500 to 2,000 liter/hour throughput gives separated solids at 40% to 50%/volume. Scale-up factors which need to be considered include heat lability of the product, sterilization, and containment requirements. The mechanical work in the centrifuge can cause temperature rises of 3 to 20°C in the liquid and solid discharges, thus necessitating cooling. The machine and its mechanical seals must be designed to allow for sterilization prior to disassembly for maintenance. Containment, when needed, necessitates design of the centrifuge system to prevent contamination of the surrounding atmosphere with materials being processed by the centrifuge. These and other factors are reviewed by Erikson of Alfa-Laval, Inc., and illustrate some of the challenges faced when proceeding from the bench to the production scale (*7*).

Another approach proposed for clarifying fermentation broths is the use of biological processing aids (abbreviated as BPA). These have been recently introduced by Supelco (Bellefonte, PA) and are also referred to as submicron-sized polymeric particles (SSPP). In this approach, 0.1 to 1 micron size polymeric particles are added to a broth containing cells or cell debris. Since many types of cells have a negatively charged surface, quaternary amine type SSPP (i.e., BPA) can be used to adsorb cells and form flocs at pH 7.2 (*8*). It is reported that *E. coli* cells, when adsorbed on the quaternary amine form of styrene-divinylbenzene particles sediment under gravity in less than 15 minutes. *Saccharomyces cerevisiae* cells have a settling time of 5 minutes when adsorbed on the BPA, as compared to 24 hour in the absence of the BPA (*8*). Desorption is accomplished by changing the pH or ionic strength of the bulk solution. This approach, in combination with centrifugation or filtration can enhance throughputs for removal of insoluble materials. The use of an SSPP or BPA must be optimized for each type of separation. An excess of SSPP retards flocculation and sedimentation by stabilizing the suspension of cells or cell debris. For *E. coli* the desired concentration of SSPP was found to be 0.2 to 0.6 g/g dry cell weight, while for *S. cerevisiae* this was 0.02 to 0.2 g. The SSPP is also capable of adsorbing proteins. Consequently, careful control of solution pH and ionic strength is probably required if cells are to be removed from a broth containing an extracellular protein product.

Filtration techniques include: non-cake forming, cross flow filtration; or cake forming methods such as constant pressure, constant rate, and centrifugal filtration. The principles of cake-forming filtration are discussed by McCabe and Smith (*4*) who give generalized correlations. Since biomass materials are usually compressible solids, the cake resistance, α, is a function of pressure. In a constant pressure filtration, the time, t, required to collect the total volume of filtrate, V, at a constant pressure drop, Δp, is given by (*4*):

$$t = \frac{\mu}{g_c(-\Delta p)} \left[\frac{c\,\alpha}{2} \left(\frac{V}{A}\right)^2 + R_m \frac{V}{A} \right] \qquad (3)$$

where μ is the viscosity of the filtrate; g_c is Newton's-law conversion factor; c is the mass deposited on the filter per unit volume of filtrate; α is specific cake resistance; A is filter area; and R_m is filter medium resistance. Estimates of R_m and α for the fermentation broth to be filtered is obtained by measuring V and t at selected pressure drops, and then fitting the data to Equation 4:

$$\frac{dt}{dV} = \frac{c\,\alpha\mu}{A^2(-\Delta p)g_c} V + \frac{R_m \mu}{A(-\Delta p)g_c} \qquad (4)$$

to obtain values of α and R_m. An empirical equation for α as a function of Δp can then be obtained. These values of α can also be used in calculations for continuous rotary drum filtration where R_m is usually small, or in continuous rate filtration. The necessary equations and procedures are clearly presented in standard chemical engineering texts (*4,9*). What appears needed, at this point, are general tables and correlations which give values of α and μ for a number of "typical" fermentation broths to aid the process engineer in devising preliminary process flowsheets.

Microbial cells can also be concentrated for purposes of cell recovery using microfiltration membranes having pores with sizes of 0.2 microns or lower. Such membranes can either be inorganic (*10*) or organic (*11*) based. A comparison of a pleated sheet microfilter based on an acrylic polymer, a tubular polypropylene microfilter, and a hollow-fiber, polysulfane ultrafilter with respect to yeast cells showed the tubular and hollow-fiber units to be best suited for this use. These membranes, having 0.2 and 0.007 micron pore sizes, respectively, gave quantitative cell recoveries while the pleated sheet microfilter (0.2 micron pore size) did not. Flux of the tubular microfilter was about 100 liters$\bullet m^{-2}\bullet h^{-1}$ at yeast cell concentrations of 100 g/liter while the hollow fiber had a flux of 40 at a cell concentration of 250 g/liter (*11*). These authors (*11*) conclude that almost any microfilter will do if the goal is to simply clarify a spent fermentation medium. If cell recovery is the goal, straight through module designs with unobstructed channels and small pore asymmetric membranes are recommended. Since tubular- and hollow-fiber modules are available for large-scale applications, this type of separation has potential for scale-up and commercial use.

Centrifugation and filtration may be carried out in a sequential manner, according to Erikson (*7*). Thus, for separation of bacterial amylases from the microbial cells, centrifugation can reduce cell concentration from ca. 10^7 cells/ml to ca. 10^4 cells/ml in the clarified product. This reduction reduces the load on a subsequent sterile filtration so that the filter elements can treat up to 50,000 liters of product before being replaced as compared to 4,000 liters if centrifugation is not used. Similarly, combinations of centrifugation with micro-filtration can be envisioned for purposes of process optimization.

Product Concentration

A direct way of concentrating the desired product in a clarified broth is to use a membrane which allows water and molecules smaller than the product to pass, while rejecting the product itself. A good overview of this topic is given by Michaels (*12*). Key practical problems often mentioned are: (1) concentration polarization in which buildup of the concentration of the rejected solute against the membrane causes a decrease in flux; and (2) fouling in which components adsorb on the membrane and decrease effective membrane porosity, and thereby, the flux. Research to overcome some of these limitations include: operation of the membrane in a cross-flow configuration to reduce the depth of the boundary layer of concentrated solute next to the membrane; and development of new membrane chemistry which resists adsorption of components which can foul the membrane. Despite these problems, membranes are in use in industry, not only for water treatment but also for processing of pharmaceuticals. Examples include cell concentration of *E. coli* cells to 10^{14} cells/L; interferon recovery; and pyrogen removal (*13*).

Major practical operational features have been summarized for the Iopor (Dorr-Oliver), ultrafiltration system (*13*). Transmembrane pressure affects the flux according to:

$$J = \kappa(\Delta P)^M \tag{5}$$

where J is the flux; κ is a proportionality constant; and ΔP is the pressure drop across the membrane and M is the slope of the line of flux data plotted as a function of ΔP. A concentration gradient is established between the membrane surface and the bulk solution as the permeate is pushed through the membrane, and solvent is removed from the rejected species at the membrane surface. Mass transfer through this boundary layer limits flux. If a high enough crossflow velocity can be established, the formation of the boundary layer is limited and the operational flux J is proportional to the expression:

$$J \; \alpha \; \frac{(v)^b (D)^d}{(\nu)^a} \log (C_w / C_b) \tag{6}$$

where v is linear velocity across (parallel to) the membrane; D is diffusivity; ν is kinematic viscosity; C_w is concentration of rejected component at the membrane surface; and C_b concentration of rejected component in bulk solution. The benefit of increased flux attained at high crossflow velocities may be offset by increased pumping costs (*13*). An alternate configuration encompasses membranes in a hollow fibers of 10 to 20 micron diameter, assembled together in a tubular bundle. In this case, high flowrates are also attainable with the same effect as described above. Examples of such systems are Amicon, laboratory scale, hollow fibers, and Rhomicon hollow fiber systems for process scale separations.

When the product molecule is found in a mixture containing both lower and higher molecular weight components, or if a membrane approach is otherwise unsuitable, sorption offers another method by which the product may be concentrated. An example of concentration of a recombinant protein using a composite ion exchange media is given by MacGregor et al. (*14*). In this case, a proprietary recombinant protein was sorbed on DEAE cellulose in a Zeta-Prep cartridge (Cuno, Meridien, CT) from a clarified,"crystal clear" feed. The cartridge consists of a spiral wound ion exchange paper. Flow is radial with liquid passing from the outside to the inside of the cartridge. Scale-up of 240-fold was achieved by a ratio approach (see Table I), in which length is added to a standard diameter cartridge and/or multiple cartridges are assembled in parallel to increase volumetric throughput. If the cartridge radius stays the same, flux and compression of the media should also be the same since increases in throughput are attained while keeping the linear fluid velocity (through the cartridge radius) constant.

Table I
Example of Scale by Ratio for
Concentration of a Recombinant Protein

Parameter	Laboratory Scale	Pilot Scale	Scale-up Factor
Volume of Ion Exchange Media (liter)	0.08	19.2	240x
Adsorption Step			
Influent Protein Conc. (g/liter)	0.15	0.15	NA^{+}
Effluent Protein Conc. (g/liter)	NG^{++}	NG^{++}	NA^{+}
Influent volume (liter/cycle)	10	NG^{++}	NA^{+}
Loading (mg/ml)*	12.5	12.5	NA^{+}
Flowrate (liter/min)	<0.05	6 to 8	>160X
Desorption Step			
Influent Salt Conc. (M)	0.3	0.3	NA^{+}
Effluent Protein Conc. (g/liter)	3.2	1.5	NA^{+}
Effluent volume (liter/cycle)	0.3	75.	250X
Recovery of Activity (%)	85-100	85-100	NA^{+}
Specific Activity Increase	4 to 8X	4 to 8X	NA^{+}

NG^{++} denotes not given; NA^{+} denotes not applicable;
* as mg recombinant protein per ml ion exchange media.

The Zeta-Prep unit is an example of a unit operation in which the equilibrium step during the loading step is favorable so that the sorbed solute present in the feed solution goes *on* the matrix and does not desorb. Elution requires a change in conditions (pH, ionic strength, or polarity) so that the solute comes *off*. *On-off* chromatography consists of: load; wash to displace the feed solution from the column void volume; elute solutes one-by-one; wash and then repeat. This cycle is typical for ion exchange, affinity chromatography, and adsorption (*15,16*). During the loading step a constant pattern profile of solute breakthrough should result. This profile should not increase in width for longer columns. Equations for constant pattern systems for fixed beds of cylindrical geometry are based on the concept of a mass transfer zone (*15-17*). The width of the pattern over which the solute concentration changes inside the bed is called the length of the mass transfer zone, L_{MTZ}.

A lumped parameter mass transfer expression is given by the expression (*16*):

$$\rho_B(1-\epsilon)\frac{\partial q_i}{\partial t} = -k_M a_p (c_i^* - c_i) \tag{7}$$

The length of the mass transfer zone for a single solute is given by Equation 8:

$$L_{MTZ} = \frac{u_{sh}\rho_B(1-\epsilon)}{k_M a_p}\frac{q_{feed}}{c_{feed}} q_s \tag{8}$$

where q_s is a function of the feed concentration and equilibrium expression. The well known Langmuir and Freundlich isotherm equations given in Table II, can be used. It should be noted that the sorbent weight, as defined here, includes the weight of fluid inside the sorbent's pores.

The pattern or shock wave velocity u_{sh} is the velocity at which the stoichiometric center moves in a column having a void fraction ϵ:

$$u_{sh} = \frac{v}{1 + \frac{1-\epsilon}{\epsilon}\rho_B\frac{\Delta q}{\Delta c}} \tag{9}$$

For large molecules such as proteins pore diffusion will usually control. The mass transfer coefficient, k_m, together with the interfacial area, a_p, can then be estimated by (*17*):

$$k_M a_p = \frac{60 D_p\ m\rho_B}{d_p^2} \tag{10}$$

where m $= \Delta q / \Delta c$, D_p is the diffusivity in the pores, and d_p is the particle diameter.

Table II.
Typical Adsorption Equilibria Equations

Type	Equation*	Comment
Langmuir	$q_s = \dfrac{k'cQ}{1 + k'c}$	Single Component Equation
	$q_s = \dfrac{k' k_1 c_1}{1 + \sum_{i=1}^{m} k_i c_i}$	Multicomponent Equation
Linear	$q_s = k' cQ$	k' c<1; Commonly used isotherm in chromatography. Applicable to dilute solutions.
Freundlich	$q_s = kc^{1/n}$	Empirical. For fitting a non-linear isotherm.

*Nomenclature:

Q,k',k,n	empirical constants determined for experimental data
c	concentration of solute in mobile phase
q	concentration of solute in solid phase

For *symmetric* constant patterns the fraction of the bed used, x, is given by:

$$x = 1 - 0.5\,(L_{MTZ}/L) \tag{11}$$

As L_{MTZ}/L decreases, a sharper breakthrough profile results and the fractional bed use increases. For L_{MTZ}/L of less than 0.33, the improvement of bed use on increasing the bed length or decreasing L_{MTZ} becomes marginal. If L_{MTZ} is small, beds with a relatively short path length can be used. This includes the radial flow geometry described above.

In a radial flow system, v is a function of radial distance:

$$v = \frac{r_{ref}\, v_o}{r} \tag{12}$$

where v_o is the eluent interstitial, velocity measured at the radius r_{ref}. In this case, Equation 9 becomes:

$$u_{sh} = \frac{r_{ref}\, v_o / r}{1 + \dfrac{1-\epsilon}{\epsilon}\, \rho_B\, \dfrac{\Delta q}{\Delta c}} \tag{13}$$

If flow proceeds from the outside to the inside of the radial flow cartridge, $r_{ref} = r_o$ where r_o is the inside radius. If the flow is in the opposite direction r_{ref} = R, where R is the outside radius.

For a given separation with pore diffusion control, L_{MTZ} is proportional to d_p^2 as indicated by combining Equations 8 to 10:

$$L_{MTZ} \alpha v d_p^2 \quad (14)$$

The mass transfer zone is measured by passing the entire mass transfer zone through the column and measuring the time required for the zone to exit the column. This time, t_{MTZ}, is then related to the length of the mass transfer zone in the column by

$$L_{MTZ} = t_{MTZ} u_{sh} \quad (15)$$

Decreasing effective particle diameter can decrease L_{MTZ} and increase the fractional bed use although pressure drop eventually limits how small of a particle size can be used.

Column Staging. An alternative to increasing the column length or decreasing the particle diameter when L_{MTZ} is large, is to use beds in series as illustrated in Figure 1. In this example, the feed first enters column A which is already partially loaded. For a long mass transfer zone, breakthrough occurs long before column A is saturated. Thus, the effluent from column A is sent to column B which is initially clean. Once column A is saturated, it is washed with a bed volume of wash water or solvent to remove feed solution from the interstitial void volume. Then the system is switched to part 2 of the cycle where column A is eluted and then washed. Column B becomes the lead column in the loading step while clean column C is the trailing column. When column B is saturated, the columns are switched to part 3.

If $L > L_{MTZ}$, two columns are sufficient for the loading step. Otherwise more than two columns are required for this step. If elution of the solutes in a series of steps or with a gradient takes longer than one part of the loading step, a four column system with two columns being eluted can be used. Concepts involving other combinations with more than four columns can also be envisioned.

The beds-in-series system has several advantages. Complete saturation of the loading column can be obtained with $L = L_{MTZ}$ in each of the two columns. For example, if a single column with $L = 2L_{MTZ}$ were used, Equation 11 shows that the fractional bed use would be only 75%. Thus less adsorbent is required when the two beds are in series. The beds in series system is also useful for maintaining smooth operation when the sorbent is changed over. When the capacity of a column drops below an economical level, the column is removed from service. While the exhausted column is dumped and repacked, the remaining columns continue to do the desired separation.

Purification

Purification encompasses many different areas. In the context of this chapter, purification will refer to fractionation of a soluble product from other soluble components, after insoluble materials have been removed and the product has been concentrated and partially purified. Purification processes include adsorption and chromatography, as well as affinity chromatography.

Reviews of theoretical considerations (*18*), as well as the chemical and biochemical basis of affinity chromatography (*19*), have recently been published. Affinity chromatography is used in the biotechnology industry for purification of high value proteins and polypeptides, although specific details are yet to be published.

Sorption is a common technique used in protein purification. Commercially available ion exchange resins are potentially applicable for purification of proteins and polypeptides such as lysozyme, cytochrome C, hemoglobin, asparaginase, pepsinogen, and pepsin. A volatile, aqueous acid (acetic acid), base (ammonia) or buffer (pyridine-acetate) for elution the proteins is suggested. In this way a purified component could be obtained with volatile eluents which could be removed by direct lypholization (*20*).

The adsorption cycle will result in a breakthrough curve. The equation for this case, based on local linear equilibrium with significant longitudinal dispersion, is (*21,22*):

$$D \frac{\partial^2 x_A}{\partial z^2} = v \frac{\partial x_A}{\partial z} + \frac{\partial x_A}{\partial t} \left(1 + \left(\frac{1-\epsilon}{\epsilon}\right) \frac{C_{ASO}}{C_A} \right) \tag{16}$$

with the boundary conditions of:

$$at\ z = O,\ x_A = 1 \text{ for } t > O \tag{17}$$

$$at\ z = \infty,\ x_A = 0 \text{ for } t > O \tag{18}$$

and the initial condition of

$$at\ t = O,\ x_A = O \text{ for } z > O \tag{19}$$

The solution for Equations 16 to 19 is (*21,22*):

$$x_A = \frac{C_A}{C_{AO}} = \frac{1}{2} \left(1 + erf\ [Pe_z^{1/2} \frac{(V - \bar{V})}{2(V\bar{V})^{1/2}}] \right) \tag{20}$$

where the Peclet number Pe_z, for a column of length, L, at a superficial velocity, v is:

$$Pe_z = \frac{Lv}{D_{am} + E_A} = \frac{Lv}{D} \quad (21)$$

The axial dispersion coefficient, D, represents the sum of diffusivity, D_{am}, and eddy diffusivity, E_A. The volume of solution, $\bar{V}$, required to saturate a column of length z in the absence of dispersion and assuming linear equilibrium applies, is:

$$\bar{V} = Sz\ [\epsilon + (1-\epsilon)(\frac{C_{ASO}}{C_{AO}})]$$

$$= V_o + KV_s \quad (22)$$

where Sz is column volume for a column of length z, and cross-sectional area, S; V_o = Szϵ is the column void volume (based on the volume between particles); K = C_{ASO}/C_{AO} is the distribution coefficient for concentration of solute A in the stationary phase in equilibrium with that in the feed solution (i.e., linear equilibrium assumption); and $V_s = Sz\ (1-\epsilon)$ is the volume of the stationary phase.

The utility of Equation 20 is that it reflects a "best-case" analysis for adsorption breakthrough for a given system where a single component is adsorbed, and linear equilibrium applies. Hence, a breakthrough curve which results from an adsorption run will allow determination of $\bar{V}$ (which corresponds to the inflection point) and D (by parameter estimation of curve fitting to obtain Pe_z). Subsequent calculations can then use values of these parameters to calculate the elution profile as a function column length, L, and superficial velocity, v.

In its most simple form, chromatography has the advantage of fractionating multicomponent mixtures into pure components using a single eluent. Unlike an adsorption system, an eluent of a different composition is not required to regenerate the stationary phase between injections of the sample to be purified. In comparison, gradient chromatography entails desorption of solutes from the stationary phase by changing pH, ionic strength, and/or hydrophobic character of the eluent during a chromatographic separation. An example for cellulase enzyme purification is given in Figure 2 (*23*). After the last component elutes, the column is re-equilibrated using the starting buffer before the next separation is carried out. In this respect, gradient chromatography resembles an adsorption system, since a change in eluent condition is required to prepare the stationary phase for the next run. Equations for gradient chromatography have been recently described (*24*).

A major disadvantage of chromatography is that the component being purified will be diluted as it passes through the chromatography

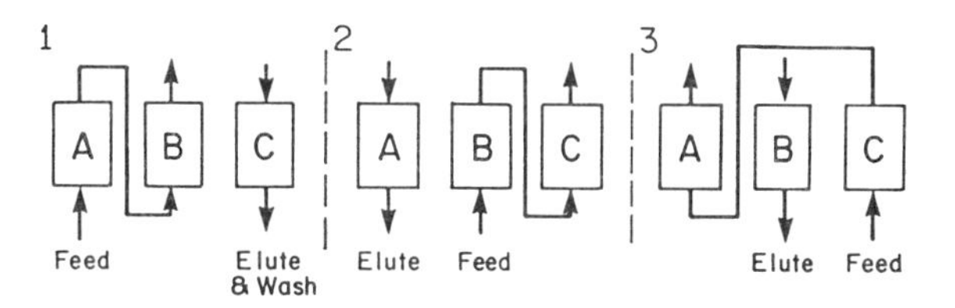

Figure 1. Schematic representation of multiple column system.

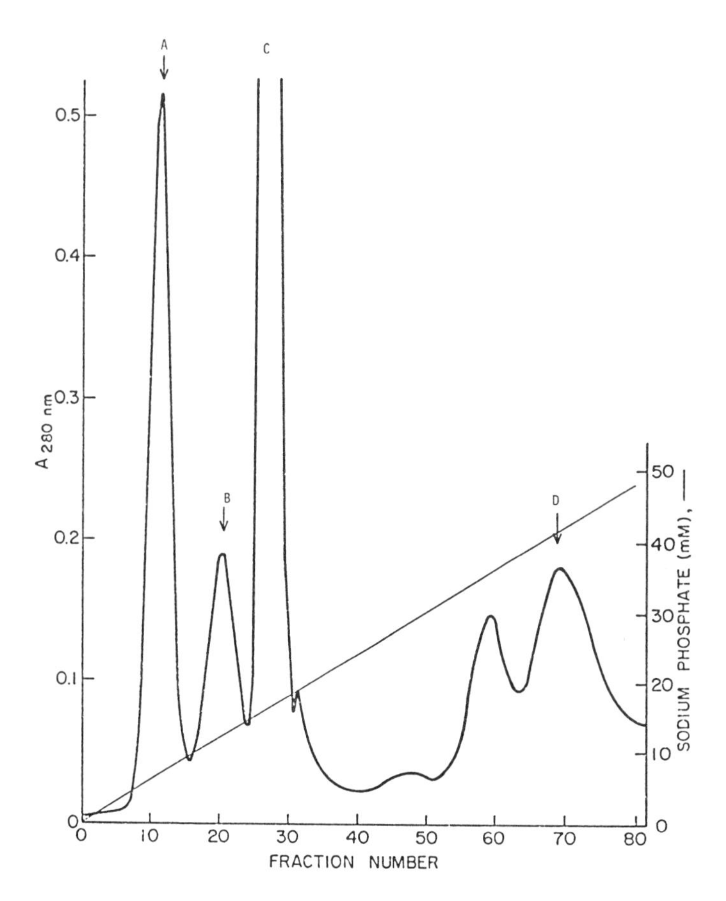

Figure 2. DEAE Sepharose chromatography of a high molecular weight endoglucanase. A, B, C, and D, denote protein peaks having enzyme activity (Reproduced from ref. 21. Copyright 1952 American Chemical Society).

system. An example is given by ion exclusion chromatography for the separation of glucose from sulfuric acid (*25*). In this case the electrolyte (H_2SO_4) is excluded from the resin phase of a strong cation exchanger (H+ form) according to the Donnan equilibrium principle. The electrolyte travels through the void space at close to the interstitial velocity of the eluent (distilled water), while the non-electrolyte glucose distributes itself between the fluid and resin phases. The presence of acid has little impact on glucose distribution in this case. Hence, the acid and glucose are treated as non-interfering components. Figure 3 illustrates concentration profiles which result for a column temperature of 55° C for IR-118H cation exchange resin (Rohm and Haas Co.) packed in a 2.54 cm i.d. by 61 cm long column. Dilution effects are obvious, with an increase in sample size resulting in decreased dilution. In this system, glucose exhibits linear equilibria with respect to the stationary phase (*25*). If local equilibrium and negligible end effects are assumed, the ratio of the outlet concentration, C_A, to initial feed concentration, C_{AO}, as a function of eluent volume can be estimated if linear equilibrium applies. The principle of superposition, applied to the analysis of such a linear column process, defines the concentration profile for an adsorption process as (*22*):

$$x_A = C_A / C_{AO} = f(z, V) \quad (23)$$

For a feed pulse of volume, V^o, followed by eluent at the same, constant flow rate (i.e., isocratic conditions), the effluent profile (i.e., chromatogram of a single component) is given by:

$$x_A^o = f(L, V) - f(L, V - V^o) \quad (24)$$

As the sample volume V^o approaches zero, taking the limit as V^o --> 0 of Equation 24 gives:

$$x_A^o = V^o \left(\frac{\partial f(L,V)}{\partial V} \right) \quad (25)$$

The equation for an elution profile with a large injection volume, V^o (for example in Figure 3 where $V^o > 0.1\ V_o$), follows from Equations 20 and 24 to give:

$$\frac{C_A}{C_{AO}} = \frac{1}{2}\left[1 + erf\ \frac{Pe_z^{1/2}(V - \bar{V})}{2(V\bar{V})^{1/2}} \right]$$

$$- \frac{1}{2}\left[1 + erf\ \left(\frac{Pe_z^{1/2}(V - V^o - \bar{V})}{2[(V - V^o)\bar{V}]^{1/2}} \right) \right] \quad (26)$$

The fit of this equation to several glucose elution profiles is illustrated

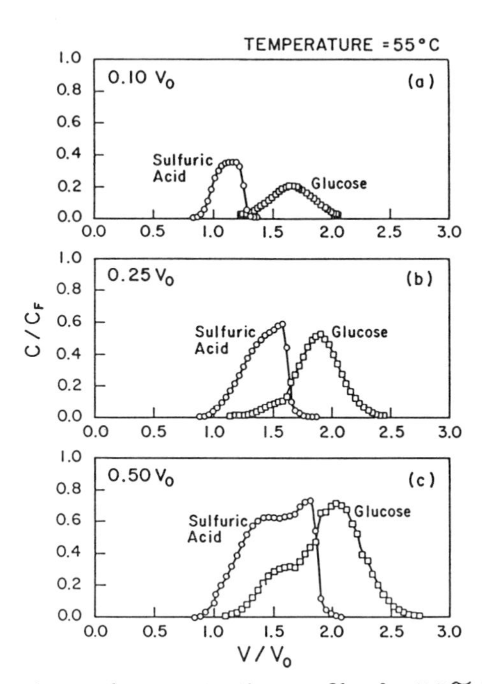

Figure 3. Experimental concentration profiles for 7.7% H_2SO_4 and 1.0% glucose, at 55° C for separation over IR118H with water as eluent (Reproduced with permission from ref. 25. Copyright 1987 Elsevier)

in Figure 4 (*25*). For an analytical scale column where V^o is small ($<$ 0.01 V_o), Equations 20 and 25 give:

$$x_A^o = \frac{1}{2}\left(\frac{V^o}{\bar{V}}\right)\left(\frac{Pe_z}{\pi}\right)^{1/2}\exp[-Pe_z(V-\bar{V})^2/4V\bar{V}] \tag{27}$$

Equations 20, 26, and 27 are of practical use in size exclusion chromatography of various proteins, since linear equilibria and local equilibrium conditions apply. Sephadex G-50, which is used in processing insulin, and other types of dextran and agarose gels (*26*) are examples of chromatographic supports for which we believe these equations should apply. It should be noted that one characteristic of this type of system is that the maximum volume in which a sample elutes will be less than the volume of the empty column in which the stationary phase is packed (*27*).

The utility of Equations 20, 26, and 27 lies in their ability to predict elution profiles as a function of eluent velocity, column length, and sample volume. It should be noted that resolution between two components is not given by these equations, which treat components of a mixture on an individual basis, i.e., it is assumed that the presence of one component does not affect the adsorption and dispersion of the other. The other major assumption of Equations 20, 26, and 27 is that dispersive tendencies which cause peak broadening are primarily due to diffusional effects in the mobile phase. In fact, for a given particle size the perceived peak broadening (i.e., dispersion) can be attributed to other factors including adsorption/desorption kinetics, diffusion effects in the stationary phase, and interchannel effects as well as longitudinal (mobile phase) diffusion and dispersion. These effects are described elsewhere based on a random walk model (*28*), and lead to the definition of plate height, plate count, and resolution which is usually associated with liquid chromatography.

The definition of plate height, H, is difficult to describe in terms of a physical concept. We have chosen to simply define it as the measure of peak broadening (variance) which occurs as a pulse of feed travels a distance L through a liquid chromatography column. Consequently, for a Gaussian peak (Figure 5) resulting from passage of a feed pulse through a uniformly packed column, the plate height H, is:

$$H = \sigma^2/L \tag{28}$$

If a column is divided into a number of theoretical plates, the general approach of Glueckauf as described by Giddings (*28*) gives a material balance for a component traveling through a distance equivalent to H. The result of the diffusional treatment is the same as the plate height approach, i.e.:

$$\sigma = \sqrt{HL} \tag{29}$$

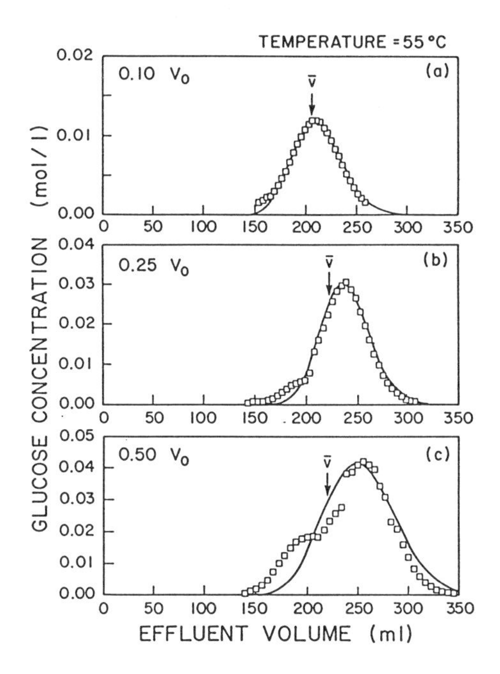

Figure 4. Curve fits obtained using Equation 26 with values of Pe_z of (a) 138; (b) 200; and (c) 109. Sample sizes as indicated. Elution at 55° C (Reproduced with permission from ref. 25. Copyright 1987 Elsevier)

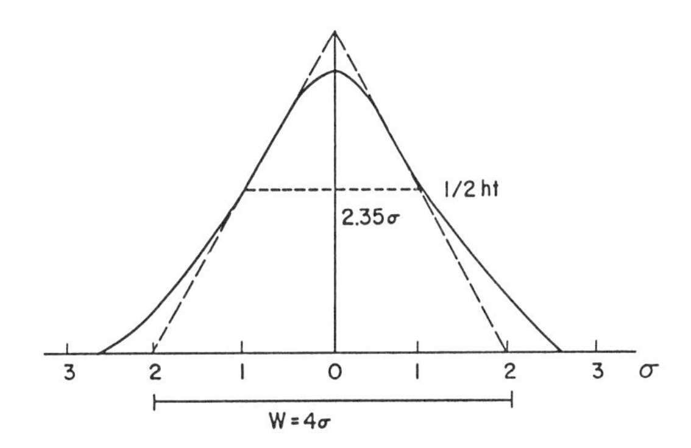

Figure 5. Schematic representation of a Gaussian peak. (Reproduced with permission from ref. 30. Copyright 1984 J. Wiley and Sons)

Here L denotes the distance migrated by the center of the solute peak. In a one-dimensional random walk model a solute's movement is treated in terms of a molecule moving backward and forward in discrete steps, as determined by chance. For a large number of molecules starting together, the resulting standard deviation, σ, is given by:

$$\sigma = \ell \sqrt{n} \tag{30}$$

for a step length, ℓ, and number of steps, n. For two sets of random walks in which each molecule takes n_1, and n_2 steps, respectively, Giddings (*28*) shows that

$$\sigma^2 = \sigma_1^2 + \sigma_2^2$$

For any number, j, of simultaneous random processes, this is:

$$\sigma^2 = \sigma_1^2 + \sigma_2^2 + \sigma_3^2 \cdots \sigma_j^2 \tag{31}$$

Einstein's equation relates this random walk to diffusional processes:

$$\sigma^2 = 2Dt_D \tag{32}$$

where D is the diffusion coefficient and t_D is the time over which diffusion occurs.

These relations form the basis for the concept of plate height. The plate count, N, is:

$$N = L/H = L^2/\sigma^2 \tag{33}$$

Since $\sigma =$ w/4 (see Figure 5), the plate count can be estimated from a chromatographic peak by simple measuring retention time, t_r, and peak width (in terms of time) at the base, t_w, or at half height, $t_{w,1/2}$, to calculate, N, where:

$$N = 16(t_r/t_w)^2 \tag{34}$$

or

$$N = 5.54(t_r/t_{w,1/2})^2 \tag{35}$$

respectively.

The distribution of a single solute is given by the capacity factor k' (*29*):

$$k' = K\frac{V_{s,x}}{V_{m,x}} = \frac{V - V_o}{V_o} = \frac{t_r - t_o}{t_r} \tag{36}$$

where $V_{s,x}$ and $V_{m,x}$ are volumes of the sample in the stationary and mobile phases, respectively, and V_o is the void volume as measured by an excluded solute. The capacity factor for a single component is quickly obtained by measuring the retention volume, t_r (time between sample injection and elution of the peak maximum) and t_o (time between injection and elution of the peak maximum of an excluded component such as blue dextran) for a given flowrate. The resolution, R_s, of two components in analytical chromatography is given by thee empirical expression (*29*):

$$R_s = \frac{t_{r,2}-t_{r,1}}{1/2(t_{w,1}+t_{w,2})} \cong \frac{1}{4}(\alpha-1)N^{1/2}(\frac{k'}{1+k'}) \tag{37}$$

where α is the separation factor ($= k_2'/k_1'$); k' is an average capacity factor $k'(=(k_1' + k_2')/2)$; N is an average plate count ($=(N_1+N_2)/2$); and the subscripts 1 and 2 refer to components (peaks) 1 and 2, respectively. The definition of R_s is empirical and applies to analytical chromatography where the sample volume is quite small (*29*). Application of Equation 37 to process or preparative scale chromatography introduces further approximations as reviewed elsewhere (*27,30*).

The various processes which impact, H, and therefore, N and R_s, are thoroughly discussed by Giddings (*28*). Some of the plate height contributions due to diffusional effects are summarized in Tables III and IV. The dispersion associated with the flow exchange (eddy diffusion) is given by

$$H_f = 2\gamma_i d_p \tag{38}$$

Thus, the overall plate height, for example, which reflects diffusion between fast and slow flow streamlines (H_D) and eddy diffusion effects (H_f) is (*28*):

$$H = \frac{1}{1/H_f + 1/H_D} = \sum_i \frac{1}{1/2\gamma_i d_p + D_m/\omega_i v d_p^2} \tag{39}$$

The parameters in Table III and Equations 39 and 40 are defined as follows: γ is an obstructive factor; D_m is the solute's diffusion coefficient in the mobile phase; D_s is the diffusion coefficient of the solute in the stationary phase; R is the fraction of the solute in the mobile phase; q is a configuration or shape factor; d is a distance over which diffusion occurs in a stagnant liquid; and ω_i ($=\omega_\alpha^2\omega_\beta^2/2$) is given in Table IV. The involved derivation and explanations for these parameters is given in an excellent discussion by Giddings (*28*).

The number of parameters which can impact plate height are large. Hence, prediction of plate height from molecular properties given the current paucity of data in the literature, is difficult. Nonetheless, the definition of the plate height is useful in defining the regime which predominates the dispersion process (i.e., diffusive, diffusive + flow, or flow mechanisms). Thus, when the effects illustrated in Equation 39 are combined with contributions due to longitudinal diffusion and adsorption-desorption kinetics, an equation of familiar form is obtained:

Table III
Plate Height Contributions Due to Diffusional Effects

Effect	Plate Height Contribution
Longitudinal Molecular Diffusion	$\frac{2\gamma D_m}{v}$
Longitudinal Diffusion in Stationary Phase	$\frac{2\gamma_s D_s}{v} \frac{1-R}{R}$
Adsorption/Desorption	$2\,R\,(1-R)\,vt_d$
Diffusion in Stationary Phase	$qR\,(1-R)d^2v/D_s$ where q is a configuration factor
Diffusion between fast and slow stream paths in a packed bed (see Table IV for values of ω_i)	$\frac{\omega_\alpha^2\,\omega_B^2}{2} \frac{d_p^2 v}{D_m} = \omega_i \frac{d_p^2 v}{D_m}$

$$H = \sum \frac{1}{1/A + 1/C_m v} + \frac{B}{v} + Cv \qquad (40)$$

where the parameter A combines constant terms; B combines longitudinal molecular diffusion effects; and C combines velocity dependent terms. The effect of particle diameter, d_p, can be used to define a reduced plate height:

$$h = \frac{H}{d_p} \qquad (41)$$

and reduced velocity:

$$\nu = \frac{v}{(D_m/d_p)} \qquad (42)$$

Rearrangement of Equation 40 in terms of reduced parameters gives:

$$h = \sum \frac{1}{1/2\gamma_i + 1/\omega_i \nu} + \Omega\nu + \frac{2\gamma}{\nu}\,(1+\beta_s) \qquad (43)$$

In Equation 43, $2\,\gamma/\nu$ reflects longitudinal molecular diffusion in the mobile phase while $2\,\beta_s/\nu$ reflects diffusion in the stationary phase. The $\Omega\nu$ term reflects kinetic or mass transfer effects, while the first term is a "coupling" term of the type described in Equation 39. The

Table IV
Values of ω_i

i	Parameter	Magnitude of Order Values for ω_i
1	Flow in channels between particles	0.01
2	Stationary mobile phase trapped inside particles	0.1
3	Combined effects of diffusion involving large and small channels (short range interchannel effect)	0.5
4	Long-range interchannel effect	2
5	Wall effects and change in uniformity of packing across column diameter ($m_x = D_c/d_p$)	0.002 m_x

form of Equation 43 is illustrated schematically in Figure 6, which shows that a range of reduced velocity could be expected which minimizes the plate height h. Assuming that a minimum plate height, h, is desirable, most practical processes would be expected to fall within a limited range of velocities. Thus, for example, a process which falls in the range where longitudinal dispersion controls, would be amenable to modeling using Equation 26. If a higher flow rate is used where the reduced velocity corresponds to point B (kinetic effects are important), Equation 26 might still be applicable, even though not derived for this condition, since the plate height is still the same. In the case where plate height is not a strong function of ν, one equation might be applicable to empirically model elution profiles which are obtained at the values of ν between points A and B (illustrated in Figure 6) since h changes little in value.

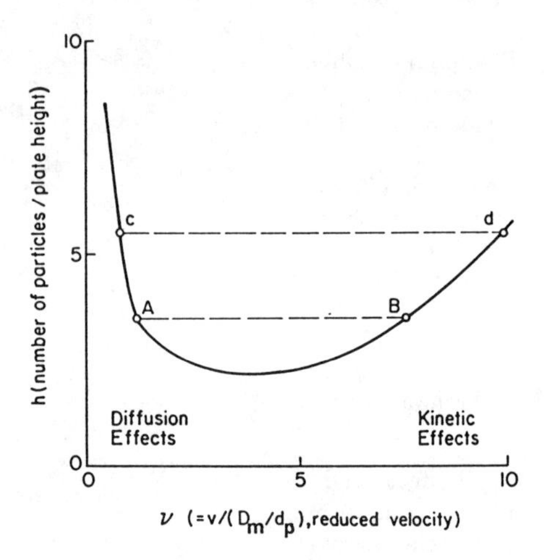

Figure 6. Schematic diagram of general shape of curve for reduced plate height, h, as a function of reduced velocity, ν based on description of Giddings (*28*). Symmetry and exact values will vary from one support to the next.

The approach suggested above is empirical, but nonetheless, useful. Once a stationary phase has been identified which will give the desired separation, a preliminary scale-up calculation may be desirable. If the separation is one in which linear and local equilibria assumptions apply, possible steps for carrying out a first estimate of chromatography column length and volume for cost estimate purposes is as follows. We assume that column operation will be at conditions which give close to minimum values of h. A feed solution of a mixture to be separated into two components ("A" and "B") is assumed as well.

1. Choose the particle size to be used on the process scale.

2. Choose a column diameter which minimizes wall effects ($D_c \approx 40$ to $200\ d_p$) (*16,30*).

3. Pack a column which is long enough so that end effects will be a small portion of total dispersion effects. This is the infinite column assumption. A column length/diameter ratio (L/D_c) of greater than 10 is recommended for bench scale columns. In practice, large scale columns (diameters of 10 cm and higher) may have an L/D_c significantly less than 10 if uniform feed distribution is attainable. In this case, the same calculation procedure is still used as a first approximation. The use of proper end fittings and packing techniques is essential (*27,30,31*).

4. Select an appropriate operational temperature (usually between 4° and 80° C, depending on the molecule) and run samples of the individual components A and B at different flow rates over the range of interest but within predetermined pressure drop limits. Pressure drop may be calculated as given elsewhere (*31*). Estimate N for component A and B using Equation 35. Then calculate H from Equation 33. Estimate k' from Equation 36 for A and B. Calculate R_s from Equation 37 and check by injecting a mixture of A and B to see if resolution obtained is consistent with that determined for individual components. If the observed and calculated resolutions differ by 25% or more, the possibility of interference effects (the presence of one solute affects separation of the other) must be considered. Sample size effect on resolution can be examined as reviewed elsewhere (*30*). Comments on the effect of interference affects on the form of the equations is given by Lightfoot et al. (*22*).

5. Using values of H from step 4, calculate and plot the reduced plate height, h (Equation 41) as a function of reduced velocity ν (Equation 42) for individual components A and B. Values of diffusivities for proteins are in references described by Bernstein et al. (*32*). If A and B exhibit plate height curves having significantly different shapes, the choice of ν (i.e., flowrate) may not correspond to the minimum value of h.

6. If the graph from step (5) indicates longitudinal diffusion to be controlling at the value of ν corresponding to the chosen flowrate, use Equation 26. Otherwise, go to step (7) below. Equation 26 will give an elution profile for component A and then for B, if the sample size is large ($V^o > 0.1\ V_o$). For a small sample Equation 27 may be used. Parameter estimation should give a value of Pe_z which can be approximately equivalent to 2N (*22,25*). The effect of sample size and column length on the elution profile can then be estimated using Equations 21 and 26. Column length will affect Pe_z (Equation 21) while sample size for a given column length will affect the elution profile expressed as C_A/C_{Ao} as a function of V (Equation 26). Flow velocity effects can also be estimated, although this would only have a small affect on h at close to optimum conditions. Plots of resulting individual curves for A and B will give an estimate of overlap of peaks (assuming non-interference).

7. If the graph from step (5) indicates that kinetic, or kinetic and diffusion effects are controlling, the appropriate model should be developed. If a quick estimate is desired, a run at a corresponding low flow rate (for example point c in Figure 6) can be carried out and the calculations in step (6) completed. The effects at the corresponding higher flow rate (point d, Figure 6) can then be estimated, using the value of the Pe_z obtained at the lower flow rate. This is a strictly empirical approach and is not safe for extrapolation purposes, since Equations 26 and 27 are derived for conditions consistent with point c, and not for point d.

8. From calculations in step (6) or (7), specify the maximum sample size which will give a specified product recovery for a fixed purity. This is done by substituting selected values of V^o into Equation 26, and calculating C_A/C_{Ao} as a function of V for each sample volume size. Express sample volume in terms of fractional void volume, V_o.

Once the sample volume for a given size column and column length corresponding to a given plate height are estimated, an estimate of cost based on column (stationary phase) volume, stability of the stationary phase, and column loadings may be generated using procedures outlined elsewhere (*27,30*). It should be noted that the calculations above are for linear chromatography, and as such, do not apply to situations where there are high solute concentrations and loadings beyond the range where a linear equilibrium relationship exists, or where local equilibrium is not applicable. There is now a need to develop workable approaches for analyzing and developing correlations for non-linear systems, particularly as transfer of new biotechnology processes to large scale production facilities continues to evolve.

Conclusions

A number of different separation techniques were presented to illustrate the diversity of processes required to recover and purify molecules of biological origin. Where possible, experimental data on a pilot scale should be used to specify and generate a conceptual process for a given purification. In many cases, this data is not initially available and first estimates of scale-up parameters may need to be carried out in areas for which there is a limited base of experience. Consequently, approximations and semi-empirical correlations are sometimes used to obtain first estimates of equipment size, and operational cost for a specified throughput. There appears to be a need for developing data bases of physical and chemical properties for selected bioseparation processes. This, combined with development of more systematic design strategies will aid in the development of new separations processes which are economically as well as technically viable.

Nomenclature

a	proportionality constants in Equation 6
a_p	interfacial area ($cm^2\ cm^{-3}$)
A	area of a filter (ft^2)
A,B,C	constants in Equation 40
b	width of centrifuge basket (ft)
b	constant in Equation 6
c	mass deposited per volume filtrate ($lb \bullet ft^{-3}$); or concentration ($g \bullet liter^{-1}$). Appropriate dimensions indicated in text
C_b	concentration of retained solids in bulk solution (M)
c_i^*	concentration of solute in fluid which is in equilibrium with a solid having a solute concentration of q_i ($g \bullet liter^{-1}$)

C_w	concentration of retained solids at membrane surface (M)
C_A	outlet concentration of component A (moles $liter^{-1}$)
C_{AO}	inlet concentration for component A (moles $liter^{-1}$)
C_{ASO}	concentration of component A on the stationary phase in equilibrium with the feed concentration C_{AO} ($moles \bullet liter^{-1}$)
d	constant in Equation 6
d	distance over which diffusion occurs in a stagnant liquid (cm)
d_p	particle diameter (mm)
D	axial dispersion coefficient (cm^2 $\min^{-1}$)
D_c	column diameter (mM)
D_m	solute's diffusion in mobile phase ($cm^2 \bullet \min^{-1}$)
D_s	diffusion coefficient for solute in mobile phase ($cm^2 \bullet \min^{-1}$)
D_p	diffusivity of solute in pores ($cm^2 \bullet \min^{-1}$)
D_{pc}	particle diameter eliminated by centrifuge (ft)
d,e	proportionality constants in Equation 6
g_c	Newton's law conversion factor (32.174 ft-lb$\bullet lb_f^{-1} sec^{-2}$)
h	reduced plate height in Equation 41
H	plate height (cm)
H_D	plate height which reflects eddy diffusivity effects
H_f	plate height which reflects flow effects
J	flux (liter $\bullet m^{-2} h^{-1}$)
k	proportionality constant in Equation 5
k'	capacity factor (dimensionless)
k_M	mass transfer coefficient as in Equation 10
K	distribution coefficient as in Equations 22 and 36
L	distance migrated (cm)
L_{MTZ}	length of mass transfer zone (m)
m	number of processing steps
m_x	D_c / d_p
M	empirical parameter representing slope of line of log J vs log ΔP in Equation 5
N	plate count (dimensionless)
p	pressure (bar)
Pe_z	Peclet number
q	configuration or shape factor in Table III
q_c	volumetric flowrate corresponding to cut diameter ($ft^3 sec^{-1}$)
q_i	amount of solute i adsorbed on sorbent (g solute/g sorbent)
q_s	concentration of solute on sorbent due to solute adsorbed on the adsorbent as well as solute in the pores of the adsorbent (g solute/g sorbent); sorbent weight includes weight of fluid inside pores.
r_{ref}	radius of reference
r_1, r_2	inner and outer radii of cake in centrifuge (ft)
r_o	inner radius of a radial flow cartridge (m)
R	outer radius (m)
R_m	filter medium resistance (ft^{-1})
R_s	resolution

S	cross sectional area of a column, cm^2
t_d	mean desorption time (min) in Table III
t_r	retention time (min)
$t_{w,1/2}$	peak width at half height (min)
u_{sh}	shock wave velocity ($cm \bullet min^{-1}$)
v	linear velocity ($m \bullet sec^{-1}$ in Equation 6; $cm \bullet min^{-1}$ in other equations).
v	flow velocity in Table III and Equation 40
$\underline{V}$	volume of filtrate in Equation 3 (liter)
V	volume of solution required to saturate a column (liter)
V^o	sample volume (liter)
V_o	void volume (liter)
V_s	volume displaced by the stationary phase (liter)
$V_{m,x}$	volume of sample in mobile phase (liter)
$V_{s,x}$	volume of sample in stationary phase (liter)
w	peak width (cm)
x_A	fractional concentration of A ($= C_A/C_{AO}$)
z	length of column (cm)
ΔP	transmembrane pressure drop (bar)
ρ_B	density of fluid passing through the column (i.e., bulk phase)
μ	viscosity ($lb \bullet ft^{-1} sec^{-1}$)
ϵ	ratio of volume of fluid surrounding the stationary phase, to the volume of an empty column; i.e., external void fraction
α	specific cake resistance ($ft \bullet lb^{-1}$)
ρ	density
ρ_b	density of particle ($lbs \bullet ft^{-3}$)
ρ_B	bulk density of a solid (stationary phase) including fluid in the pores
ω	constant
ω	angular velocity (radians sec^{-1})
η	fractional recovery of a single component
σ^2	variance
γ_i	obstructive factor in Equation 38 and Table III
ν	kinematic viscosity in Equation 6
ν	reduced velocity in Equation 42
Ω	lumped parameter in Equation 43

Acknowledgment

The material in this work was supported by NSF Grant ECE 8613167. Support of activities (for MRL) in equipment aspects of chromatography design by ARTISAN Industries is also acknowledged.

Literature Cited

1. Belter, P. A. "Recovery Processes-Past, Present and Future," Paper 82-54, 184th ACS Meeting, MBT Division, Kansas City, MO, 1982.

2. Hacking, A. J. *Economic Aspects of Biotechnology*; Cambridge University Press: Cambridge, 1986; p. 127.

3. Atkinson, B; Mavituna, F. *Biochemical Engineering and Biotechnology Handbook*; The Nature Press: Surrey, England, 1983; p. 938-950.

4. McCabe, W. L.; Smith, J. C. *Unit Operations of Chemical Engineering*; McGraw-Hill: New York, 1967; p. 875-931.

5. Boss, F. C. in *Fermentation and Biochemical Engineering Handbook*; Vogel, H. C., Ed.; Noyes Publications: Park Ridge, NJ, 1983; p. 296-316.

6. Wang, D.I.C.; Cooney, C. L.; Demain, A. L.; Dunnill, P.; Humphrey, A. E.; Lilly, M. D. *Fermentation and Enzyme Technology*; Wiley-Interscience, NY, 1979; 261-267.

7. Erikson, R. A. *Chemical Engineering Progress*, 1984, *80*(12), 51-54.

8. Kim, C. W.; and Rha, C-K. *Enzyme Microb. Technol.*, 1987, *9*, 57-59.

9. Smith, J. C.; Ambler, C. M.; Bullock, H. L.; Dahlstrom, D. A.; Dale, L. A.; Emeet, R. C.; Gurnham, F. G. in *Perry's Chemical Engineers' Handbook, 4th Edition*; Perry, R. H.; Chilton, C. H.; Kirkpatrick, S. D.; McGraw-Hill, NY, 1963; p. 19-42-19-100.

10. Moto, M.; Lafforgue, C.; Strehaiano, P.; Goma, G.; *Bioprocess Eng.*; 1987, *2*, 65-68.

11. Patel, P. N.; Mehaia, M. A.; Cheryan, M.; *J. Biotechnol.*, 1987, *5*, 1-16.

12. Michaels, A. S.; *Chem. Tech.*, 1981, 36.

13. O'Sullivan, T. J.; Epstein, A. C.; Korchin, S. R.; Beaton, N. C. *Chem. Eng. Progress*, 1984, 80, 68-25.

14. MacGregor, W. C.; Szesko, D. P.; Mandara, R. M.; Vishva, R. R. *Biotechnology*, 1986, *4*, 526-527.

15. Ruthven, D. M. *Principles of Adsorption and Adsorption Processes*; John Wiley and Sons, NY, 1984.

16. Wankat, P. C. *Large-Scale Adsorption and Chromatography*; CRC Press, Boca Raton, FL, 1986.

17. Sherwood, T. K.; Pigford, R. L.; and Wilke, C. R. *Mass Transfer*; McGraw-Hill, NY, 1975.

18. Arnold, F. H.; Blanch, H. W.; Wilke, C. R. *The Chemical Engineering Journal*, 1985, *30*, B9-B23.

19. Absolom, D. R.; *Sep. Purif. Methods*, 1981, *10*(2), 239.

20. Pollio, F. X.; Kunin, R. *Chem. Eng. Symp. Ser.* *67*(108), 66-74.

21. Lapidus, L; Amundson, N. R.; *J. Phys. Chem.*, 1952, *56*, 986-989.

22. Lightfoot, E. N.; Sanchez-Palma, R. J.; Edwards, D. O.; in *New Chemical Engineering Separation Techniques*; Schoen, H. M.; Interscience, Wiley and Sons, NY, 1962; 100-181.

23. Gong, C. S.; Ladisch, M. R.; Tsao, G. T.; *Adv. Chem. Ser. No. 181*; Brown, R. and Surasek, L., eds., Washington, DC, 1979; 261-288.

24. Gibbs, S. J.; Lightfoot, E. N.; *Ind. Eng. Chem. Fundam.*, 1986, *25*, 490-498.

25. Neuman, R. P.; Rudge, S. R.; Ladisch, M. R.; *Reactive Polymers*, 1987, *5*, 55-61.

26. Jansen, J-C.; Hedman, P.; in *Advances in Biochemical Engineering, 25*; Feichter, A., ed.; Springer-Verlag, Berlin, 1982; 44-99.

27. Rudge, S. R.; Ladisch, M. R.; in *Separation, Recovery, and Purification in Biotechnology*, Asenjo, J. and Hong, J., eds., Am. Chem. Soc., Washington, DC, 1986; 122-152.

28. Giddings, J. C.; *Dynamics of Chromatography*; Marcel Dekker, NY, 1965, 13-94.

29. Snyder, L. R.; Kirkland, J. J.; *Introduction to Modern Chromatography*, 2nd ed.; Wiley, NY, 1962.

30. Ladisch, M. R.; Voloch, M.; Jacobson, B. J.; *Biotechnol. Bioeng. Symp. Ser. No. 14*; 1984; 525-540.

31. Ladisch, M. R.; Tsao, G. T.; *J. Chromatogr.*, 1978, *166*, 85-100.

32. Bernstein, H.; Yang, V. C.; Langer, R.; *Biotechnol. Bioeng.*, 1987, *30*((2), 196-207.

RECEIVED September 16, 1987

Chapter 8

Mammalian Cells as Factories

Randall W. Swartz

Tufts University Biotechnology Engineering Center, Medford, MA 02155 and Swartz Associates, Biotechnology Consultants, 15 Manchester Road, Winchester, MA 08190

It now appears that many of the products of recombinant DNA technology (including TPA, MAB's and EPO) will be produced in mammalian cells. Several technologies exist for preparative mammalian cell culture which differ depending on whether the cells are suspended or attached and with regard to the degree to which the cells are maintained in a growing versus a non-growing state. In preparative culture costs depend strongly upon the requirement for serum or substitute growth factors (which is linked to the relationship between qp, product formation rate per cell, and μ, the specific growth rate) and to labor. Alternative technologies will be examined from the perspective of product unit cost and suitability for specific culture and product characteristics.

It now appears that many of the most important diagnostic and therapeutic products of biotechnology in the late 1980's and early 1990's, including tissue type plasminogen activator (TPA,) prourokinase (KPA,) erythropoetin (EPO,) monoclonal antibody (MAB,) and possibly even some lymphokines and interferons, will be produced in mammalian or similar eukaryotic cell systems.

Estimates of the $ volume of these new products are quite varied. The prediction of explosive growth at some future time from relatively small volumes today is a common property. While there is agreement on the explosive growth, the "experts" agree that no one can accurately predict when it will occur. Table I presents an estimate of these market volumes in 3 to 5 years.

TABLE I. Biotechnology Markets (Adapted from Boston Biomedical Consultants and other sources.)

	SALES (million$)
rDNA Therapeutic	2600
Immunodiagnostic	2300
In Vivo Diagnostic	300
MAB Therapeutic	700
TOTALS	6000

0097-6156/88/0362-0102$06.00/0

Mammalian Cells as Factories

Mammalian cells must be cultured to produce viral vaccines, protein products, and more cells. Included are polio vaccine, vaccinia (cow pox) virus being used as a vector for antigens of a variety of viruses (HIV,malaria, and other parasitic diseases.) Numerous animal virus vaccines are prepared also. Vaccine preparation is the earlier of the applications listed above and experience with bovine foot and mouth disease and polio vaccine has contributed much technology.The cells are useful in some cases including the culture of skin for transplantation, the activation and proliferation of certain anticancer T-cell populations and the use of cells (e.g.;islet cells) as implants to produce hormones. (see Table II.)

Even as late as 1982 and 1983, commercial organizations whose focus is pharmacologically active proteins produced using recombinant DNA technology ("rDNA" or "biotechnology" companies) anticipated a much larger role for microbial expression systems than now seems likely, at least for first generation products. Mammalian cells have become increasingly important for the preparation of therapeutic proteins, and this is so in spite of the high costs associated with mammalian cell production systems (several hundred or several thousand dollars per gram range for the purified protein.)

The protein products which have sparked all the interest are diverse. They include medical and agricultural diagnostic applications which require monoclonal antibodies, experimental therapeutics; notably plasminogen activators (tissue type and prourokinase -KPA) which dissolve thrombin clots and erythropoetin which ammeliorates certain anemias. The focus of process development has been on these new products, and particular attention has been paid to hybridoma cultivation since so many groups are preparing monoclonal antibodies (MAB).

Mammalian cells are chosen in preference to bacteria for production only in those cases where the much cheaper microbial expression was not feasible. With many larger proteins (>20,000 MW) which have a number of disulfide bonds (>1) it is difficult or impossible to return the insoluble molecule isolated from a bacterium to its active state. Further bacteria could not perform the often critical additional modifications such as removal of the so called signal sequence and addition of multisubunit sugars, (glysosylation.) Mammalian cells are able to do these and other so called post-translational modifications. (Table III.)

There are other recombinant production systems which use bacteria (E.coli) and yeast as production systems. It is much less expensive to produce certain proteins in E.coli. but in general this host/vector system faithfully reproduces only the linear sequence of amino acids which make up the protein, and this as an insoluble mass inside the bacterium (an inclusion body.) To the extent that a given product can be prepared from this material in a cost effective way, this may be the preferred system. This is the case with certain smaller molecules such as IL2, several interferons, and certain animal growth hormones where correct folding of the molecule in vitro is feasible. In other cases, also, generally small proteins (< 20,000 MW) having only one or two disulfide bonds, and with peptides or proteins whose use (diagnostic antigens or components of vaccines) does not require a particular conformation, these systems are excellent. (See Figure 1.)

Therapeutic and diagnostic enzymes require correct conformation (or folding) in order to be active and may require other chemical modifications or additions for activity, proper pharmacokinetic behaviour (e.g.; clearance), specificity and normal antigenic response. The further processing required may include addition of sugars(glycosylation), formation of correct disulfide bonds, activation or modification by clipping, and other "post-translational modification. E.coli cannot do this, mammalian

Table II. PRODUCTS OF MAMMALIAN CELL CULTURE

ARTIFICIAL SKIN, NERVES, VEINS, ETC.
CELLS FOR IMMUNOTHERAPY
HUMAN GROWTH HORMONE
MIXED ALPHA INTERFERON
TPA, UK, PUK
EPO
AHF
VACCINES
LYMPHOKINES
MONOCLONAL ANTIBODIES

Table III. WHY MAMMALIAN CELLS ?

BECAUSE THEY DO THINGS BACTERIA DON'T !
CORRECT FOLDING
CLIPPING
GLYCOSYLATION
GAMMA CARBOXYLATION
SUBUNIT ASSEMBLY
SECRETION FOR EASIER ISOLATION

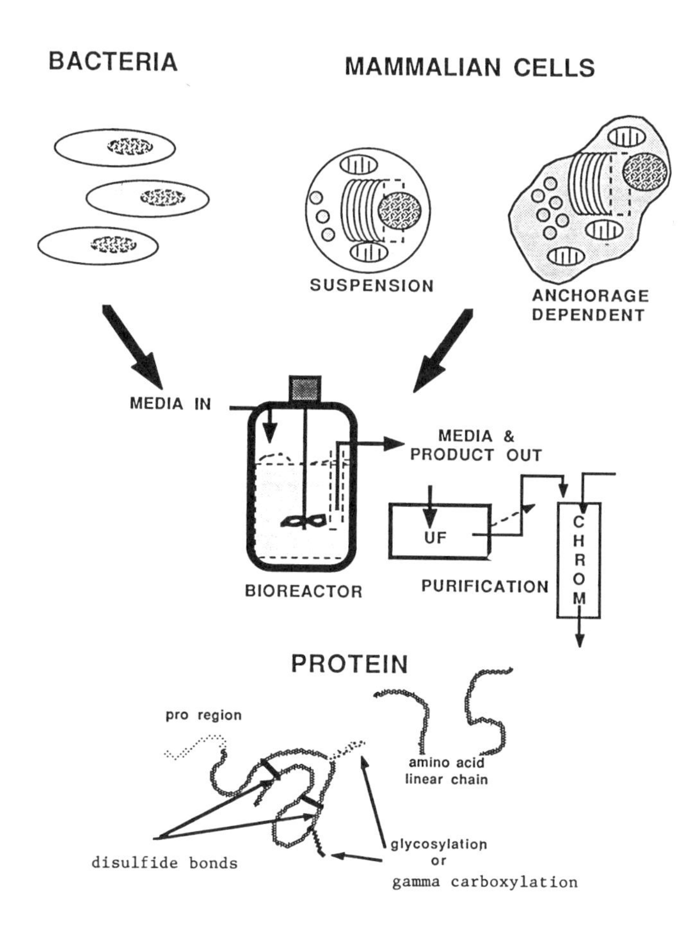

Figure 1. Recombinant production systems that use bacteria for production.

cells and yeast,(so called eukaryotic cells) can. Also, the products of mammalian and some yeast systems present the product as individual, intact, soluble molecules outside the cells, albeit in quite low concentrations. Product in the broth or supernatant may be readily separated from the hundreds of cellular proteins, making the purification step much simpler. In recent months great progress has been made in the development of yeast for secretion of glycosylated, active enzymes. These successes will in part displace mammalian cell culture due to their much higher volumetric production rate due both to higher cell density and higher production rate per cell.

<u>Issues for commercial drug development</u>

Regulatory issues are diverse.These include drug safety, efficacy, homogeneity, identity, purity, and various issues relating to our confidence that each batch of material consistently performs like the batches tested to obtain approval. The details of expression system , cultivation system, product purification and formulation all influence regulatory compliance in each area. Cost of manufacturing and our ability to produce the required amount of product is determined by the performance of the expression system and the cell cultivation and product purification process.

<u>Production Technology</u>

MAB's have been the initial focus of preparative mammalian cell culture even though these are produced with reasonable efficiency in the ascites fluid of mice. The reason for this interest has been the need to produce really pure antibody, free of mouse IGG, etc. for linking of MAB to other enzymes (ELISA) or molecules (cytotoxic drugs or imaging agents, etc.) and concerns about other immunogenic or disease producing agents (viruses, etc.) which might remain with material targeted for therapeutic use. Technologies similar to those described below for MAB are being optimized for production of TPA, Factor VIII, hepatitis B vaccine, etc.

Thus production technology is critical to successful product introduction and affects profitability. In fact, because of the high cost of manufacturing and drug testing of therapeutic proteins, technology influences whether or not a particular product can be profitably commercialized. If the amount of TPA or EPO required per patient were as large as the dose of many antibiotics, they would not be commercialized.

Initially the technology for cultivating the cells was adapted from those used in research. Scaleup was modeled after vaccine production and microbial fermentation. The usual techniques were roller bottles, cell factories and small fermentors with low shear agitators appropriate to the fragile mammalian cells.

The process begins with a cell line which produces the product in a suitable form. The preparation of MAB in quantity is facilitated by the relatively high productivity of some hybridomas. Hybridoma productivity varies over a wide range due to intrinsic cellular factors. In turbidistat culture (cells grown continuously at their maximum rate) S. Fazekas de St. Groth (J. Immun. Meth. 57,(1983) 121-136.) showed that for one group of hybridomas, each growing at its μmax, the product formation rate per cell varied between 8 and 130 μg/106 cells/day. Formation of other products was made possible by the cloning (or isolation) of the gene coding for the synthesis of each product and the development of suitable mammalian expression systems. Today's production rates are near the low end of the hybridoma range between 1 and 30 μg/106cells/day. Great potential exists for continued development of recombinant host/ vectors. Beyond this, we can suggest that many of the systems now in use have elements which will be useful in hybrid vectors of today and tomorrow.

There are several MAMMALIAN HOST / VECTOR systems. All were initially developed for research purposes. A few have had properties favoring further development as commercial production systems. Characteristics of an ideal host/ vector for preparative use can be deduced from the commercial development issues discussed above. Properties I see as important are tabulated below in Table IV:

Maintaining the host/ vector or the hybridoma in a rapidly growing or synthesizing state requires both a medium and a bioreactor. The cell is grown in a suitable mixture of nutrients and protein "growth factors" contained in a bioreactor" which contains the cells and media, regulates the physical environment, and prevents invasion by other organisms. Media to date has been largely designed for growth of cells in research labs and is unsuitable for high density cultivation. One major area for process improvement is the design of media for high density culture. Adjustment of the media composition can affect protein products in other ways than yield alone. It has been shown that feeding certain sugars involved directly or indirectly in gylcosylation can affect the homogeneity of glycosylation and it could be that rate could be affected. This would have potential effects on the safety and efficacy of a therapeutic protein.

Extensive use is being made of defined serum substitutes and their effect on costs and simplification of purification can be profound. While most processing still uses some reduced fraction of serum, horse, cow or FBS, more and more groups are trying to find simple cocktails of so called growth factors as substitute. As the market for these products expand and their cost drops, in part because many can be made in bacteria using recombinant techniques, this trend will accelerate. The task of serum substitution is easier with transformed hosts as these require fewer of the protein growth factors.

The Host/ Vector System Determines The Process Strategy.

Cultivation can be batch (media and cells put in at the start, no medium changes, harvest at end,) fed batch (medium replaced during cultivation to replace depleted nutrient, remove wastes, and harvest products, continuous (a variant of fed batch where medium is added and a homogeneous mix of reactor contents , including cells, are removed,) and perfusion where cells are retained in the reactor, fresh medium is added, and waste containing medium removed. If the host/ vector requires cell proliferation for product formation, a batch, fed batch or continuous system must be used. If product is formed by non-proliferating cells, perfusion is satisfactory and offers the advantage of conserving expensive growth factors, since these are primarily required for cell proliferation, and less for the resting state. Because the cells are expensive to produce, perfusion may offer an economic advantage even if the per cell rate of product formation is highest with growing cells. The process options are diagramed in Figure 2.

Alternatives For in vitro Production of Monoclonal Antibodies

The primary alternatives for in vitro production of monoclonal antibodies are two: 1) chemostat (Celltech,J.R. Birch,et.al.,presented at fall,1985 American Chemical Society,Div.of Microbial & Biochemical Technology mtg.) or turbidostat(S.F.de St. Groth, J. Immun. Methods, 57 (1983)121-136.) type suspension cultures where cells, partially spent media, waste products and desirable products are removed on a continuous or periodic basis and fresh media is added likewise and 2) perfusion or other

Table IV. THE IDEAL HOST / VECTOR SYSTEM

HOST CELL	EXPRESSION VECTOR
High level synthesis, modification and sectretion.	High level production in host at low or zero growth rate.
Differentiated to sectretory function.	High DNA copy number.
Lack virus and oncogene segments.	Very strong promoter and stable message.
Modest growth factor needs.	Stable without selection with toxic drugs.
Not affected by product.	Rapid development of strain to high production rate.
Stable on passage.	
Low protease secretion.	

immobilized cell culture in which cells, once produced, remain in the reactor,as described by R.Rupp of Damon,W. Tolbert of Monsanto, B.K. Lyderson, then with K.C. Biologicals (he is now at Hybritech), R. Dean of Verax and P. Brown of Bioresponse at the same ACS meeting. Also in class 2 are are the microporous membrane system described by S.S. Seaver and the hollow fiber U.F. membrane system described by M. Gruenberg and R. Schoenfeld and the hollow fiber U.F. membrane system described by J. Hopkinson at the 1985 Automation, Scale-up Congress in San Francisco. The perfusion of a suspension culture in a reactor such as the "spin filter" fermentor developed by Dennis Johnson of ADL would also fit in this group. The papers and presentations cited all reported work with murine hybridomas producing monoclonal antibody (MAB).

The Host Determines Whether a Surface is Required for Growth.

Mammalian cells derived from tissues generally require a solid support which simulates the normal cell-cell association in the tissue. They must be covered with a liquid as in the tissue. Circulating cell types can grow suspended directly in a liquid. Cells differ in their sensitivity to shear, the more sensitive types, especially anchorage dependent cells requiring very gentle treatment. Hardy cells should do well on a microcarrier or support providing direct contact with the medium.

The shear sensitive types require particular supports, especially when grown in agitated liquid culture. Solid spherical cellulosic or polymeric beads (such as Cytodex) provide relatively little protection. Other beads or microspheres encapsulate the cells in a semipermeable sac (Damon) or in a gel (alginate), etc. The bead developed by Verax is a quite open matrix of collagen, weighted to provide a suitable settling rate for their fluidized bed cultivation system. One benefit of the Verax technology is the physical protection afforded by the collagen matrix.

Several high density bioreactors are diagrammed below. See figure 3. Each can be used with either anchorage dependent or suspension cells. The selection of the bioreactor depends on several factors. The bioreactor must provide a satisfactory physical environment for the cells, allow for maintenance of the nutritional environment and for removal of cells and product as appropriate. It must provide for dissolution of oxygen in the medium at a rate equal to that at which the cells use oxygen. For cells relatively insensitive to shear which do not require a solid support, this may be accomplished in a conventional stirred reactor or "fermentor." If cells are shear sensitive, the stirrer or mixer must be modified to reduce shear. Marine propellers may replace flat blade stirrers because they provide more mixing for a given amount of shear. In mammalian cell culture the use of air lift type mixers is another way to reduce shear. In fermentors, where air is introduced by bubbling, shear increases the rate at which oxygen can be introduced so reducing shear reduces the available oxygen.

Previously the amount of air which could be introduced was limited because bubbles cause the medium (which contains proteins) to foam. New developments which help alleviate this are the "caged aerator" and the use of low protein content serum substitutes. The caged aerator is simply a cylinder of stainless steel mesh which allows medium and dissolved gas to pass freely and excludes the fragile cells from the aeration column. Any bubbles are broken as they pass thru the mesh. The mesh typically has a pore size of 5 μ if cells are in suspension and 80μ if on microcarriers.

The rate of oxygen supply is typically the factor limiting cell density in a mammalian bioreactor until the cells are so dense that there is no more space for the liquid which carries the oxygen and media.

An alternative approach is used for high density immobilization systems such as the ceramic matrix where there is no opportunity to directly oxygenate the culture by exposing the fluid inside the bioreactor to gas.

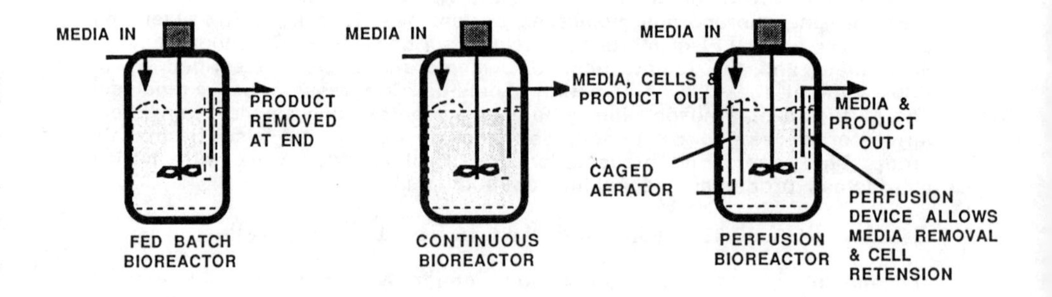

Figure 2. Process design alternatives.

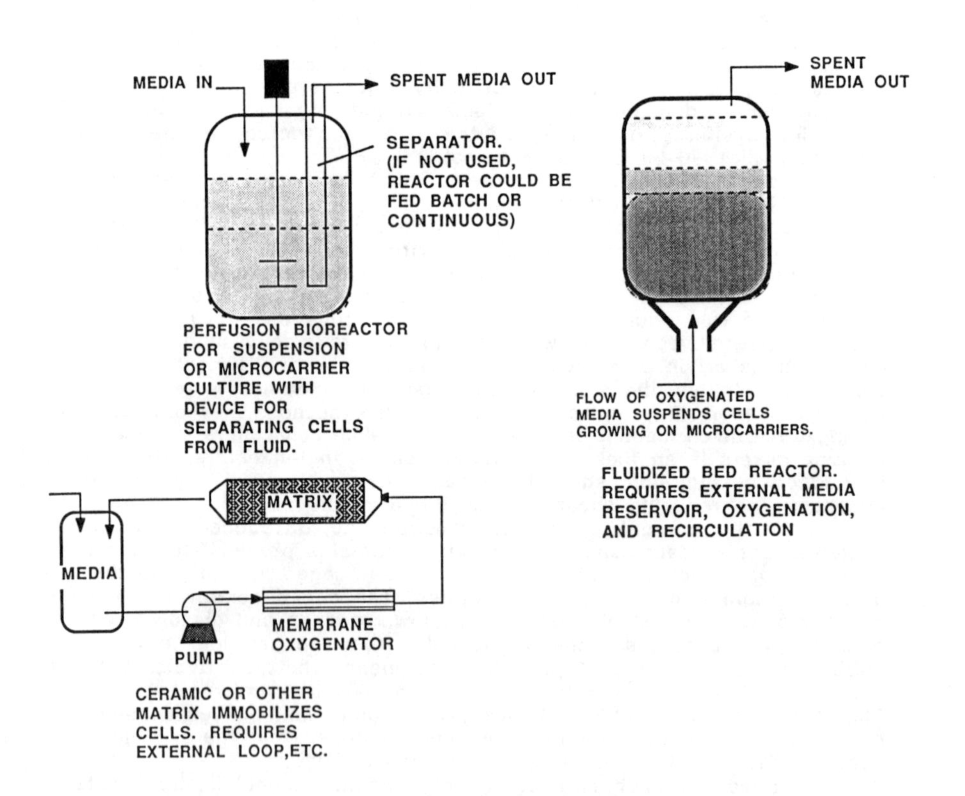

Figure 3. Mammalian cell perfusion bioreactors.

Here the gases (oxygen and carbon dioxide) are exchanged with the liquid media using a gas permeable membrane, usually constructed of microporous polypropylene (Celguard) hollow fibers, but sometimes still constructed of silicone which has a lower permeability. This is typically done in a separate hollow fiber device external to the bioreactor. Because of the relationship between the permeability of the membrane and the demand of the cells, the flow rate of liquid thru the oxygenator must be equal to one or more times the volume of the bioreactor per minute and may not stop even for a few seconds due to power, pump or connecting tubing failure.If stopped the cells will die. In some systems a coil of silicone tubing is incorporated in the fermentor and air is passed thru it, but space and concerns for cleanliness limit the amount of tubing which can be incorporated.

Reactor Productivity

One desirable characteristic of bioreactors is a "high volumetric productivity." This is the rate of product produced per volume of reactor. Recognizing that the "factory" which produces our protein product is the cell, volumetric productivity may be maximized by maximizing the product of the number of cells in the reactor times the rate at which each cell produces product. The PRODUCTION RATE/CELL is dependent on many factors, including all aspects of the physical environment and the effects of nutrient type and availability upon the host/vector system. The cell growth rate may be related to the per cell rate of product formation. Clearly the most important determining factors are the intrinsic, genetically determined, ability of the host to produce , process (modify) and export the protein of interest and the ability of the vector to divert the synthetic machinery of the cell toward the product of interest without compromising the viability of the host. These factors are determined by the genetics of the host , the vector and their interaction.

PRODUCTION RATE / LITER = [PRODUCTION RATE / CELL] X [CELLS / LITER]

The technological alternatives are presented in figure 3. This is obviously a conceptual document only and does not present the many differences in process details among alternatives. Arrows depict the direction of media flow in and out. Media out is sometimes accompanied by product and/or cells but the important factors are conservation of cells (so that expensive media components are conserved) and concentration of product relative to byproducts (for ease of purification.) Of course this assumption may not apply to all cells. The expensive medium components may in some cases be necesary for maintenance of cell viability independent of growth.

Bioreactor Systems Exhibiting High Volumetric Productivity.

The focus of this discussion will be three bioreactors providing high reactor volumetric productivity and at the same time conserving the cells (USING PERFUSION). These are the ceramic matrix technology available thru Charles River Biotechnology (Wilmington,MA) , agitated vessels equipped for withdrawal of media while retaining the cells in the reactor , and the fluidized bed reactor. Each can readily be operated as perfusion cultures.

Stirred and fluidized bed reactors with excellent overall design can do this provided some means is provided to remove the liquid while retaining the cells in the reactor. For anchorage dependent cells, New Brunswick

Scientific offers a well engineered bench scale system combining gentle agitation, caged aeration, and two alternate approaches for media removal which allow cell retension. A settling tube with screen is used with microcarrier beads providing a surface for cell attachment. For suspension culture, a recirculating stream thru a x-flow filter is satisfactory for cultivations of circa 1 month duration. Several other vendors can provide suitable vessels of this type. Verax has engineered a quite suitable fluidized bed reactor which, when used in conjunction with its weighted beads, should perform comparably to the other reactors mentioned. Each reactor type is likely to have advantages and disadvantages for particular applications.

The ceramic matrix developed by Corning and now sold by CRBS retains the cells thru entrapment and a charge based cell/ surface interaction. Cells are retained in stirred vessels using a filter of 0.1 to 5 μ pore size. The filter may be a spinning stainless steel mesh, as used by Shaul Reuveny, et. al. for high density perfusion culture of hybridomas, and by Collaborative Research in the large scale preparation of prourokinase by a transformed human kidney line. Reuveny encountered difficulty with plugging using a simple spinning cup shaped filter. On larger scales, special design features can minimize these problems and filters can be maintained inside the reactors for circa 6 months of continuous perfusion. In smaller reactors, hollow fiber and flat sheet microporous membranes have been successfully used in a pumped recirculation configuration for perfusion of small suspension reactors. A 0.1 μ cross flow filtration device was used to perfuse a device at New Brunswick Scientific Co. It remained operational and produced monoclonal antibody at high rates for four weeks.

These systems and one or two others which have been developed as systems for preparation of products incorporate reliable control systems for control of pH and dissolved oxygen tension as well as the standard controls for temperature. Addition of devices for feeding of nutrients and withdrawal of culture fluid (either with or without cells) is straightforward. Each system can produce 1/2 to 1 1/2 g MAB/ day from a benchtop device using a fairly productive cell line. Both have been used effectively with a number of recombinant host vector systems.

These approaches have been selected primarily for the production of monoclonal antibodies by hybridomas. The characteristics of hybridomas are : low sensitivity to shear, low oxygen tension in some regions of the reactor affects only those regions, product is produced by both growing and non-growing cells, although the rate per cell may differ. These cells are easy to grow and produce MAB at relatively high rates per cell compared to recombinant expression systems. Perfusion reactors have been used successfully to produce prourokinase using high density cultivation of a shear sensitive transformed human kidney cell line.

It is important to recognize that high volumetric productivity does not necessarily result in low manufacturing costs to the extent that high productivity is achieved solely thru high cell density . This is because the same nutrients must be batched and fed regardless of cell density. Note that even with MAB production by hybridomas, Celltech has achieved a cost effective production system at relatively low cell density in their air lift reactors. Lilly has also done so cultivating the cells in modified microbial fermentors at relatively low density (<5 x 106 cells/ml) and this as a batch cultivation. Naturally, in a case where media cost is a small portion of the product cost, and purification cost are volume dependent, (rather than product quantity dependent) volumetric productivity would be very significant.

The cells in the reactor must be produced and each will have similar nutrient and growth factor requirements on a per cell basis regardless of their density. Thus as we push to higher and higher density we will need to perfuse at higher and higher rates or increased media concentrations. In fact as cell density approaches the limit for a given reactor some fraction of the cells will be exposed to a less than optimal environment and pushing to higher and higher density will be counterproductive beyond a certain point since the physiological state of the average cell may be affected in a way which impedes product formation. It should be noted that recent work suggests the significance of "inhibitory factors" in serum. In some cases it may be benificial to dilute the medium with a suitable buffer and raise perfusion rate, lowering the cost of production. This assumes an efficient concentration step in purification.

The significance of this analysis is that optimization of the nutritional environment of the cells provided by the media and growth factors (or serum) and the way in which these are fed, together with the development of increasingly efficient host/vector systems are the factors which are most significant in controlling product cost. The high density systems have worked successfully and these technologies will spread and become standard. Cell densities of circa 1 to 5x 107 cells/ ml. will be common in suspension reactors and in stirred reactors using various cell immobilizing carriers and filters for perfusion.The high cell density systems will continue to be very useful but because the actual effect of their use, even for MAB production, is now more on the convenience of production than on a clear cost basis, efforts should continue to be focused on the host/vector development and on the use of optimal media and process strategy (fed batch, continuous or perfusion culture).

Some anchorage dependent transformed cell lines, such as the CHO cells used by Genentech and Genetics Institute for preparing TPA, normally grow attached to a surface but can be "adapted" to suspension culture. It is likely that genetic and certainly profound physiological changes occur as a result of this "adaptation." The extent to which rates of product formation and post translational modification of proteins are affected are experimental questions. One might wonder whether a product produced by a given host/vector in roller bottles (anchorage) would be precisely the same as one produced in suspension. If not, this could have profound effects on the pharmacological performance of the product. It may be that for certain very shear sensitive cells, or cells which require attachment but adhere only weakly, that these now "standard" approaches will be unsuitable. For recombinant production the best long run recommendation would be to alter the HOST/ VECTOR system.

The conclusion to be taken are that suitable bioreactors are available for cell cultivation at preparative scales but one must know which designs are most appropriate to a given application. The real need now is for continuing work on the development of improved host/ vector systems and on the optimal medium, feeding regimen and process strategy for a given host/ vector and product. These are largely experimental issues but require a quite sophisticated and comprehensive approach which is sensitive to product cost (mainly influenced by expression technology) and to issues of product quality.

Cell Culture Technologies

Technologies suitable for the growth of mammalian or similar cells at preparative scale may be classified based on whether the cells are grown in suspension or are attached to a substrate. More importantly from a culture physiology and a process economic perspective, there are technologies where cells are retained in the reactor in a non-growing state (or nearly so) vs. those where the cells are continually growing and dividing and some fraction of the population is discarded continuously such that each cell has only a short productive period. If the product were the cells,the latter approach would be entirely satisfactory, but as we shall see, it is generally some product of cell metabolism which is desired and the cell itself is the factory producing it. The key to understanding the comparative economics of the preparative technologies is recognizing that the cost of producing product is linked to different degrees to the cost of producing cells, depending on the degree to which product formation is tied to cell growth. One major cost element in mammalian cell culture is the medium, which can account for up to 50% of the manufacturing cost in extreme cases. The high cost of medium is primarily a result of the serum or serum substitute which provides certain "growth factors" (such as insulin, selenium, transferrin, "fetuin", phospholipids, ...) While these are not consumed as a carbohydrate or amino acid or mineral might be, they are required in quantities stoichiometrically tied to cell mass or quantity. It is the growing of the cells that is the costly item, and the cost of the cell product can be drastically reduced if we can find a way to prolong the productive life of each cell produced. Certain "growth factors" may be required for maintenance of cell viability. The significance of this requirement will be a function of cell type and vector.

The primary source of the potential advantage of the Bioresponse, CRBS, Damon, Endotronics, InVitron, Millipore, New Brunswick Scientific, Verax ,and Vista Biologicals technologies are this opportunity to retain the cells produced at relatively high cost in a productive state for extended periods. Because side by side comparisons of these technologies are not available, we cannot identify technology factors which convey an advantage to one over another within this group, trademarked or patented technology notwithstanding.

Table V , PREPARATIVE CELL CULTURE ALTERNATIVES, reviews organizations servicing this in vitro mammalian cell culture market. The list includes suppliers having unique technology or a special position today in cell culture . Absent from this table are media suppliers. The important ones today are those producing media suitable for support of increased cell density. This generally means an increase in amino acid concentration and altered buffering. Another common characteristic is the availability of a well defined serum substitute.

In their advertising, several of these commercial firms claim a per gram production cost for their systems at small scale of circa $500/gm. This seems reasonable for a relatively productive hybridoma but of course the cost will be extremely sensitive to qp , the production rate per cell per time. There is no obvious reason why the economics for the CRBS ceramic system (Opticell) should vary significantly from the hollow fiber technology of Endotronics, Invitron or Bioresponse. The Damon, Verax,or In Vitron technologies may be somewhat more capital or labor intensive, and in the Celltech systems, somewhat more serum might be used, but with the potential benefit of higher growth rate, which could be important in some systems. A significant advantage of the fluidized bed and high density suspension/ perfusion technologies is that they promote a more homogeneous cell environment than is feasible at high density using a fixed or packed bed design. The technology is developing quite rapidly and from a manufacturing cost perspective it is difficult to select a winning technology at this point. Actually among technology types (immobilized vs. continuous growth) clear winners are unlikely. The problem is more one of weeding out the clear losers which unfortunately remain in the market place. If we could confidently select based on unbiased performance data and comparative user evaluations the picture would be clear. Unfortunately such is not yet available and even systems with serious or cumbersome design flaws are adequate for laboratory, but not necessarily routine preparative use. Such problems increase the cost of using the system.

Among the service organizations the approach of Damon and InVitron among others is to offer more extensive services including enhancement of mammalian expression (Damon's based on the Tonagawa , In Vitron's

Table V. PREPARATIVE CELL CULTURE ALTERNATIVES

ORGANIZATION	SUPPLIES or SERVICES	TECHNOLOGY	OTHER SERVICES
Bio-Response	contract production	hollow fiber,	purif.,GMP facility
Celltech Ltd.	contract production	air lift, continuous flow for MAB	expression, GMP, purif.
Damon Biotech	contract production	entrapment in semiperm. beads	purif., GMP, filling finishing, expression
Endotronics	equipment&	hollow fiber	process supplies
InVitron	contract production	several	purification, GMP, and reg. affairs, expression
Charles Rivers Biotech. Svcs.	equipment, supplies, contract production	ceramic cell immobilization for perfusion	some purif.
Millipore/Waters	bioreactor in development	sheet membrane immobilization for perfusion	purification equip. & supplies
New Brunswick Scientific	equipment	suspension & μ carrier perfusion	broad line of equipment
Verax	contract development/production,	cell entrapment &fluidization	limited purif.
Vista Biologicals	contract development/production	μ carrier and suspension perfusion culture	limited purif., GMP facility,

being based on the work of Chris Simonson who did the DHFR based expression in CHO cells at Genentech until joining InVitron. Other services being offered are purification, filling and finishing and assistance with regulatory affairs such as IND filings and GMP manufacturing issues. This allows them to differentiate themselves from organizations using related technologies and to integrate in the direction of pursuing products on their own account which should be their primary focus. It should be recognized by potential clients that these organizations are quite new and their regulatory experience derives primarily from the expertise of the staff assigned to the project, rather than any depth in this area for the organization as a whole. Also, while the physical facility may be in compliance with pharmaceutical cGMP's, the potential client should assure that SOP's and validation, etc. are also complete. Similarly, while most of these organizations have purified MAB to >95% purity, few have actually taken a product thru IND approval and if purification of something other than IGG is involved, bear in mind that MAB purification is really rather straightforward compared to TPA, factor VIII, etc.. Only a few of the groups have sophisticated, preparative protein chemistry/ purification groups in place and with only one or two experts at the most advanced companies, clients should pay close attention to the team assigned to their project especially if the product is not an MAB.

Process Economics

The manufacturing cost of these products is critical to the financial success of many projects. These proteins are quite expensive to produce and may or may not be therapeutically realistic, depending on the cost of treatment. In order to provide a framework for estimating manufacturing costs, the following is presented as an example. One must bear in mind that this is NOT a "typical" example- none exist due to the wide range of expression levels and of therapeutic doses.

Technology for the preparative culture of mammalian cells is developing at an incredible pace as are the related areas of culture media, and expression system development. The driving force behind this development is reduction of the manufacturing cost of commercial products such as monoclonal antibodies, tissue plasminogen activator, factor VIII, etc. In order to evaluate alternative technologies, we can examine the major contributions to product cost and then examine the extent to which these technologies affect costs. Table II presents in column A, an analysis of production costs for an anchorage dependent cell line producing a recombinant product at 150 mg/ liter. Product formation is dependent upon the specific growth rate of the culture being maintained near its maximum level (cell doubling each 24-48 hrs.). It is further assumed that calf or horse serum at the 5% level is adequate to sustain this growth. Expression technology which decouples growth and product formation rate results in a substantially lower serum requirement and thereby a lower cost. This strong coupling of growth rate and product formation kinetics with cost results from the disproportionately large contribution of even these relatively low growth factor (low percentage of less expensive serum) costs to total product cost. Obviously expression level is most strongly coupled to cost and a proportionate increase in volumetric production rate would be feasible regardless of reactor type. This would result in a proportionate decrease in those costs associated with cell culture.

This sort of analysis is essential at an early stage of any project where mammalian cell culture will be used as the production system. This analysis has been simplified for illustrative purposes and the details have been omitted. A full analysis would analyze the purification and formulation steps in a more precise way and include an analysis of the sensitivity of costs to key assumptions.

Fermentor Productivity Calculation

If we assume that a BPV expression system might yield circa 150 mg / liter vessel capacity / week or circa 50% of that purified, and that approximately 3 liters of media (containing circa 5% serum, type uncertain) would be used in semi continuous culture mode during that week, then a 100 liter vessel would produce circa 7.5g of product per week. This is considered in case "A", below. Continued development of this or some other mammalian expression system (such as the antibody promoter/ enhancer system being commercialized by Damon) might allow production levels to eventually approach the 500 mg / liter / day levels achieved in optimal MAB production in small scale perfusion vessels (using some variant of cross flow filtration , e.g.; "spin filter",to separate cells and media) at a density of 2x108 cells/ml. In this later case, 2 liters of media might be used per liter of reactor per day, however the serum requirement in such perfusion culture is circa 2%, substantially less than with the chemostat cultures referred to in the former case. This is because the serum (growth factor) requirement is in the main proportional to net cell growth rather than product formation per se. Preparative purification will be no worse than 50% yield . Serum costs, the main disposable item, are assumed to be substantially reduced thru reduction of use levels and substitution of calf or horse, etc for FBS. (In bulk, FBS would be circa $3-$6 / liter of medium at the 5% level ($60-$120/literFBS, calf or horse would be $30-$60/liter depending on quantity.) We have assumed our medium is $8 / liter total cost for the continuous culture or repeated fed batch("splitting cells") process. If we assume the same basic expression level and the 2 liter per liter per day media use at 2% horse serum for a perfusion process, the total medium cost in column "A" could be as low as $4 per liter or $4000000 for materials and supplies. In this example, even assuming favorable serum costs, serum remains a major cost component. The impact of perfusion versus continuous culture (a savings of $300/gm) would obviously change with yield (it would be less on a per gm basis for an MAB produced at 500mg/liter/day) and total serum cost would increase in smaller quantities.

Estimation of MAB production cost at the 10 kg level is not relevant for most uses, however it can give us a sense of the long run cost potential for such products as TPA, etc. assuming that expression technology approaches that achieved by hybridomas. Column "B", Table VI B, presents such a case.

Fermentor Size Calculation

See Table VI A & VI B. Assume 10 kg of product are required annually.
Column A: Assuming the 150 mg / liter / week projected for the BPV system and a 50% purification yield :
kg / yr x (1 yr / 50 wks) x (liter - wk)/150 mg x 1/.5 mg /mg x 1/ .7 eff. x mg / 10-6 kg = 3800 liters.

Column B: Assuming 500 mg/liter/day in perfusion culture, 10 kg/yr x (1 yr / 50 wks) x (liter - wk)/3500 mg x 1/.5 mg /mg x 1/ .7 eff. x mg / 10-6 kg = 163 liters.

TABLE VI, A. MAJOR PRODUCTION EQUIPMENT:

	A.	B.
fermentors:	2x 2000 l. 2x 400 l. 2x 80 l. 2x 20 l.	2x 100 l. 2x 20 l. 2x 5 l.
spinner bottle facility:	yes	yes
chromatography:	2x 25 l. gel disk stacks prep HPLC U.F. systems waste treatment support tanks/ etc.	2x 25 l. gel disk stacks prep HPLC U.F. systems waste treatment support tanks/ etc.

No QC, tech. service, analytical or filling/ finishing areas included. An allowance for these is included below.

CAPITAL ESTIMATE:	$15 MM	$7 MM

TABLE VI, B. ANNUAL OPERATING BUDGET:

	A	B
salaries/ benefitsdirect)	$ 750000.00 (24 direct)	$ 500000.00 (16 direct)
overhead	$ 450000.00	$ 300000.00 (60% of S/B)
equipment depreciation	$ 1500000.00	$ 700000.00
facilities	$ 400000.00	$ 300000.00
utilities	$ 300000.00	$ 100000.00 (stm., elec., H2O)
maintenance	$ 750000.00	$ 350000.00 (5% of capital)
miscellaneous	$ 400000.00	$ 200000.00
supplies and materials	$ 7000000.00	$ 500000.00 ($4/l. + $1mm.)
TOTAL	$11,550,000.00	$ 3,300,000.00
	$ 1,600000.00	$ 980,000.00 (35% cont. on non- media part)
TOTAL	$ 13,150,000.00	$ 4,280,000.00
Unit cost at bulk:	cost/ 10,000 gm $1315 / gm ($1015)	cost/ 10,000 gm $ 428 / gm

Costs associated with Final Product Preparation:

Capital required:	$ 5,000,000.00
Operating :	$ 1,500,000.00

Cost at final:	$1465/gm ($1165)	$ 578/gm

These estimates are for manufacturing costs and exclude r&d, licensing, marketing, distribution, and profit. It is critical that they be viewed as examples for illustrative purposes only.

These estimates are for manufacturing costs and exclude r&d, licensing, marketing, distribution, and profit. It is critical that they be viewed as examples for illustrative purposes only. One conclusion from this analysis is that while a therapy requiring even 100 mg would be feasible, a therapy requiring 10 gm of product per patient treatment would need to be both effective and offer very significant benefits over alternative therapies in order to justify its cost. Remember that these are manufacturing costs. . We do not necessarily propose to use conventional fermentors, other options such as ceramic or modified hollow fiber units may prove advantageous. The operating cost of such systems would not be widely different from fermentors although in particular cases there may be significant differences in capital, utility or labor costs. This could be particularly true for production of smaller quantities of material where capital, labor and facility and utility cost could be significantly affected in the long run. To represent a real advantage, "immobilized" culture systems, whether based on perfusion reactors, on hollow fibers,on ceramics,or on nitrocellulose, collagen, alginate, or alginate formed semipermeable lysine beads require that product formation rate per cell (qp) be a relatively weak function of cell growth rate (μ). Actually in perfusion, immobilization on ceramic, and even in fiber reactors, there is some net growth primarily replacing cells which die or sluff off, etc. If cell growth strongly enhances product formation rate in the recombinant system, the economics are not favored because high levels of growth factors are required to continuously produce the new cells.

Alternatives

In selecting a production system, one first selects a host/ vector system according to availability and ideally based on process economic criteria. Next cultivation conditions and product formation kinetics should be optimized. Finally a selection of bioreactor type should be made.

The current reality is that a given organization typically has only one or two host/ vectors available (there is a significant investment in developing even one.) Media has generally not been optimized- in fact it is wasted in most perfusion experiments where very high media thruputs are used. Work to reduce serum use has begun as has media optimization in the more advanced groups.

In selecting bioreactor type, for the near term, we can rely on immobilization approaches as satisfactory for most moderate scale hybridoma cultures. Hollow fiber,suspension/ perfusion, ceramic(CRBS), alginate/lysine(Damon), and fluidized bed technologies all are likely to work if competently implemented. All could be implemented in most cell culture or fermentation laboratories. Among the technologies within a given class, there is little basis for selecting one or another on theoretical grounds. Unfortunately some of the systems being sold in these categories have shortcomings in such areas as aseptic design and even improperly designed bioreactors and oxygenation systems. Even the best are somewhat cumbersome, but all will improve with time. The published information on these systems does not include side by side comparisons. It is obvious that certain types would be inappropriate for certain host types and certain product formation kinetic behavior. This discussion must be viewed as a preliminary analysis. Ideally each organization would have available at least one of each basic type (continuous suspension or μcarrier vs. immobilized perfusion) culture technology, but resources are not unlimited for startup groups. A review of several biotech companies shows that with only one or two exceptions each is strong in only a single technology. It would be a serious error to select a perfusion system if formation of the target product is strongly growth associated. It would likewise be a mistake to purchase an "experimental" system. Fortunately, several of the systems available today for preparative cell culture have benefited from several years of improvement and are quite well engineered. Each of the three perfusion systems described as having high volumetric productivity are likely to give quite satisfactory performance for MAB preparation. Among these and a conventional stirred reactor, with or without microcarriers, a suitable system should be found and a process optimized .

Options

For a company planning to produce such products, the alternatives to the major technology development effort outlined are limited. One such alternative is to contract for a large pharmaceutical firm or a smaller contract manufacturing group to handle the process development, scaleup, preparation and even licensing and marketing. Table VII, while very approximate, illustrates the impact of such a decision on future profits. Obviously the investment at risk is also reduced, but it is important to make a careful strategic decision before contracting for production services.

TABLE VII. BUSINESS REASONS FOR IN HOUSE VS. CONTRACT MANUFACTURING.

COST COMPONENT	SALES, PROFITS OR COST COMPONENT (million$) INTERNAL	CONTRACT HOUSE	VENTURE PARTNER
MARKETING & SALES	60	60	60
OVERHEAD(10%)	30	30	30
(10%)	30	30	30
LICENSING	30	30	30
MANUFACTURING			
COST	90	90	90
PROFIT		20	20
SHARE OF PROFITS		10	10
SHARE OF PROFITS			
FOR OTHER CONTRIBUTIONS			15
TOTAL PROFITS	60	30	15

Clearly the details of the relationship can drastically influence profits.

RECEIVED August 31, 1987

POLYMER

Chapter 9

Polymers in Biotechnology

Raphael M. Ottenbrite

Department of Chemistry and Massey Cancer Center, Virginia Commonwealth University, Richmond, VA 23284

Macromolecules comprise many of the natural materials found in living matter. The most important and abundant are the proteins, nucleic acids and polysaccharides. These and other basic biomolecules were selected during the course of biological evolution for their capacity to perform specific functions. Consequently, specific biomacromolecules have evolved that function as membranes for compartmentalization, as enzymes and coenzymes for reaction catalysis, as polynucleic acids for memory and protein replication, and as polysaccharides for connective tissue, energy and structural components.

The elucidation of the complex function, structure, and modes of reaction of these biomacromolecules has been a major goal of many scientists. Recently it was discovered that some synthetic polymeric materials have physicochemical properties that are compatible with biological systems. Subsequently, polymer chemists in collaboration with basic scientists are developing new synthetic materials with attention to synthesis, macromolecular design, structure-reactivity relationships, molecular weights, solubilities and other physicochemical properties related to physiological activity and toxicity. Similarly, advances in the development of synthetic polymeric materials for artificial skin, sutures, prosthetic devices and surgical implants are being made by polymer chemists in collaboration with scientists in surgery, pathology and immunology.

Consequently, polymers and basic polymer science itself, are now playing an important role in the area of medical treatment. Although polymers have been used for a considerable length of time for such biomedical applications as prosthetic devices, cosmetic implants, and enteric coatings, only recently have their importance been recognized for drugs and for drug administration. Therefore, drug research has moved beyond the biochemistry laboratories and is securing its own position in polymer chemistry. The reason is that the requirements for any materials utilized for biological application are exceptionally demanding; they are required to be extremely pure and have specific physicochemical properties. Characteristics

0097–6156/88/0362–0122$06.00/0

that are very important in these materials are, biocompatability, durability, nontoxicity, and biodegradability.

Polymer chemists are now playing a unique and important function in developing new monomers, polymers and copolymers, evaluating the physicochemical properties, and determining how these properties relate to biological systems which are in need of medical assistance. Normally, unless topically active, a drug is administered by means of ingestion, injection, or by infusion. In all cases, the reagent enters the body and is transported to the diseased site in time dependent concentrations. A drug is generally distributed throughout an organism via the aqueous phase of the blood plasma or lipid phases of the body, reaching the tissues of an organ at a rate determined by blood flow through that organ and by the rate of passage of the drug molecule across the capillary bed and into the tissue cells. Most of the drug is excreted, held in various fluid compartments, or localized in subcellular areas such as macromolecular surfaces and in fat deposits by adsorptive or partition processes. Even within the target tissue, cellular fractionation studies reveal that most drug molecules are associated with structures not involved with the specific drug activity.

Consequently, polymer scientists have become involved in developing new and improved methods of drug delivery as well as specificity. The most difficult problem in drug administration is getting a sufficient quantity of agent to the inflicted site at a proper dosage. Ingestion and injection methods often require repeated administration wherein the whole body becomes infused with a high concentration of drug that slowly dissipates. The optimum therapeutic levels are only realized when the ascending or descending concentrations pass through this "therapeutic window". Periods of over-medication often results in acute side effects, with no therapeutic value. The ultimate answer is to achieve site specific drugs which will be delivered at both an effective concentration and at an optimal rate.

Research is now directed toward the development of synthetic polymers which will be utilized for controlled drug release and transport. A number of techniques have been devised to achieve zero-order drug delivery in an effort to avoid the problems inherent with the present methods of administration.

The polymer systems that are being explored for pharmaceutical applications include; (a) polymer drugs: polymers or copolymers that are physiologically active themselves, (b) drug-carrying polymers: polymers that have active drugs bound to a parent polymer backbone; these may be drugs perminently or semiperminently attached to the polymer backbone or pendent to it, (c) time-release drug polymers: polymers used for controlled release of the drug; the drug may be encapsulated with water soluble polymer coatings which dissolve at different rates and release the drug at various times, or the drug may be imbedded into a polymer matrix from which it diffuses at specific rates, (d) site-specific drugs: polymers that have special chemical groups attached to the polymer carrying the drug; these groups can combine with specific receptor sites on protein, cell surfaces or in the lipid areas to achieve specific binding and drug delivery.

Although synthetic polymers already play a role in biological systems, it is evident that our utilization of these substances for these purposes is only in its infancy. In order to make polymers more effective, much more has to be understood about toxicities, specificity, structure-relationships, molecular weight and other physicochemical properties. New polymers are required that will compliment the delicate balance of the biological systems. Further research into polymer stability and biocompatibility for implants and pharmacokinetics of polymers for controlled drug release and longer duration release are needed. Thus, we see before us an exciting and unique challenge to be met which will require the resources of several scientific disciplines, but with outstanding rewards to be achieved.

RECEIVED September 19, 1987

Chapter 10

Polymers in Biological Systems

Raphael M. Ottenbrite

Department of Chemistry and Massey Cancer Center, Virginia Commonwealth University, Richmond, VA 23284

The application of polymers in biological systems has become an important feature of medical care. The most common uses have been for implants and prosthetic devices; however, recent interest has been to use polymers as drugs and drug delivery systems. Polymeric drugs were the first to receive attention, and now the primary interest focuses on new modes of drug administration. Research in this area is highly interdisciplinary and involves not only polymer chemists but health scientists and physicians.

Most of the organic matter in living cells consists of macromolecules; these include not only proteins and nucleic acids but also other polymeric substances such as starch and collagen. Since the macromolecules in all living organisms are made from only a relatively few, simple, building-block molecules, it appears probable that these basic biomolecules were selected during the course of biological evolution for their capacity to serve several functions. Various nucleotides, for example, are not only utilized as building blocks of nucleic acids by the body but also as coenzymes and as energy-carrying molecules. These primordial molecules may be regarded as the precursors or ancestors of all other biomolecules; they are the first alphabet of living matter. Other macromolecules of living organisms evolved from these simple, low-molecular-weight substances which formed an ordered hierarchy of complex mesomeric structures. From this pool of complex macromolecular biological systems evolved highly specific polymeric substances that function as membranes for compartmentalization, polynucleic acids for memory and replication, amino acid combinations for a variety of proteinaceous functionalities, and polysaccharides for energy storage, structure, and connective tissue applications.

The elucidation of the complexity of both function and structure of these biological macromolecules has been a principal goal of many scientists. Recently, polymer chemists have become involved in the development of separation and purification techniques aimed at the

0097–6156/88/0362–0125$07.50/0

identification of the function and structure of these compounds. In addition, polymer chemists have been exploring synthetic methods of preparing both natural and synthetic macromolecular analogs for biological use. The evaluation and modification of these materials has led to the preparation of an array of polymeric compounds which have been applied to biological systems (1-3).

This chapter, "Polymers in Biological Systems," was chosen to introduce the pertinent research in this interdisciplinary field. The economic impact of polymers in biological systems is illustrated in Table I. The present market is in excess of 15 billion dollars

Table I. Polymers in Biological Systems

Biopolymer	Billions 1985	Billions 1990
Pharmaceuticals	0.3	2.5
- Delivery Systems		
- Controlled Release		
Agriculture	0.2	1-3
- Hormones		
- Immunostimulants		
- Pesticides		
- Nutrients		
Prosthetic Devices	7.0	12-15
- Contact Lenses		
- Dental		
- Joints		
- Catheters		
Artificial	2.5	3.5
- Kidney		
- Heart		
- Skin		
General Medical Care	5.0	10-20
Total Market (Billions)	15.0	28-48

with a projected growth to about 50 billion in the next five years. The topics that will be discussed include: basic studies of prosthetic device implants; biocompatibility of polymers; drug administration (by polymeric drugs, polymeric drug carriers and polymeric drug delivery devices). The purpose of this chapter is to give researchers a comprehensive overview of current knowledge, developments, and trends in this rapidly expanding and exciting field.

Polymeric Implants

Biomaterials used for prosthetic devices are defined as substances which are used in the human body for a prolonged period of time without applicable changes in their properties. Although biomaterials range from strong metals to brittle ceramics, polymers, with their wide range of properties including strength and hardness are presently the major source of biomaterials. Listed in Table II are some physical properties of polymer materials presently being used. Many devices such as catheters, shunts, cannulae, orthopedic

Table II. Variable Polymer Properties

Property	Polymer	Use
Hard	Methylmethacrylate	Replacement Parts, Teeth and Bone
Soft	Silicone	Can be Made to Feel Like Human Tissue
Tensile Strength	Kevlar, Polyethylene	Reinforcement of Prosthetics
Transparent	Methylmethacrylate, Polyacetals	Contact Eye Lens
Biodegradable	Poly(Galactic Acid), Orthoester Polymers	Sutures Controlled Drug Delivery

devices, arterial grafts, and heart valves have successfully been fabricated and have been successfully implanted through surgical means and have functioned for many years (4). The ultimate goal in the development of biomaterials is to implant independently functioning prosthetic organs such as kidneys, lungs, and hearts into a living host (5).

The replication of living tissue, with the ability to respond and repair itself after physical trauma or debilitating diseases, is a major goal of prosthetic research. Presently, one of the greatest problems with surgical implants is the rejection of the foreign biomaterial by the living organism (6). Even the most chemically inert materials can cause blood clotting or tissue encapsulation. Furthermore, transplants such as heart valves and arterial grafts must be designed to function for more than 30 years. Consequently, meticulous research is continuing for materials or composites that will maximize the desired properties and minimize the undesired side effects listed in Table III.

Table III. Criteria for Biomaterials

Biopolymer Must Have	Biopolymers Must Not
High Purity	Cause thrombosis
Chemical, physical, and mechanical properties to meet proposed function	Destroy cellular elements
	Alter plasma protein
Easy fabricability	Destroy enzymes
High stability	Deplete electrolytes
Sterilizability	Cause immune responses
	Cause cancer
	Produce toxic and allergic reactions

Sutures are the most commonly implanted biomaterials in humans. Thirty years ago cotton thread, synthetic nylon, polyester fibers, and metal clamps were commonly used for this purpose. These materials and the first absorbable suture, cat gut, which absorbs in 40 days, cause significant inflammation of the wound and impede the natural healing process. Consequently, there has been considerable interest in the development and subsequent application of absorbable sutures composed of biocompatible materials. The two principal absorbable materials now used are polyglycolic acid provided by American Cyanamide (120 days) and poly(glycollic acid-co-lactic acid) by Ethicon Incorporated (Vicryl - 90 days).

The application of these two products as sutures is based on their ability to be readily hydrolyzed in a physiological environment and the resulting hydrolysis products are readily metabolized without causing any physiological reaction. Continued research in this area has brought about new products with better degradation properties and biocompatibility (7).

The second most commonly known biological implants are contact lenses (8). These were initially developed from polymethyl(methacrylate) which was found to be compatible with the eye based on the observed tolerance of windshield fragments imbedded in the eyes of some World War II pilots. The technology to develop contact lenses has evolved over the last 30 years to the point that there are over 40 different types of gas permeable hard contact lenses on the market with variations in color, oxygen permeability, shape and patency (Table IV). Soft contact lenses that have been developed for comfort and extended wear represent another large segment of this industry. Permanent intraocular lenses are now being developed for replacement of natural lenses lost due to injury, cateracts, and other eye diseases (9).

The utilization of synthetic or modified natural polymers for prosthetic devices is still in its beginning phases. Some of the present applications of this important area of surgical medicine are shown in Figure 1 (10).

Biological Compatibility of Biopolymers. Tissue and blood compatibility are essential requirements of any biopolymer introduced into a biological system. The activity of a biopolymer in an organism is directly related to its structure and its solubility in the tissue or tissue fluids. The biocompatibility of an implanted material or prosthetic device is a dynamic two-way process that involves the time dependent effects of the host on the material and the effects of the material on the host. The implantation of any material whether it is a natural polymer or a synthetic material triggers a series of complex biological mechanisms which include inflammation chemotaxis and phagocytoses (11).

Inflammation is the host's response to injury or the presence of a foreign material. The intensity and duration of the response is determined by a variety of biological mediators responding to the size and nature of the implant, the trauma due to implantation, and the site of implantation. Biological systems seem to be much more tolerant of implants that come in contact with tissue alone rather than with blood. Foreign body reaction within tissues can produce inflammation, tissue encapsulation or cell-change response in the surrounding tissues. In addition, reactions remote from the site, such as carcinogenic and antileukotactic responses, can occur. Within blood, however, foreign object recognition is much more sensitive (12). Initially, there is a rapid adsorption of plasma protein. This may be followed by a number of responses such as platelet adhesion and aggregation, thrombin formation, and a fibrinogen activation that could result in blood clot formations.

Implanted polymers are generally subjected to severe biological environments which can cause the polymer to degrade and thus changing the properties of the polymer. More important, these degradation byproducts may be much more sensitive to the surrounding tissue than the polymer itself. Furthermore, the leaching of small molecules such as monomer, initiator, oligomers, plasticizers, additives, emulsifiers from implants, and products of hydrolytic or enzymatic reactions on the polymer material can cause serious tissue reactions. These nocuous materials are usually transported by diffusion and can cause inflammation, edema, antigenic reactions, and pyrogenicity. Chronic inflammation, local tissue damage and neurosis can lead to serious infections. Mechanical irregularities or polymer fragments caused by abrasions have produced fibrous tissue growth as well as tissue encapsulation.

In blood, it is extremely important to clarify the influence of adsorbable plasma proteins on an implant; the subsequent platelet adhesion and aggregation when designing new nonthrombogenic polymers

Table IV. Hard Gas Permeable Lens Materials

Company	Name Lens/Material	Polymer
A.I.T.	Medicon	Sil/Acry
BH/HC	Cabcurve	CAB
BH/HC	GP II	CAB
Biocontacts (Canada)	Revlens	But/Acry
Bio Products	Oxy-PMMA	Sil/Acry
Corneal Contact	Alberta (SM38)	Sil/Acry
Danker	Meso	CAB
Dow Corning	Silicon	Silicone
Frigitronics	Opus III	N.A.
Frigitronics	Saturn II	N.A.
Fused Kontact	Sil O_2 Flex	Sil/Acry
Glasflex	Dioxyflex	Sil/Acry
Glasflex	Electrocab	CAB
Neefe Optical	Siloxycon 14	Sil/Acry
Neefe Optical	Bioflex	CAB
Ocular Technology	OTC V	Sil/Acry
Ocular Technology	OTC VII	Sil/Acry
Optacryl	Optacryl 95	Sil/Acry
	Optacryl 60	Sil/Acry
	Optacryl K	Sil/Acry
Paragon	Paragon 95	Sil/Acry
Paragon	Paraperm O_2	Sil/Acry
Paragon	ParaCab II	CAB
Polymer Technology	Boston I	Sil/Acry
Polymer Technology	Boston II	Sil/Acry
Rynco	Rx 56	CAB
Syntex	Polycon I	Sil/Acry
Syntex	Polycon II	Sil/Acry
Toyo	Menicon O_2	Silicone
Wesley Jessen	Airlens	Alkyl Styrene
GBF	BFB 5	N.A.
Bio Medic Polymers	Oxycon	Sil/Acry
Bio Medic Polymers	Oxycon 32	Sil/Acry

Sil/Acry = Silicone/Acrylate; But/Acry = Butyl Acrylate

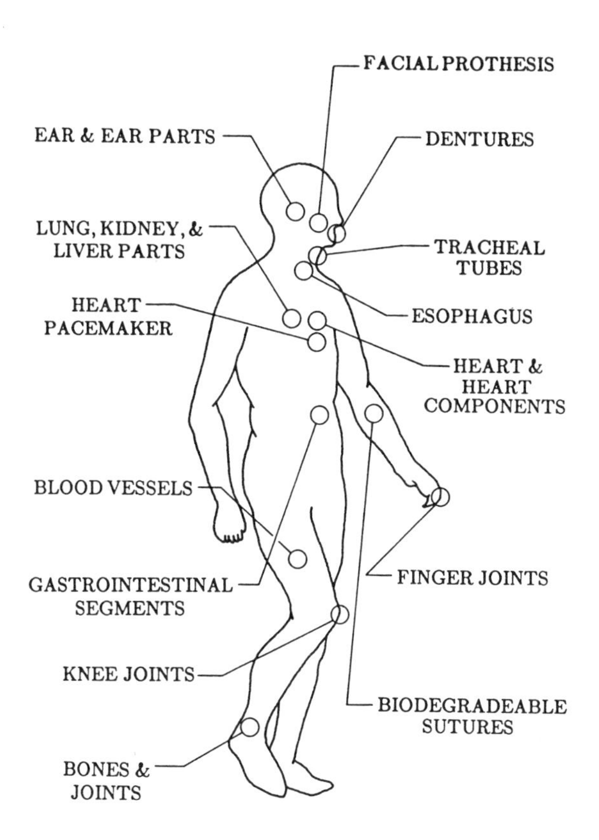

Figure 1. Applications of synthetic or modified natural polymers for prosthetic devices.

(12). One of the best known methods of improving blood compatibility is by modifying the surface of the implant with substances that interfere with thrombin and fibrin activation. Heparin, heparinoids, and mucopolysaccharides are among the most effective water soluble polymeric anticoagulants. Heparin is a naturally occurring sulfated polysaccharide and is found in trace amounts in most mammalian tissue. Heparinoids are biopolymers prepared by the partial degradation of natural polysaccharides that are sulfonated; these modified materials exhibit promising anticoagulation properties. Mucopolysaccharides are naturally occurring carbohydrates similar to heparin, but have fewer sulfonate groups and have weaker anticoagulant activity (13).

Continual systemic heparinization is a difficult procedure because of problem with controlling the dosage and the possibility of initiating hemorrhaging. A number of years ago, however, it was found that when a polymeric material was treated with heparin that the resulting surface was much more compatible in a plasma environment, and that coagulation was decreased. Heparinization has now become a standard method for treating devices that are to be placed in contact with blood. The principle method of heparin attachment has been by electrostatic bonding, but a considerable amount of effort has been made to covalently bond heparin to polymer surfaces. Heparinization has also been accomplished by adding heparin to the bulk polymer; the heparin then slowly diffuses out of the polymer matrix effecting body compatibility of the polymer device (14).

Nonthrombogenic properties have also been exhibited by materials other than the heparin-types. Block polymers of polyurethane (DuPont Co.), polyether polyurethanes (Stanford Research International), polyurethane with silicone (Avoco Corp.) and fluorocarbon polymers have all shown appreciable anticlotting activity. A recent unique approach to controlling rejection of implants involves the attachment of clot-lysing agents such as streptokinase and urokinase to the polymeric material. The theory is that if a clot does begin to form on the surface of an implant, it is immediately lysed by the attached enzyme. Another approach has been to replicate the vascular endothelium which has a macrophase structure composed of hydrophilic and hydrophobic microdomains. Segmented-synthetic polymers of this type have been prepared and are being evaluated with promising results (15). Various types of polyionic complexes prepared by combining polyanions with polycations are also being evaluated for "polymer-platelet" interaction. All of these new techniques have produced some interesting results (16).

Drug Administration

Drugs are generally distributed throughout the body in the aqueous phase of the blood plasma. Unless they are topically active, most drugs must first enter the blood system. They reach the tissues of each organ at a rate determined by the blood flow through that organ and by the rapidity of the passage of the drug molecule across the capillary bed and into the tissue cells of that particular organ. Most drugs are retained within the blood plasma due to solubility or binding to cells and substances such as plasma proteins and do not diffuse freely out of the plasma. Consequently, the amount of drug present in the tissues at the site of activity is only a small portion of the total drug present in the body. The majority of the drug remains in various fluid compartments or is localized in subcellular particles, at macromolecular surfaces, and in fat depots by adsorptive or partition processes. Even with the target tissue, cellular fractionation and radioautographic studies reveal that most drug molecules are associated with structures that have nothing to do with the specific drug effect.

Consequently, one of the most difficult problems in drug administration is getting the agent in sufficient quantity to the desired site for the required period of time. With conventional delivery systems, such as oral or injection methods, it is necessary to administer repeatedly, and the entire body thus becomes infused with the drug. Furthermore, if drug administration interval is increased, there will be periods when an insufficient amount of the drug is present, and the disease or infestation can reoccur. Alternatively, reapplication can lead to a build-up of the agent in the body to the point where toxic levels are exceeded, as shown in Diagram I. Thus the therapeutic utility of many drugs is often limited by their short in vivo half-lives, lack of specificity, acute toxicity, and undesirable side effects.

Attention, therefore, has been directed to the development of therapeutic systems for the controlled administration of pharmacologically active agents that utilize synthetic polymeric materials as predictable barriers for the regulated release and transport of drugs. A number of techniques have been devised to alleviate many of the problems inherent with the repeated dosages associated with oral and injection methods. The methods presently being explored include: (a) polymeric drugs - these are polymers that are physiologically active either as polymers themselves or are copolymers of active monomers; (b) polymeric drug-carriers - these are polymers that have active drugs covalently attached to the polymer backbone; (c) polymeric prodrugs - these are polymers that have pendent drug molecules that are only active after being released from the polymer; (d) polymeric drug delivery devices - these involve encapsulation of a drug into a polymer depot from which the drug is released and is disseminated by diffusion *in vivo*. These include osmotic pumps, microcapsules, liposomes and dermal patches.

Polymeric Drugs. Polymeric drugs are macromolecules that elicit an efficaceous biological activity on their own. Current developments in the areas of biochemistry and molecular biology have resulted in the elucidation of the structure and function of many polymer molecules essential to the overall metabolism of living organisms. Consequently, there has been significant development in research regarding the synthesis and the properties of physiologically active polymers. It is now possible to prepare macromolecules with predetermined structures, discrete molecular weights, and with specific functional groups. The major reasons for the development of polymeric drugs are that they exhibit delayed action, prolongation of activity, decreased rate of drug metabolism, and drug excretion.

The biological activity demonstrated by these polymeric compounds may be the result of the polymer itself, such as in the case with polyanions and polycations (*16,17*) which are made from inactive

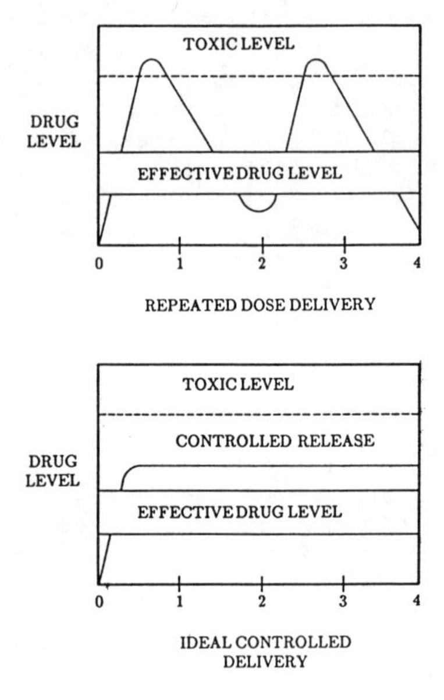
TOXIC LEVEL
DRUG
LEVEL
EFFECTIVE DRUG LEVEL
0
1
2
3
4
REPEATED DOSE DELIVERY
TOXIC LEVEL
CONTROLLED RELEASE
DRUG
LEVEL
EFFECTIVE DRUG LEVEL
0
1
2
3
4
IDEAL CONTROLLED
DELIVERY

DIAGRAM 1

monomers or may be the result of polymerization or copolymerization of an actively known monomeric drug. Urea formalde-hyde copolymer, for example, known by the trade name "Anafelx", has been used as an antibacterial and antifungal agent. Other polymeric drugs currently being used include; poly-N-oxides, which inhibit silicotic fibrosis and significantly increases the self-purification of the lungs from quartz dust, and copolymers of formaldehyde with sulfapyridine or sulfanilamide which possess antimalarial activity (18). Listed in Table V are several active polymeric drugs and their medical applications.

Table V. Medicinal Applications of Some Polymeric Drugs

Antibacterial agents	Quaternary Ammonium polymers Polyanionic polymers Polypeptides
Antifungal agents	Polyanionic polymers Urea-formaldehyde
Antiviral agents	Polyanionic polymers Urea-formaldehyde
Antitumor agents	Polyanionic polymers Quaternary Ammonium polymers
Anticoagulants	Polyanionic polymers Heperin Heparinoids
Antisilicosis	Poly-N-oxides
Plasma extenders	Polyvinylprolidones Dextran Gelatin
Hemostatics	Carboxymethylcellulose Polyanions

Many active drug monomers can lose their activity in the polymeric form, but others exhibit enhanced activity. It was found, for example, that most amino acids exhibit no antibacterial activity in the monomer form but became very active when polymerized and administered as the cationic salts (Table VI). The most notable is L-lysine, which has no apparent antibacterial activity as the monomer but is very active against E. coli and S. aureus as the polyamide. Donaruma (19), who has prepared a large number of polymers and

Table VI. Antibacterial Activity of Poly-Basic Amino Acid Polymers

	E. Coli µg/mL	S. Aureus µg/mL
L-lysine·HCl	-	-
Di-L-lysine·HCl	450	-
Poly-L-lysine·HCl	2.5	1
Poly-DL-lysine·HCl	5	3
DL-Ornithene·HCl	-	-
Poly-DL-ornithene·HCl	10	5
DL-arginine	-	-
Poly-DL-arginine	10	5

copolymers by utilizing different active drugs as monomers, has concluded that polymerization or copolymerization of drugs can enhance, decrease, or have no effect on the activity elicited by the resultant material. He also found that the incorporation of a drug into a polymer chain could affect toxicity. The polymerization of active drug monomers has in many cases decreased toxicities, but it has also been shown to enhance or create new toxicities.

In evaluating structure-activity relationships in polymer drug systems, two of the most important variables that apparently regulate biological activity are molecular structure and stereochemical configurations. These two features are manifested in most polymer systems by molecular weight, copolymer properties, chain coiling and/or folding, cross linking, tacticity, and electrolytic character (20). With drug copolymers it has been shown that the nature of the nondrug comonomer is very important and that biological activity is not just related to the drug comonomer alone. It appears, therefore, that certain properties characteristic of polymer systems may correlate with biological activity. Consequently, there exists a great need for more detailed structure-reactivity studies of polymeric drugs to serve as a basis to facilitate future drug design.

Polymeric drugs have been applied most extensively in the treatment of cancer. Synthetic polyanions have shown excellent potential for this pathology. For example, poly(vinyl-sulfonate) inhibits L1210 leukemia, Krebs 2 carcinoma and Ehrlich ascites. Many carboxylic acid polymers elicit antitumor activity but are also very toxic. The most widely studied polyanionic polymer is poly(divinyl-ether-co-maleic acid) known as pyran or MVE-2. This polymer has a wide range of biological activity including, antitumor, antiviral, antifungal and antibacterial activities. More recently, we have developed a polymer, poly(maleic anhydride-alt-cyclohexyl-3,5 dioxepin) which exhibits little toxicity and has produced the first murine survivors to Lewis lung carcinoma (21).

Since nucleic acids and enzymes play such a large role in the replication of cell materials for mitosis, a considerable amount of research has been conducted in this area to control virus replication. On the molecular level, analogs of nucleic acids are capable of forming complexes with adenine, cytosine, uracil, thymine, and guanine. Through complexation, these nucleic acid analogs are potential inhibitors of biosynthesis and require nucleic acids as templates. The polyvinyl analogs of nucleic acids are one of the few polymers that have been tested in living systems to investigate their bioeffects. The most thoroughly investigated polymer is poly[vinyl adenine, which has been reported as being effective against viral leukemia, chemically induced leukemia, and infections caused by other viruses (22). The inhibition of viruses by the complexation of nucleic acids with their polymer analogs is apparently virus specific. For example, poly(9-vinyladenine) inhibits viral replication through the reverse transcriptase step, while poly(9-vinylpurine) is ineffective.

Other antineoplastic polymers include the aziridine alkylating funtionalized polymers cyclophosphoramides, mustard type alkylating functionalized polymers, and conjugates of methotrexate, adriamycin and cis-platinum (23). More recently, smaller macromolecules have been found to be active such as muramyl dipeptide (24) and neocarzino-statin conjugated to poly(styrene-co-maleic anhydride) (25).

Biopolymeric Drug Carriers or Conjugate Polymeric Drugs. The utilization of macromolecules to serve as drug carriers is a technology that holds a great potential for drug administration. The principle reason is that many drug companies have drugs that are now nearing the end of their patent protection period. However, by involving these drugs in new controlled release or delivery methods of administration, new patents can be obtained.

Most medications are micromolecular in size and, as such, are relatively free to diffuse throughout the biological system. Consequently, drugs have been inherently difficult to administer in a localized, concentrated mode within the primary target tissues and organs. Since polymers diffuse slowly and are often adsorbed at interfaces, the attachment of pharmaceutical moieties produce a biopolymer with distinct pharmacological activity. The main reason for the development of these "polymeric-drug carriers" is to obtain desirable properties such as sustained therapy, slow drug release, prolonged activity, and drug latenation, as well as decreased drug metabolism and excretion (Table VII).

A model for pharmacologically active polymer-drug carriers has been developed by Ringsdorf and others (20), similar to that shown in Diagram II. In this schematic representation four different groups are attached to a biostable or biodegradable polymer backbone. One group is the pharmacon or drug, the second is a spacing group, the third is a transport system, and the fourth is a group to solubilize the entire biopolymer system. The pharmacon is the agent that elicits the physiological response in the living system; it can be attached permanently by a stable bond between the drug and the

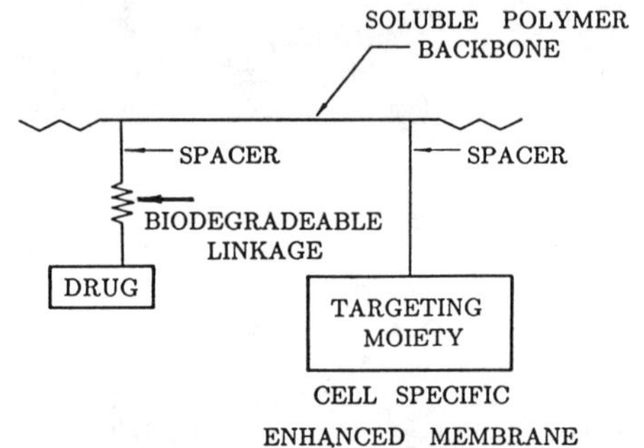

MODEL FOR POLYMERIC DRUG CARRIER

DIAGRAM 2

Table VII. Characteristics of Polymeric Drug Carriers

Depot effects	Slower diffusion of drug Slower absorption of drug Slower elimination of drug Nonadsorbability for topical application
Pharmokinetic	Sustained release of drug Variable drug metabolism routes
Body distribution	Localization of drug Inhibition of adsorption in some areas due to molecular weight Cell-specific interactions and uptake of drug Protein binding
Pharmacological activity	Variable activity from none to an increase Variable toxicity Incorporation of drug combinations on polymer backbone Reduction of side effects such as nausea and irritation common with large doses of drugs

polymer, or it can be temporarily attached and removed by hydrolysis or by an enzymatic process. The transport systems for these soluble polymer-drug carriers can be made specific for certain tissue cells by the use of homing or "targeting" devices such as pH-sensitive groups, receptor-active components, i.e. antibody-antigen, or they may be made nonspecific. Solubilizing groups, such as carboxylates, quaternary amines and sulfonates, are added to increase the hydrophilicity and solubility of the whole macromolecular system in an aqueous media, while large alkyl groups adjust the hydrophobicity and solubility in lipid regions. Another important feature in a polymer-drug carrier is to move the pharmacon away from the polymer backbone of other groups so that there is a minimal structural interference with the pharmacological action of the drug.

The method of attachment of a drug to the polymer is dependent upon the ultimate use of the adduct. Further, the chemical reaction conditions for the attachment of the drug to the polymer should not adversely affect the biological activity of the drug. Temporary attachment of a pharmacon is necessary if the drug is active only in the free form. If the drug is only active after being cleaved from the polymer chain then it is called a prodrug. This is usually the case with agents that function intracellularly. This form of attachment usually involves a hydrolyzable bond such as an anhydride, ester, acetal or orthoester. Permanent attachment of the drug moiety is generally used when the drug exhibits activity in the attached form. The pharmacon is usually attached away from the polymer chain and other pendent groups (Diagram II) by means of a spacer moiety which allows for drug-receptor interaction. For example, catecholamines were ineffective when bonded directly to polyacrylic acid but do affect heart rates and muscle contractions when attached away from the backbone of the macromolecular carrier. Similarly, isoproterenol was found to elicit a pharmacologic response only when coupled to a polymer by pendant azo groups but not when directly attached to the polymer chain.

Drug targeting to a specific biological site is an enormous advantage in drug delivery since only those sites involved are affected by the drug and not the whole body which can have serious side effects. Ideally, a targetable drug carrier is captured by the target cell to achieve optimum drug delivery while minimizing deposition elsewhere in the host.

Fluid-phase uptake of macromolecules by cells in general is a slow process and most administered macromolecules are cleaved from the host before any significant uptake takes place. However, if the macromolecule contains a moiety that is compatible with a receptor on a specific cell surface then the macromolecule is held to the cell surface and the uptake by that cell is tremendously enhanced. This allows the targeted drug carrier maximum opportunity for specific-cell capture with minimum deposition elsewhere in the host. Kopecek and Duncan (26) have successfully developed this type of cell-specific targeting to hepatocytes, with galactosamine; T-lymphocytes, with anti-T cell antibodies; and mouse leukemia cells, with fucosylamine.

The ultimate fate of drugs and drug metabolites is a major concern for all drugs; if they are not cleared in a reasonable time, they could promote undesirable side effects. Polymeric drug carriers are usually nonbiodegradable and because of their size, they could accumulate in the host with the potential of future deleterious effects. Consequently, the basic polymeric system used by Kopecek and Duncan is N-(2-hydroxypropyl) methacrylamide, which was originally developed and used as a plasma extender. It has been established to be nontoxic and well tolerated in the host as well as being nonimmunogenic. To further enhance the elimination of this macromolecular drug carrier, Kopecek has cross-linked smaller segments of N-(2-hydroxylpropyl)methacrylamide with biologically hydrolyzable peptides until the optimum macromolecular size for drug delivery was achieved (27). After intravenous administration, *in vivo* experiments showed that these peptide cross-links are cleaved to produce smaller components that are easily excreted from the kidneys.

The specificity of polymer pharmacokinetics is due largely to their molecular weight, which hinders their transport across compartmental barriers. However, by controlling the molecular weight of the polymeric carrier it is possible to regulate whether the drug passes through the blood-brain barrier, is excreted by the kidney, or is accumulated in the lymph node, spleen, liver, or other organs. The application of fundamental macromolecular transport theory to biopolymers and biological tissues has been successfully applied to the design, fabrication, and prediction of in vivo performance of controlled drug delivery systems. Consequently, by using only the variability of molecular weight, polymeric drug-carriers have been developed to perform much more specifically than the drug alone. When one considers the many variables such as composition of the polymer chain, structure, polyelectrolytic character, and solubility that can effect polymer behavior the chemist has a great deal to consider during the development of new polymeric drug carriers.

The full spectrum of applications for this targeted drug carrier system has only been implied. The utilization of the total concept, which includes a biocompatible drug-carrier with selective cell targeting, controlled-drug release and biodegradability, will provide one of the most potent drug administration systems in the future.

Drug Delivery Devices. Among the first drug delivery devices used by the pharmaceutical industry involved encapsulation methods. Enteric-coated pills were originally introduced to resist the action of gastric fluids. They were designed to disintegrate and dissolve after passage into the intestine. A major purpose of these coatings was to prevent nausea and vomiting induced by drugs that caused local gastric irritation, to achieve a high local concentration of a drug intended to act in the intestine or lower bowel region, to produce a delayed drug effect, or to deliver a drug to the intestine for optimal absorption there. Enteric coatings are usually composed of fats, fatty acids, waxes, shellac, or cellulose acetate phthalates that quickly dissolve once ingested orally.

Recently, more complex laminated coatings have been introduced are are referred to as "sustained release" medications (28) (Figure 2). The drug may be applied in soluble form to the outside polymer layer of a tablet containing a less soluble but porous core. More of the same drug is trapped inside the core. This is slowly dissolved by intestinal fluid, which gains access through pores in the polymer core matrix. Another variation of this method involves incorporation of the drug with coatings subject to dissolution at different rates, so that the total drug dose is released over a long period. Several variables affect the rate of release of drugs, and these are used in the development of sustained-release preparations. The size of the tablet and concentration of drug determine the surface area for solution and the rate at which the drug will dissolve. The pore size of the inert matrix, its resistance to sloughing, the presence of water-soluble substances in the matrix, and the intrinsic solubility of the drug in an aqueous medium all affect the release rates.

Rose and Nelson and others (29), have developed an implantable osmotic pump for drug delivery. This pump consists of three parts; a drug chamber, a salt solution chamber, and a water environment or water chamber (Figure 3). The drug and the salt chambers are separated from an in vivo aqueous environment by a semipermeable membrane. The water diffuses into the salt area expanding this chamber, which in turn exerts pressure on the drug-separating latex diaphragm causing the drug to be pushed or "pumped out" of the device through an orifice at a controlled rate into the biological system. Evaluations of this system indicate that similar rates of delivery are encountered for in vitro as for rats and mice in vivo.

A diffusion controlled drug delivery from a polymer matrix is

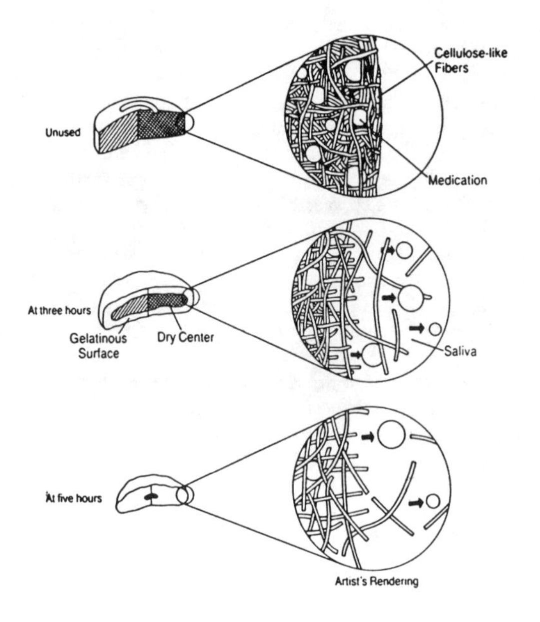

Figure 2. Oral controlled release medications.

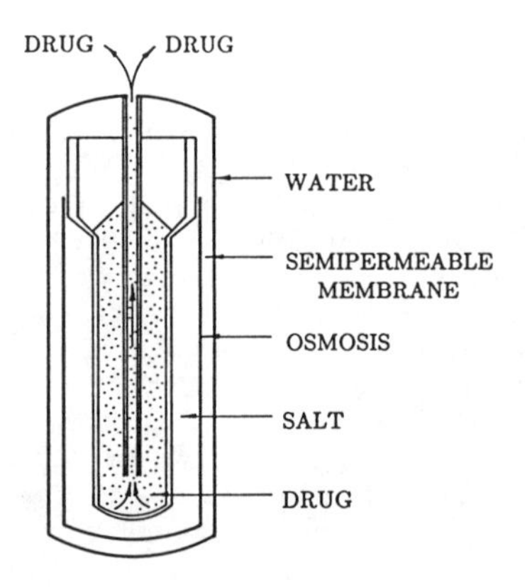

Figure 3. Implantable osmotic pump for drug delivery.

called a monolithic system. In this method the drug is encapsulated as a homogeneous dispersion, or it is dissolved throughout a rate-controlling matrix such as a gel (Figure 4). The drug diffuses at a given rate to the surface and then escapes into the biological environment. This process was first used for commercial flea collars such as Shell No Pest Strip and Hercon dispensers. These allow the release of pesticides by rate-controlled diffusion from the polymer matrix.

The membrane reservoir system is another diffusion-controlled device that involves the dispensing of a drug through a rate-controlled membrane (Figure 5). Two major purposes can be served by this technique; a long constant intravenous infusion and a prolonged local release or drug action to tissues in a specific area. This method avoids many toxic effects that could occur if the drug were introduced into the general circulation of the body. For example, Ocursert, an ocular therapeutic system is a device that delivers pilocarpine, an opthalamic drug for the relief of glaucoma (30). The device is placed under the eyelid where it continuously releases drug through rate-controlling membranes (Figure 6). This process obviates the need and complication of eye drops.

It has been found that insulin can be released with near zero order kinetics from implant devices. Glucose sensitive membranes which increase their permeability in the presence of glucose are being developed (31). Other microporous membranes, containing amine groups and entrapped glucose oxidase, have been found to alter insulin permeability in response to external glucose concentration (32). A cellulose membrane for self-regulated insulin delivery based on the principle of competitive and complementary binding behavior of concanavalin A with glucose and glycosylated insulin is also being developed.

Diabetic rats have been shown to exhibit normal blood sugar for a 30-day period with one pellet. Similarly, progesterone has been incorporated into a Silastic depot which is then directly inserted into the uterus of rabbits. The subsequent slow release of very small quantities of this hormone affected local contraceptive action without significant absorption and without suppressed ovulation.

Subdermal implants of microcapsules of drugs in a polymer matrix is another important device used for drug delivery. Originally these devices were composed of nondegradable materials such as silicon rubbers and were subdermally inserted via a trocar. A continuous and constant release of antifertility steroids were successfully delivered for over one year. After the device is surgically removed, a rapid return to fertility was observed.

The first implants were in the form of pellets and the drug release rates were dependent upon the surface area of the pellet. Subsequently, studies of controlled release from bioerodable polymers such as poly(lactic acid) and poly(glycolic acid) have been investigated based on the success of these materials as absorbable sutures (Figure 7). The drug is released as the implant begins to erode or break down. These devices are carefully designed so that a constant rate of release is obtained. Recently better bioerodable (biodegradable) polymers have been developed such as poly(esters) (33) poly-(orthoesters) (34) and poly(anhydrides) (35). The main advantage of these devices over others is that time related delivery can range

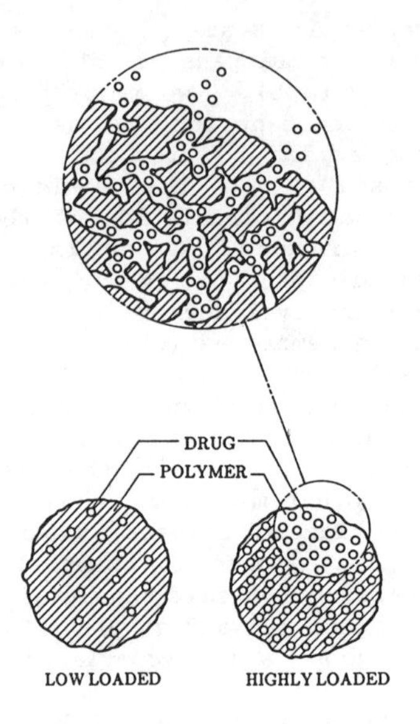

Figure 4. Diffusion through polymer matrix.

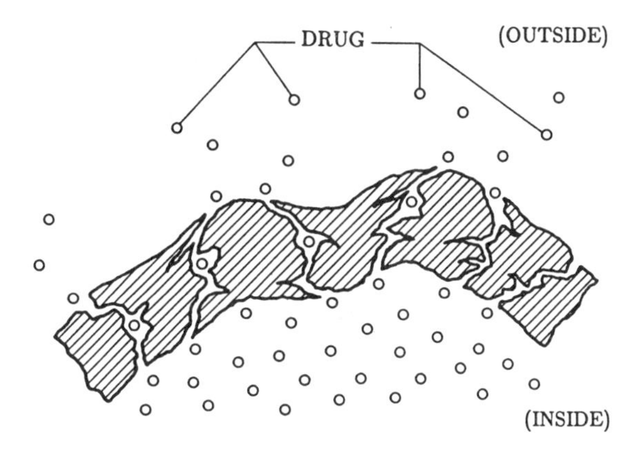

Figure 5. Diffusion through polymer membrane.

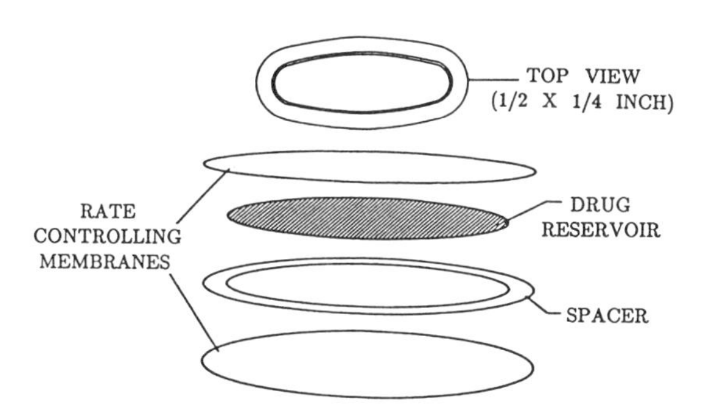

Figure 6. Pilocarpine ocusert system

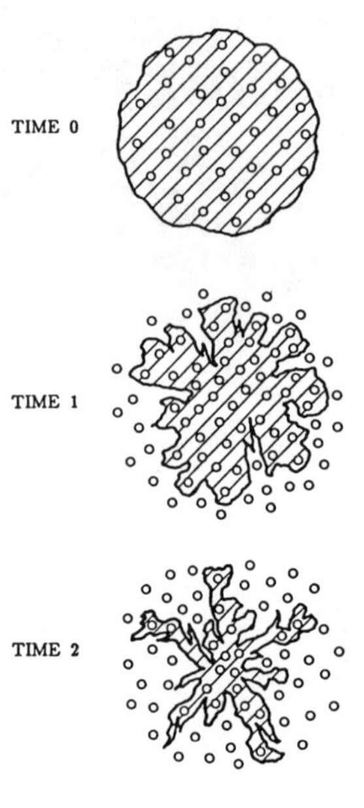

Figure 7. Biodegradable surface eroding system.

from a few months to several years and they do not require surgical removal.

Transdermal drug delivery may well become the preferred method of administration for 25% of the prescription medication in the near future (36). The skin is a multilayered organ and some years ago it was thought to be impermeable to chemical intrusion. However, the observed effects from exposure to toxic substances has lead to the concept of utilizing the skin's permeability for drug delivery. The first commercial product was Transderm-Scop which delivers scopolamine transdermally for the prevention of motion sickness (Figure 8). The rate of delivery is controlled by the diffusion properties of scopolamine through a membrane.

A similar device, Transderm-Nitro, is used for delivery of nitroglycerin for heart ailments. Two monolithic systems, Nitro-Dur and Nitro-disc, are also on the market (Table VIII). In these latter

Table VIII. Transdermal Delivery of Nitroglycerin

Approved Transdermal Products

Key Pharmaceuticals	Nitroglycerin	Monolith Angina
Searle Pharmaceuticals	Nitroglycerin	Monolith Angina
Ciba-Geigy	Nitroglycerin	Membrane Angina
Ciba-Geigy Sickness	Scopolamine	Membrane Motion
Boehringer-Ingleheim Hypertension	Clonidine	Membrane

systems, the rate of release depends on the concentration differences between the encapsulation matrix and the skin; consequently, the device releases whatever the skin is capable of transporting. Nitroglycerin, which has a wide therapeutic index, has good skin permeability while scopolamine binds to skin proteins, and these sites need to be saturated before effective administration occurs. Several new drugs are presently seeking FDA approval for transdermal administration (Table IX). It is evident that the utilization of transdermal systems will receive considerable attention in the near future. The principle reason is that many pharmaceutical houses have profitable drugs that are now coming to the end of their patent protection period. However, by incorporating these drugs into new controlled drug release or delivery methods of administration, patents can be extended without having to repeat many of the costly procedures that are associated with FDA compliancy. Therefore, extended market protection, profitability, and improved efficacy are major incentives. The advantage of transdermal delivery is that drugs can be delivered over a long period of time at a constant rate and that administration can be aborted by simply removing the transdermal patch.

Table IX. Drugs in Research and Development

Category	Drug
Hormone	Estradiol Estradiol esters
Cardiovascular	Isosorbide dinitrate Timolol Propranolol
Analgesic	Salicylate
Antihistamine	Chlorpheniramine
Cholinergic	Physostigmine

Liposomes were first used as model biomembranes and are presently being evaluated as drug carriers for improved drug therapy. Liposomes are bilayer molecular assemblies of molecules that have long hydrophobic carbon chains with a hydrophilic head (Figure 9). These bilayer vesicles are of interest as drug delivery systems since water soluble drugs can be encapsulated in the central aqueous core, and nonpolar drugs can be situated in the hydrophobic bilayer area of the liposome. This potential application received extensive initial interest which soon dimmed because of two major drawbacks; first, these assemblies are not very stable and "leaked" or release encapsulated material very rapidly; and secondly, they are taken up by the liver. Consequently, the drugs were released into this metabolitically active area and large amounts of lipid material (37) were being deposited in the liver. These difficulties stimulated research which solved several pharmaceutical problems such as stability, sterility, and scale up. Now practical production of liposome drug formulations for human use are available.

The liposome appears to be a specifically ideal system for drug delivery to the lung. These bilayer systems do not appear to irritate the lung or invoke foreign-body response. Research has shown that guinea pigs after aeresol administration of antihistamine in a liposome formulation resisted a challenge by histamine. Furthermore, these liposome-encapsulated bronchodiadators produced significantly lower heart rates (37). Related to this, Sunamoto (38) has effectively delivered antibacterials to the alveolar cells by coating the liposome surface with polysaccharides. The polysaccharides not only stabilize the liposomes, but target them to the alveolar cells. This liposomal system provides for an effective treatment of Legionnaire's disease.

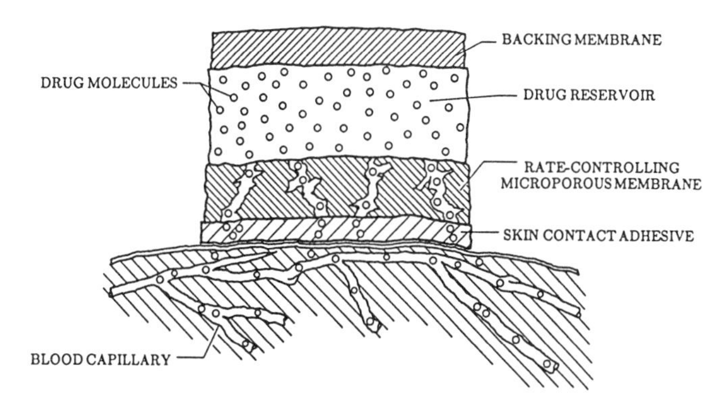

Figure 8. Transdermal therapeutic system.

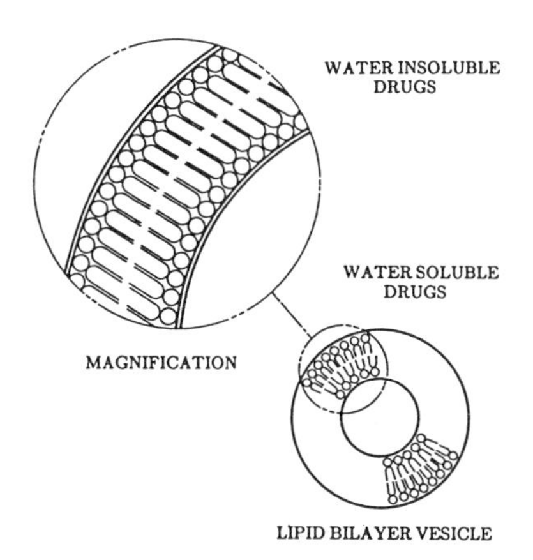

Figure 9. Liposome.

Site specific delivery of liposomes is the present research thrust in this area. The ability to specifically destabilize the lipid assembly has led to the design of heat-sensitive, light-sensitive and pH-sensitive liposomes while the chemical approach to obtaining site-specific delivery involves the attachment of targeting moieties such monoclonal antibodies, specific carbohydrates, hormones and proteins.

The application of polymeric drugs and drug delivery systems should become more popular in the near future for two reasons. One, they provide for a more effective method of delivering medications; this is from the viewpoint of time and type of delivery, dosage levels, and the biological demand for the medication. Secondly, new FDA policies with regard to extending the protection period of a drug by improving its efficacy through new administrative modalities will stimulate drug companies to market these systems. The two drug delivery systems that probably have the greatest commercial potential are the transdermal systems for external applications and the erodable membranes or "self-destructing" devices that slowly disappear during or subsequent to drug delivery for internal application. Other devices will also play an important role but more for specific applications.

Literature Cited

1. Anderson, J. M. and S. W. Kim, Eds.; Advances in Drug Delivery Systems; Elsevier: New York, 1986.
2. Tirrell, D., L.G. Donaruma, and A. B. Turer, Eds.; Macromolecules as Drugs and Carriers for Biologically Active Materials; New York Academy of Science Press: New York, 1985.
3. Shalaby, S., A. Hoffman, B. Ratner and T. Horbett, Eds.; Polymers as Biomaterials; Plenum Press: N.Y., 1983.
4. Szycher, M., Ed.; Biocompatible Polymers, Metals and Composites; Technomics Publishing Co.: Lancaster, Pa., 1983.
5. Kambic, E., S. Murabayashi, and Y. Nose. Chem. Eng. News, 1986, 64 (15), p 31-42.
6. Johnson, H.J., S.J. Northrup and Seagraves. J. Biomed. Mater. Res., 1985, 19(5) 489.
7. Stillman, R.M. and Z. Sophie. Arch. Surg., 1985, 120(11) 1281.
8. Refojo, M.F. Curr. Eye Res., 1985, 4(6), 719.
9. McCaffery, and F.W. Lusby. J. Cataract. Refract. Surg., 1986, 12(3), 278.
10. Iwata, H., H. Ameyiya, T. Matsuda, Y. Matsuo, H. Takano, T. Akutsu. Artif. Organs, 1985, 9(3), 299.
11. Jacker, H.J., R. Meter and S. Grutter. Pharmazie, 1985, 40(7), 472.
12. Rosen, J.M.. Ann Plast. Surg., 1986, 16(1), 82.
13. Noishiki, Y. and T. Miyata. J. Biomat. Res., 1986, 20, 337.
14. Jozefowicz, M. and J. Jozefowicz. Asaio J., 1985, 8(4), 218.
15. Toshihiro, A. Oyo Butxuri, 1984, 53, 357.
16. Ottenbrite, R.M., L.G. Donaruma, and O. Vogl, Eds.; Anionic Polymeric Drugs; Wiley-Interscience: New York, 1980.
17. Ottenbrite, R.M. and G.B. Butler. Synthesis and Characterization of Interferon Drugs; Marcel Dekker: New York, 1984.

18. Donaruma, L.G. and O. Vogl, Eds.; Polymeric Drugs; Academic Press: New York, 1978.
19. Donaruma, L.G., et al. Polymer Preprints, 1979, 20, 346.
20. Ringsdorf, H.. J. Polym. Sci., 1975, 51, 135.
21. Kaplan, A., K. Kuus and R.M. Ottenbrite. Ann. N.Y. Acad. Sci., 1985, 446, 169.
22. Levy, H.B. J. Bioact. Comp. Polym., 1986, 1, 348.
23. Gebelein, C. and C. Carraher, Eds.; Polymer Materials in Medicine; Plenum Press: New York, 1985.
24. Petering, D. In Anticancer and Interferon Agents; Ottenbrite, R.M. and G.B. Butler, Eds.; Dekker Co.: New York, 1984.
25. Meada, H.. J. Med. Chem., 28, 455 (1985).
26. Duncan, R., H. Cable, P. Rejmanova, J. Kopecek and J. Lloyd. Biochim. Biophys. Acta.; 1984, 799, 1.
27. Kopecek, J.. Biomaterials; 1984, 5, 19.
28. Decoursin, J.W.. Special Issue on Drug Delivery Systems, Pharm. Tech.; 1985, 29.
29. Theewes, F. and S.I. Yum. Ann. Biomed. Eng., 1976, 4, 343.
30. Chandrasekaran, S.K., S.K. Benson and Urquhart. In Controlled Release Drug Delivery Systems; Robinson, J.R. Ed.; Marcel Dekker: New York, 1978.
31. Adbin, G., T.A. Horbett and B.D. Ratner. In Advances in Drug Delivery Systems; Andersen, J.M. and S.W. King, Eds.; Elsevier: N.Y., 1986.
32. Young, S.Y., S.W. Kim, D.L. Lomberg and J.C. McReain. In Advances in Drug Delivery Systems; Andersen, J.M. and S.W. King, Eds,; Elsevier: New York, 1986.
33. Bailey, W.J. Ann. N.Y. Acad. Sci.; 1985, 446, 42.
34. Heller, J., B.V. Fritzinger, S.Y. Ng and D.W. Penhale. J. Controlled Release; 1985, 1, 233.
35. Leong, K.W., B.C. Brott and R. Langer. J. Biomed. Mat. Res.; 1985, 941.
36. Good, W.G.. In Drug Delivery Systems; Pharm. Tech.; 1985, 40.
37. Mufson, D. Special Issue on Drug Delivery Systems; Pharm. Tech., 1985, 16.
38. Sunamoto, J., M. Goto, T. Iida, H. Kara, and A. Tomonaga. In Receptor-Medicated Targeting of Drugs; Gregoradis, G., G. Poste, J. Senior and A. Trouet, Eds.; Plenum Press: New York, 1985, p 359.

RECEIVED September 29, 1987

Chapter 11

Polymer Chemistry and Liposome Technology

David A. Tirrell

Polymer Science and Engineering Department, University of Massachusetts, Amherst, MA 01003

Polymer chemistry has a great deal to offer in the construction of synthetic liposomal membranes for use in biology and medicine. This chapter explores the preparation and properties of polymeric liposomes, with particular emphasis on the use of controlled polyelectrolyte adsorption to manipulate liposomal membrane properties.

Nature has combined the disciplines of polymer chemistry and colloid science to remarkable effect in her design of the bilayer membranes that surround cells and subcellular organelles. These membranes do many intriguing things: they recognize one another, they respond to the binding of drugs and hormones, and they control the flow of mass, information and energy within and between cells. It would be natural, then, for us to select such structures as we try to intervene in biological processes through the controlled delivery of drugs, markers and genetic material. Following Bangham's discovery (1) of the barrier properties of lipid vesicles, or liposomes, there has grown an enormous literature devoted to liposomal delivery (2-6). What has emerged more recently is the realization that polymer chemistry has much to offer in the technological development of liposomal delivery systems.

Research on liposomal delivery systems has been motivated in large part by the notion that the material to be delivered can be entrapped in the liposomal interior, and that the entrapped material will remain inside until the liposome is opened at its target. Implementation of this idea requires the formulation of liposomal systems that are resistant to leakage of entrapped substances, resistant to attack by foreign agents used in the preparation of the delivery system (e.g. surfactants and organic solvents), stable toward attack by endogenous agents (e.g. enzymes or plasma proteins) and capable of tissue localization or selective rupture at the target. The following discussion will provide an overview of the present state of the union between polymer chemistry and liposome technology. We will address first the methods of preparation of

0097-6156/88/0362-0152$06.00/0

polymeric liposomal systems, and then proceed to an evaluation of the contributions that polymer chemistry can make to the solution of the problems outlined above.

Preparative Methods

The major approaches to the preparation of polymeric liposomal systems may be distinguished on the basis of the locus of polymerization. Most effort to date has been directed toward the preparation of phospholipid analogues that are intrinsically polymerizable through reactions of attached vinyl, butadienyl, diacetylenic, isocyano, thiol or disulfide functions. This approach produces liposomes in which the primary permeability barrier - the bilayer itself - is polymeric. The second approach invokes the attachment of extrinsic macromolecules to the lipid bilayer, in a fashion analogous to the assembly of biological membranes.

Polymerizable Lipids. The preparation of polymerizable lipids has been extensively reviewed (7-10), so the present discussion will be brief. It is useful to distinguish two portions of amphiphilic, membrane-forming lipids: a polar headgroup that forms the interface between the membrane and its environment, and the hydrocarbon chains that provide the primary barrier to the escape of entrapped solutes. It is important in the design of polymerized liposomes that one keep in mind these distinct functions, and that the locus of polymerization be selected in such a way that the desired functional properties are preserved or enhanced. Polymerization through functional groups appended to the lipid headgroup would be expected to preserve the packing properties of the hydrocarbon chains (assuming that the geometric requirements of the polymerization are consistent with those of chain packing), but would alter the surface recognition behavior of the membrane. On the other hand, polymerization in the hydrocarbon core - either in the middle of the chain or at its end - offers the prospect of an unperturbed lipid surface, but compromises in a profound way the packing of the hydrocarbon chains. It is important to point out, however, that such changes in packing may be advantageous, in that the barrier properties of the bilayer may be significantly enhanced by polymerization. An example of reduced solute leakage from polymerized liposomes will be discussed in the following section.

Attachment of Extrinsic Macromolecules. Attachment of synthetic polymers to the liposomal surface has been accomplished by at least four different routes (Scheme I). The simplest route - but also that least amenable to control - involves the adsorption of unmodified polymer chains (11). In adopting this route, one must walk a fine line between systems in which the polymer-bilayer interactions are too weak to maintain a substantial surface concentration of chains, and systems in which those interactions are strong enough to cause vesicle reorganization and lysis. A second route - with distinct advantages and well established biological precedent - is the use of hydrophobic "anchoring groups" to hold in place even weakly bound polymer chains. Suitable anchors are single or double chain surfactants (12,13) or cholesterol (14), and in principle such anchors

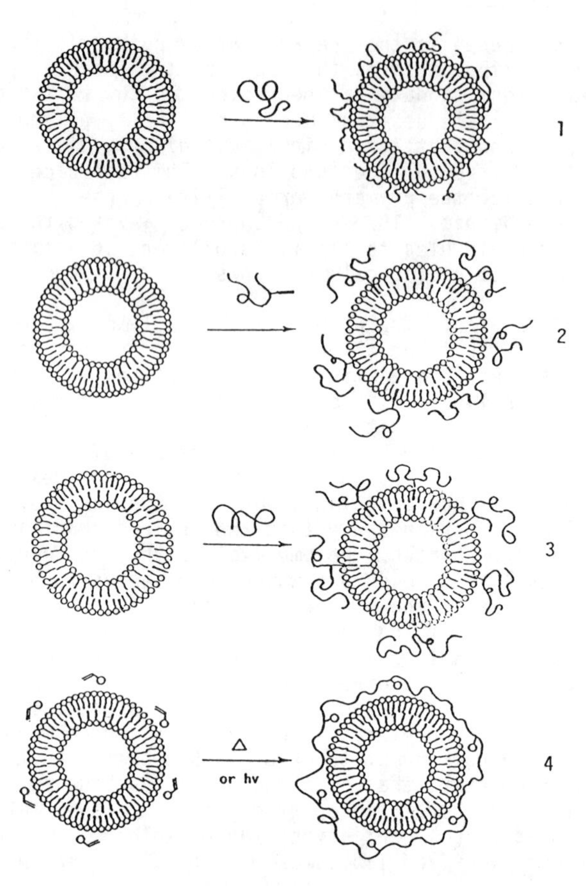

Scheme I.

may be introduced first either to the polymer chain (Scheme I, route 2) or to the vesicle surface (Scheme I, route 3). The biological analogy of course is the widespread occurrence of hydrophobic peptide sequences in membrane-bound proteins, which serve a similar anchoring function. Finally, both Regen (15) and Ringsdorf (16) have suggested that the formation of liposomes from surfactants bearing polymerizable counterions might be followed by polymerization within the double layer to provide a surface coating of polyelectrolyte chains (Scheme I, route 4).

The reader should also be aware of a very substantial, related body of work that addresses the anchoring of extrinsic, naturally occurring macromolecules to liposomal surfaces (2-6,17). Because this work has already been extensively reviewed, the present chapter will not discuss this very interesting and important subject.

Properties of Polymeric Liposomes

Liposomes from Polymerizable Lipids. Again the discussion will be brief, and the reader is directed to the several excellent reviews of polymerized vesicles (3). We discuss here only one very recent example of the preparation and properties of such a system. This example has much to recommend it, and is described by the original authors as, "the polymerizable lipid of choice for a wide variety of mechanistic and practical applications."

Regen and coworkers described in 1986 the preparation and polymerization of 1,2-bis[12(lipoyloxy)dodecanoyl]-sn-glycero-3-phosphocholine **1** (18). Aqueous dispersions of **1** produced by

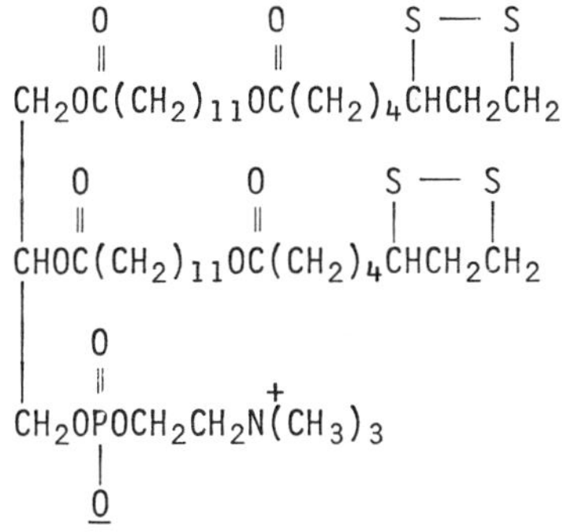

$$CH_2OC(=O)(CH_2)_{11}OC(=O)(CH_2)_4\overset{|}{C}HCH_2CH_2 \;(S{-}S \text{ ring})$$
$$CHOC(=O)(CH_2)_{11}OC(=O)(CH_2)_4\overset{|}{C}HCH_2CH_2 \;(S{-}S \text{ ring}) \qquad \mathbf{1}$$
$$CH_2OP(=O)(O^-)OCH_2CH_2\overset{+}{N}(CH_3)_3$$

injection of ethanolic solutions were shown to consist of vesicles of average diameter 270-400Å. Treatment of such dispersions with 10 mol% dithiothreitol (DTT) for 4 hr at 27°C produced complete polymerization of **1**, as shown by thin layer chromatographic analysis for unreacted monomer and by the disappearance of the characteristic UV absorption maximum of the monomer at 333 nm. Polymerization was accompanied by a small decrease in particle size as reported by dynamic light scattering, but the presence of closed vesicles was suggested by electron microscopy and confirmed by solute entrapment. The polymerized vesicles were stable to lysis even in 1% solutions of sodium dodecyl sulfate (SDS) at 60°C; in contrast, vesicles of monomeric **1** were destroyed in 0.05% SDS at room temperature. The size distribution of the polymerized vesicle dispersion was shown to be stable for at least 3 months at room temperature, and the dispersion retained 61% of its entrapped sucrose after 4 hr at 23°C, as compared to 32% retention by monomeric 1 under identical conditions.

This system indeed has much to recommend it. The polymerization conditions are remarkably mild, and appear to confer on the bilayer increased barrier properties and improved resistance to detergent lysis. In addition, the construction of the polymer chain by the formation of disulfide bonds leaves open the prospect of depolymerization in a reducing biological environment and subsequent degradation.

Liposomes Bearing Extrinsic Macromolecules. This section will consider the properties of liposomes to which extrinsic macromolecules are bound by adsorption or by hydrophobic anchors. Regen (15) and Ringsdorf (16) have discussed the consequences of polymerization of surfactant counterions (9).

Liposomes Bearing Adsorbed Chains. Much of the work to date on liposomes bearing adsorbed chains has concerned the pH-dependent adsorption of acidic polyelectrolytes on liposomes prepared from phosphatidylcholines (11,19-21). Interest in this problem has its origin in the pH-dependent conformational and solubility properties of acidic polyelectrolytes. One might expect, for example, that a poly(carboxylic acid) would be converted upon acidification from a charged, hydrophilic structure to a globular, hydrophobic coil, poorly solvated by aqueous media. The liposomal surface provides a place of refuge for the hydrophobic chain; the chain in turn alters in a profound way the geometric and thermodynamic factors that combine to determine bilayer structure. One can easily imagine that the polymer-phospholipid mixture might adopt a totally different aggregate morphology, and that in the course of the structural reorganization the barrier properties of the bilayer would be lost. Scheme II shows a reasonable working hypothesis, in which proton-driven polyelectrolyte adsorption induces a vesicle-to-micelle transition upon acidification (22).

Scheme II suggests that polyelectrolyte adsorption might provide a means of preparing pH-dependent liposomal delivery systems. Given the variations in pH that characterize certain pathological states (23) and certain subcellular organelles (24), the ability to alter membrane properties in a controlled, pH-dependent manner is a powerful methodology. Indeed, it has become clear that the cell uses subtle changes in pH to control its intracellular processing of ligands and receptors (24), and that infectious microorganisms have developed mechanisms to exploit these pH changes to gain entry to the cytoplasm (25).

Most of our work to date has concerned the interactions of poly(acrylic acid) derivatives with vesicle membranes prepared from phosphatidylcholines. In particular, poly(2-ethylacrylic acid) (PEAA, 2) is a hydrophobic poly(carboxylic acid) that undergoes a conformational transition of the kind described above (26-28). That this conformational transition occurs near neutral pH makes PEAA a candidate for use in pH-dependent liposomal delivery systems in biology and medicine.

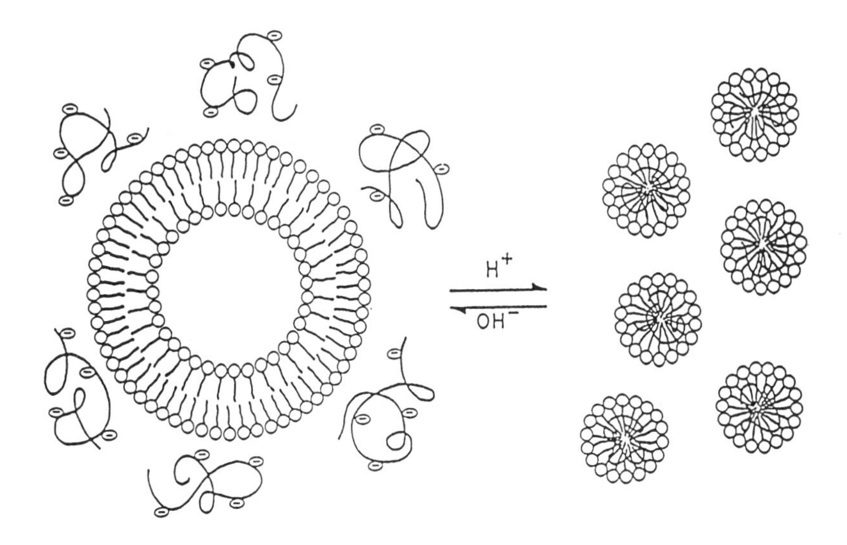

Scheme II. (Reproduced with permission from Ref. 22. Copyright 1985 Huthig and Wepf Verlag.)

$$[CH_2C(CH_2CH_3)(CO_2H)] \quad \mathbf{2}$$

Figure 1 shows that PEAA indeed functions well in this role (19). The Figure shows the efflux of the fluorescent dye carboxy-fluorescein from unilamellar vesicles of egg yolk phosphatidylcholine suspended in an aqueous solution of PEAA. Escape of the dye is very slow at pH 7.4, but essentially instantaneous upon acidification of the suspension to pH 6.5. We have since demonstrated the pH-triggered release of other substances from phosphatidylcholine vesicles by this method, and the technique should be completely general in its application to the controlled release of water soluble compounds.

Additional uses of these and related systems are suggested by the fact that many enzymes are known to yield acidic products via oxidative or hydrolytic reactions. Glucose oxidase, for example, catalyzes the oxidation of glucose to gluconic acid, which is a suitable source of H^+. A phosphatidylcholine suspension containing both glucose oxidase and PEAA should then be sensitive to glucose concentration, in that increasing glucose concentrations should lead to vesicle rupture with release of contents. Such systems might prove to be useful in self-regulated insulin delivery or in monitoring of glucose concentrations in physiologic fluids.

Figure 2 demonstrates the viability of this approach to the preparation of glucose-sensitive membranes (20). The Figure shows the results of an experiment in which dilauroyl phosphatidylcholine was suspended in an unbuffered, aqueous solution of PEAA and glucose oxidase at pH 7.4. Upon addition of glucose, the optical density of the solution was rapidly reduced, and reached a value less than 10% of the original after approximately 30 min. The reduction in optical density signals a reorganization of the bilayer that must be analogous to that induced by direct addition of H^+ as in Figure 1, so that we would anticipate quantitative release of vesicle contents under the conditions of Figure 2. This system is of interest not only from the point of view of its potential applications in diagnosis and therapeutics, but also by virtue of its analogy to the "second messenger" signalling processes so important in cell biology. That is, the signal is initiated by a rise in the concentration of glucose, but it is a second substance (H^+) that carries the message to the effector molecule (PEAA). The analogy to hormonal second messenger systems is crude, but quite real.

Liposomes Bearing Anchored Chains. We have very recently extended our work on the PEAA/phosphatidylcholine system by immobilizing PEAA on the surface of egg lecithin via a hydrophobic anchor (Maeda, M.; Kumano, A.; Tirrell, D.A. *Preprints ACS Div. Polym. Chem.*, in press). The method involves the coupling of thiolated PEAA to the maleimido phospholipid **3**, which was incorporated at a level of ca. 10% into preformed lecithin vesicles. Michael addition of the polymer-bound thiol groups to the N-alkylmaleimide functions on the vesicle surface results in immobilization of 50-60 µg of PEAA per mg of lipid. Following fractionation of the sample by size exclusion

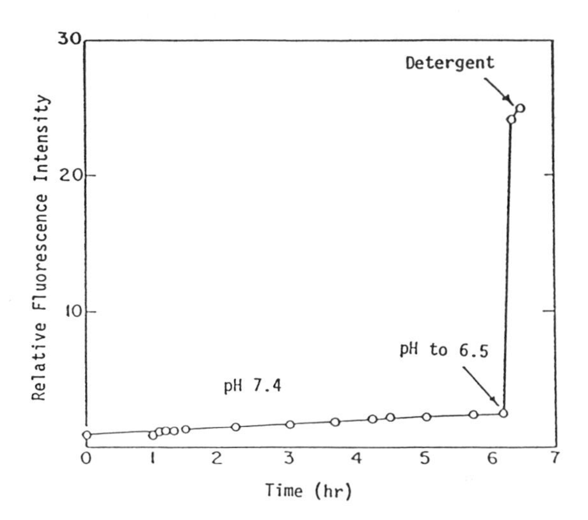

Figure 1. Efflux of carboxyfluorescein from sonicated egg yolk phosphatidylcholine vesicles suspended in 50 mM Tris-HCl, 100 mM NaCl at indicated pH. (Reproduced with permission from Ref. 19. Copyright 1985 New York Academy of Sciences.)

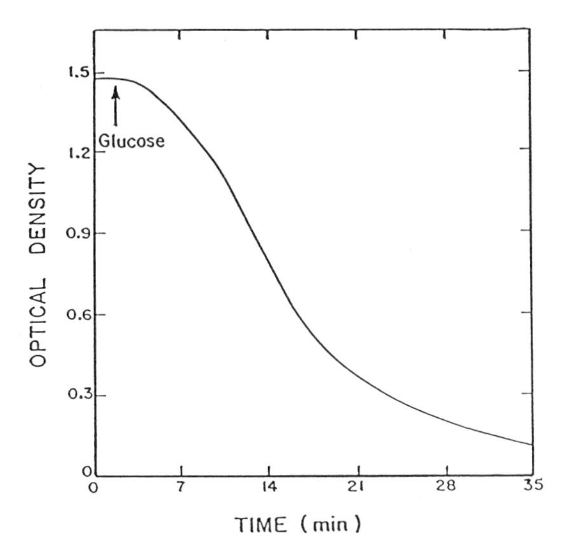

Figure 2. Optical density of a multilamellar suspension of DLPC in an aqueous solution of PEAA and glucose oxidase, prior and subsequent to addition of glucose.

chromatography, acidification of the vesicle fractions causes rapid release of vesicle contents. These results demonstrate that it is indeed possible to anchor sufficient PEAA to effect useful changes in membrane structure and function.

O O$^-$ O O

$CH_2OPOCH_2CH_2NHC$-⟨cyclohexyl⟩-CH_2-N⟨ring, two C=O⟩

CHOCOR

CH_2OCOR $R = C_{13}H_{27}$

3

Conclusions

Polymer chemistry offers a number of powerful strategies for the controlled manipulation of the properties of liposomal membranes. Polymerization within the bilayer can provide membranes of improved stability and barrier properties. Controlled adsorption or anchoring of polyelectrolytes can render the liposomal membrane sensitive to chemical and physical signals, and so can afford liposomal delivery systems susceptible to selective rupture at predetermined physiological targets or under predetermined pathological conditions.

Literature Cited

1. Bangham, A.D.; Standish, M.M.; Watkins, J.C. J. Mol. Biol. 1965, 13, 238.
2. Gregoriadis, G., ed. Liposome Technology; CRC Press: Boca Raton, 1984.
3. Ostro, M.J., ed. Liposomes; Marcel Dekker: New York, 1983.
4. Bangham, A.D., ed. Liposome Letters; Academic Press: New York, 1983.
5. Knight, C.G., ed. Liposomes: From Physical Structure to Therapeutic Applications; Elsevier: Amsterdam, 1981.
6. Gregoriadis, G.; Senior, J.; Trouet, A., eds. Targeting of Drugs; Plenum: New York, 1982.
7. Fendler, J.H. Membrane Mimetic Chemistry; Wiley: New York, 1982.
8. Tirrell, D.A.; Donaruma, L.G.; Turek, A.B., eds. Macromolecules as Drugs and as Carriers for Biologically Active Materials; N.Y. Acad. Sci.: New York, 1985.
9. Bader, H.; Dorn, K.; Hupfer, B.; Ringsdorf, H. Adv. Polym. Sci. 1985, 64, 1.
10. Fendler, J.H.; Tundo, P. Acc. Chem. Res. 1984, 17, 3.
11. Seki, K.; Tirrell, D.A. Macromolecules 1984, 17, 1692.
12. Sunamoto, J.; Iwamoto, K.; Takada, M.; Yuzuriha, T.; Katayama, K. In Recent Advances in Drug Delivery Systems; Anderson, J.M.; Kim, S.W., eds. Plenum: New York, 1984, p. 153.
13. Martin, F.J.; Papahadjopoulos, D. J. Biol. Chem. 1982, 257, 286.
14. Ottenbrite, R.M.; Sunamoto, J.; Sato, T.; Oka, M. Prepr. ACS Div. Polym. Chem. 1985, 26(1), 212.
15. Fukuda, H.; Diem, T.; Stefely, J.; Kezdy, F.J.; Regen, S.L. J. Am. Chem. Soc. 1986, 108, 2321.

16. Aliev, K.V.; Ringsdorf, H.; Schlarb, B.; Leister, K.H. Makromol. Chem. Rapid Commun. 1984, 5, 345.
17. Sunamoto, J.; Iwamoto, K. CRC Crit. Revs. Ther. Drug Carrier Systems 1986, 2, 117.
18. Sadownik, A.; Stefely, J.; Regen, S.L. J. Am. Chem. Soc. 1986, 108, 7789.
19. Tirrell, D.A.; Takigawa, D.Y.; Seki, K. Ann. N.Y. Acad. Sci. 1985, 446, 237.
20. Devlin, B.P.; Tirrell, D.A. Macromolecules 1986, 19, 2465.
21. Borden, K.A.; Eum, K.M.; Langley, K.H.; Tirrell, D.A. Macromolecules 1987, 20, 454.
22. Takigawa, D.Y.; Tirrell, D.A. Makromol. Chem. Rapid Commun. 1985, 6, 653.
23. Yatvin, M.B.; Kreutz, W.; Horwitz, B.A.; Shinitsky, M. Science 1986, 210, 1253.
24. Yamashiro, D.J.; Tycko, B.; Fluss, S.R.; Maxfield, F.R. Cell, 1984, 37, 789.
25. White, J.; Kielian, M.; Helenius, A. Quart. Revs. Biophys. 1983, 16, 151.
26. Fichtner, F.; Schonert, H. Colloid Polym. Sci. 1977, 255, 230.
27. Joyce, D.E.; Kurucsev, T. Polymer 1981, 22, 415.
28. Sugai, S., Nitta, K.; Ohno, N.; Nakano, H. Colloid Polym. Sci. 1983, 261, 159.

RECEIVED August 14, 1987

Chapter 12

Consideration of Proteins and Peptides Produced by New Technology for Use as Therapeutic Agents

Darrell T. Liu, Neil Goldman, and Frederick Gates, III

Division of Biochemistry and Biophysics, Office of Biologics Research Review, Center for Drugs and Biologics, Food and Drug Administration, 8800 Rockville Pike, Bethesda, MD 20892

Our basic philosophy in regulating biologicals produced by new technology is consistent with that used to evaluate other biologicals and may be described simply by a few concepts: sound scientific principles, flexibility, case-by-case approach, good common sense, and risk vs. benefit assessment. Several "Points to consider..." documents covering a number of specific topics have been issued which have had positive effects in facilitating the development of new drugs and biologics. There appears to be certain scientific and safety issues related to products derived from new technology that deserve special attention, e.g., alteration in molecular structure and purity of the product and impurities associated with the product.

Proteins and peptides for use as therapeutics may be isolated and purified from appropriate natural sources, prepared by rDNA technology or chemically synthesized. In this report we shall address mainly our experiences in considering production, purification, and testing of biologicals derived from rDNA technology and briefly review available documents that relate to products produced by chemical synthesis. Regardless of the source of the proteins and peptides, ultimately it is the responsibility of the manufacturers to demonstrate consistency in the safety, potency, efficacy, and purity of their products.

The Office of Biologics Research and Review (OBRR) is taking a cautious but flexible attitude in guiding these products onto the consumer market. Our basic philosophy in regulating biologicals produced by new technology is consistent with that used to evaluate other biologicals and may be described simply by a few concepts:

Sound Scientific Principles

The majority of OBRR staff involved in regulatory tasks are scientists actively engaged in basic research who, by virtue of being on the National Institutes of Health campus, maintain close interactions with world-renowned scientists present on the same campus. These close interactions have a major effect of ensuring scientific competence of OBRR staff and assuring that the decisions made on regulatory issues are based on valid scientific criteria.

Flexibility

The frontiers of new technologies are constantly being extended, and regulatory work must keep up with the advance of new scientific findings.

Case-by-Case Approach

The clinical usages of biologicals differ from one product to another. Insulin is injected 1-3 times a day for many years, and immunogenicity is to be avoided. In contrast, a vaccine is injected into a normal person 1-5 times with the primary intention of immunization. Thus, although many aspects of control apply to all products, different properties of each product may require special consideration.

Good Common Sense

As practicing scientists, OBRR regulators should be able to formulate realistic requests for scientific data from manufacturers.

Risk vs. Benefit Assessment

Regulators, academics, industry, users, and the general public, do not wish to run undue risks. However, if the benefits that can be derived from new technologies are to be realized, some potential risk cannot be avoided. It takes careful consideration of the scientific evidence, buttressed by sound judgment, to maintain appropriate risk:benefit ratios. The better each of these compounds can be defined, the better the ratio can be understood.

Products Derived From rDNA Technology

The rapid advances in molecular genetics over the past few years have allowed us to isolate and clone just about any gene from a cell's genome. These gene cloning techniques combined with our ability to express cloned genes in cells growing in culture enable us to produce useful quantitites of specific protein products for use as vaccines or as replacement therapeutics in such diverse categories as hormones, enzymes, immunomodulators, serum proteins, and viral antigens. There is an impressive list of protein products derived from cloned genes currently being used or evaluated in humans or soon to be available for clinical trials. Specific products include insulin, human growth hormone, various interferons and interleukins, albumin, clotting factors, and the surface antigens of the hepatitis and herpes viruses.

This new wave of biological products has required a reassessment of the historical concerns over product purity, potency, safety, and efficacy. This reevaluation has resulted in the development of regulatory documents, entitled "Points to Consider..." which are not guidelines *per se* that carry the force of law, but are designed to convey the current consensus of the OBRR relating to product development and testing. These "Points to Consider..." documents will remain as dated "drafts" and will never be finalized. They will be updated as needed using input from industry, academia and other regulatory agencies both from within the U.S. and abroad.

Several "Points to Consider..." documents have been issued covering a number of specific topics, including most recently interferon (July, 1982), monoclonal antibodies for *in vitro* use (March, 1982), monoclonal antibodies for *in vivo* use (July, 1983), recombinant DNA (April, 1985) and cell lines used to produce biological products (June, 1984).

This series of documents has served a number of useful purposes. Within the OBRR, they help to maintain uniformity of regulatory review, and serve as the basis for policy discussion. Externally, they provide a forum for scientist-to-scientist communication between the industry and the regulatory authority. They facilitate academic input into the regulatory process and generate international scientific consensus. They are useful in strategic planning by manufacturers, and have been especially relevant for the newly formed biotechnology firms which have had limited interactions with regulatory agencies. These flexible and evolving "Points to Consider..." documents appear to have been well received by all interested parties and have had positive effects in facilitating the development of new drugs and biologics.

In the experiences of the Office of Biologics Research and Review, there are a number of scientific and safety issues related to rDNA derived products that deserve special attention, e.g., A, alterations in molecular structure and B, purity of the product and impurities associated with the product.

Alterations in Molecular Structure

Heterogeneity in Amino Terminal Sequence. A number of mammalian proteins have been synthesized in *Escherichia coli*. While bacterial cells appear to possess many attributes for heterologous gene expression, certain problems may persist in generating products that are identical to the native protein. Most secretory or membrane proteins in eukaryotic cells are synthesized with leader peptides which are subsequently removed by specific proteases to yield native proteins. When mammalian proteins are synthesized in *E. coli* without that leader sequence, the resulting product may be homogeneous, but it is often a heterogeneous mixture of N-formyl-methionyl-protein, methionyl-protein, and protein missing one, two or more amino terminal residues. Proteins which are missing carboxyterminal residues have also been encountered. With the attachment of segments encoding bacterial or eukaryotic leader sequences to eukaryotic genes expressed in E. coli, the resulting proteins at times have been shown to have the correct amino terminal residues and are secreted into the periplasmic space (1,2).

The potential heterogeneity in amino terminal sequence is not limited to proteins expressed in E. coli, human α-interferon synthesized in the yeast Saccharomyces cerevisiae (3) was found to be a heterogeneous mixture consisting of the mature protein, protein with the intact leader peptide, and protein with the leader sequence partially removed.

Chemical and Physical Alteration. Proteins produced by rDNA in E. coli are often found deposited in insoluble inclusion bodies. The resulting extraction and purification often necessitates the use of detergents to assure solubilization, and protease inhibitors to suppress proteolytic degradation. Such agents conceivably could alter the structure of the proteins. Most detergents are known to denature proteins to some degree, and when used in conjunction with reducing agents will almost certainly cause protein denaturation. Procedures designed to remove detergents and reducing agents may not allow all of the protein molecules to restore their natural conformation. Proteolytic enzyme inhibitors such as the widely used phenylmethanesulfonyl fluoride could chemically modify side chains of amino acids in the protein. These alterations, whether in conformation or in covalent structure, may induce the recognition of these proteins by the immune system, a process that may be detrimental to the health of the recipient either by limiting the clinical efficacy of the antigenic material itself, or by stimulating cross reactive autoimmune response to the natural homologue. The ability to express specific proteins at high yields (30-50% of total cell proteins) and to have them secreted extracellularly using cloned genes with an appropriate leader sequence may greatly facilitate purification of gene products of "native" structure for clinical use.

Alteration Due to Non-glycosylation or Glycosylation. Oligosaccharides can have many more isomeric forms than peptides (4). For instance, whereas there are only six different ways in which a tripeptide can be formed from 3 different amino acids, the number of trisaccharide that can be formed from 3 different hexoses through pyranose ring structure is 1056.

Many of the hormones and plasma proteins isolated from human tissue and plasma are glycoproteins. When these human proteins are synthesized in E. coli they will not be glycosylated since E. coli can not make glycoproteins. On the other hand, proteins produced by rDNA using heterologous eukaryotic cells may have different carbohydrate composition and structure from their natural counterparts because the complement of enzymes required for the synthesis and processing of complex glycoproteins vary from one eukaryotic cell to the other (5).

For those products which are administered parenterally in large amounts or repeatedly over long periods, non-glycosylated protein may reveal new antigenic sites while glycoprotein with unnatural carbohydrate structure may be antigenic. Proteins without the appropriate carbohydrate moieties may have altered pharmacokinetics and tissue distribution and also may cause other unknown and possibly adverse effects.

Protein Folding and Pairing of Disulfide Bonds. It is generally far easier to insure and to ascertain the "intactness" of the primary structure of a protein than to establish the natural conformation and the correct pairing of disulfide bonds of a protein.

The maximum number of possible ways in which a given number of half-cystine residues can combine to form disulfide bonds (SS) upon oxidation is overwhelming (6). These numbers show, for example, that in the case of tissue plasminogen activator, the random chance of forming the correct 18 SS bonds from the available 37 half-cystine is infinitesimally small, less than one in 2×10^{20}. In the case of insulin which contains 6 half-cystines, 15 possible sets of 3 SS bonds can be made, only one of which is the native structure. How does one determine the correct paring of SS bonds in a rDNA derived protein? At present, the only realistic method is the protein-based quantitative analysis which involves peptide mapping in conjunction with compositional analysis and sequence analysis of each of the SS containing peptides. For rDNA derived insulin, this was exactly the way the 3 pair of SS bonds were established (7). The amount of protein samples required and the difficulty involved in establishing the SS bonds of a protein increases dramatically with the increase in the number of SS bonds. Moreover, even if the position of SS bonds are established for a rDNA derived protein, it will not be possible to ascertain if they are identical to the natural protein unless the positions of SS bonds for the natural protein have also been determined. In many instances this is not possible since the particular protein in question is not available in sufficient quantity to permit such an analysis.

Sufficient data (6-8) are now available to indicate, however, that the specific three-dimensional structure of a protein is dictated by the linear sequence of its amino acid residues and that the pairing of half-cystines in SS bonds in a protein is the consequence rather than a director of protein folding. This conversion from linearity to spacial organization appears to be a spontaneous process (6). The native proteins that one finds in cells are polypeptide translation products of genetic information, arranged in a form possessing maximum thermodynamic stability under physiologic conditions.

An examination of the extent to which various "derived" multichained proteins produced by specific *in vivo* cleavages of single chained proteins (e.g. chymotrypsin, insulin) and naturally occurring multichained proteins formed by disulfide bonding of two or more separately synthesized chains (e.g. immunoglobulins) undergo reversible denaturation suggests that interruption of a single chained protein is generally not conducive to proper folding. Some multichained proteins, however, that are made up of identical or genetically related subunits may be reversibly denatured. Thus, whereas chymotrypsinogen, proinsulin, and immunoglobulins can be reversibly denatured, chymotrypsin and insulin are thermodynamically unstable and informationally insufficient to undergo spontaneous refolding of the molecules and the formation of the native pairs of disulfide bonds.

In the context of rDNA derived proteins, it would seem to be advantageous and, in some instances, necessary to synthesize human proteins containing multiple disulfide bonds in eukaryotic cells that can oxidatively form disulfide bonds. When such proteins are synthesized in cells (2) that lack the ability to oxidatively form disulfide bonds intracellularly (e.g. *E. coli*) (2), the resulting proteins are often found in the inclusion bodies of cytoplasm where their half-cystines remain in the reduced form. During the isolation and purification of such intracellular proteins, steps should be taken to preserve native protein conformation and avoid unwanted proteolytic cleavages so as to favor the spontaneous generation of native protein conformation.

Purity of the Product and Impurities Associated with the Product

In regard to the assessment of the "purity" of the final product and the testing for "impurities" in the product, the consensus view is that the sensitivity of the test method and the degree of product purity should be appropriate to the product's intended use; products which are given repeatedly or in large doses will require higher purity than those given only a few times or at low doses. Tests used to support claims of purity or absence of contaminants should be capable of accurately detecting and quantitating the expected substances.

The primary reasons for concern over the "purity" of the rDNA derived product are the safety questions involved in using certain types of cell substrates for producing biologics. Concern has been expressed over the use of continuous cell lines for the production of some biologics because continuous cell lines may have biological, biochemical, and genetic abnormalities. In particular, they may contain viruses, host cell components that may be antigenic in humans, and potentially oncogenic DNA. It is important, therefore, to assure that biological products derived from continuous cell lines are highly purified by employing methods for the extensive removal and/or inactivation of cellular DNA, host cell antigens, and adventitious agents.

Residual Cellular DNA (9). In considering potential risk from residual DNA, it should be pointed out that a number of different products are given by different routes, in different amounts, and according to different schedules. For those products that are given repeatedly in large doses it is especially important that procedures for production demonstrate that no unnecessary DNA molecules will be in the final product. It is not clear how much residual DNA contamination would be necessary to induce changes in normal cellular processes. Until more information on the determination of the biological activities of DNA becomes available, a level of unwanted DNA in the picogram range per dose appears reasonable to measure and to achieve by conventional purification techniques.

Pyrogenic and Immunogenic Contaminants (10). Contaminants may come from component of cell substrates (cell wall, cell membrane, mucopeptide, lipopolysaccharide, etc.), media constituents, affinity column components or may be derived from chemical modification of the proteins used in the manufacturing process. These residual contaminants may be a potential source of risk because they may be pyrogenic or recognized as antigens by the recipient of the product, and they may have direct undesirable biological effects. It is important to monitor the production process and the final product to make sure that their levels are in an acceptable range.

Recent experience with recombinant products has emphasized that biological pharmaceuticals may be pyrogenic in humans despite having passed the *Limulus* Amebocyte Lysate test and the rabbit pyrogen test. This phenomenon appears to be due to nonendotoxin contaminants which demonstrate a marked species dependence in their pyrogenicity. To attempt to predict whether the human subject will experience a pyrogenic response, tests have been used in which human peripheral blood mononuclear cells are cultured *in vitro* in the presence of the product. The supernatant fluid from the treated cells is then injected into rabbits (11). A fever in the rabbits indicate that the substance had stimulated the cultured human mononuclear cells to produce leukocytic pyrogen, the protein mediator believed to be involved in the *in vivo* febrile response. We have successfully used this test to detect the presence of pyrogenic materials in several preparations which have been negative on both *Limulus* Amebocyte Lysate and rabbit pyrogen tests, but which were pyrogenic in humans.

The possibility of developing a humoral or cellular immune response to minor contaminants should be carefully assessed in both preclinical and clinical studies particularly in products to be administered chronically. Reliable and sensitive tests such as Western Blots are needed to assay for trace contamination present in separate production batches. Sensitive techniques such as radioimmunoassay and ELISA can be used to measure the induction of specific antibodies in recipients in response to microbial or cellular constituents likely to contaminate the final product. In addition, periodic skin testing can be performed with the product to rule out development of delayed type hypersensitivity.

Viral Contamination (12). Techniques available to test for the presence of viral agents in the cell substrate or in biological products are neither simple nor straightforward in their applications. This issue is further complicated by the difficulty in defining "viral agents." Risks imposed by viral contamination range from known viruses with predictable patterns of replication to as-yet-unknown infectious agents (e.g. potential cause of Creutzfeldt-Jakob disease) which cannot be recognized with currently available technology. Between these two extremes are "visible" viruses such as type B retroviruses whose presence can be demonstrated by electron microscopy but for which there is no sensitive in vitro assay for infectivity; latent proviral components such as viral DNA sequences integrated into the genome; and "unconventional agents" such as scrapie which has been identified as a transmissible pathogenic agent of unknown life cycle. Given the present state of science, precise measures for the elimination of risks from scrapie-like agents and as-yet-unknown viruses cannot be rationally proposed.

Because many pathogenic agents cannot be detected with great sensitivity and assurance, characterization of the final product alone is not sufficient to optimally assure safety. Initially, therefore, it is necessary to fully characterize the individual components used in the manufacturing process, especially the cell substrate. In all cases the cell line should be established and evaluated using a master cell bank system. Cells should be cultured to their furthest proposed passage level to look for induction of latent viruses and instabilities of genotype or phenotype. Judgment on the suitability of a cell line and its properties may be made on a case-by-case basis considering both the cell line and the potential clinical use of proposed products derived from it. In any case, however, in the initial assessment of the proposed purification techniques, demonstration of the elimination or inactivation of a deliberately introduced viral contaminant whose infectivity can be measured by a sensitive in vitro method, should provide some assurance that the techniques employed may be effective against unknown viruses of a similar nature.

A final point is that it is important to maintain surveillance of possible untoward effects because no matter how thorough the testing of products may be, there can always be unforeseen events. A long-term surveillance should be established and maintained relevant both to the cell lines themselves and to the proposed biological products.

The previously discussed considerations and tests to determine identity, safety, and purity will give some level of confidence that the rDNA product corresponds to the expected product. However, it should be remembered that structure and function of proteins may now be modified by means of genetic engineering thus it is possible to improve a desirable biological activity while eliminating undesirable toxic side effects associated with the natural product. Such a product is not identical with the natural product and may be observed to be immunogenic. Is such an altered product acceptable? The answer to that question is not a simple "yes" or "no" but will depend on a careful assessment of the new benefits of this product as compared to the risks identifiable during its preclinical and clinical evaluations. Requirements for long-term animal testing, including tests for carcinogenicity, teratogenicity, and effects on fertility, will depend upon the availability of animal models and should be based upon the intended use of the product, its mode of action and metabolic fate. Specific preclinical toxicity evaluations are best addressed on a case-by-case basis. Clinical trials will be necessary for all products derived from rDNA technology to evaluate their safety and efficacy.

Synthetic Peptides

Synthetic peptides are being introduced in increasing variety and frequency for human use as hormones and growth factors (13). In some instances, it appears to be commercially more attractive to synthesize peptides of up to 30-40 residues than to obtain them from biological sources or by rDNA technology. Examples include calcitonin, pentagastrin, and tetracosa ACTH peptide. In addition, new peptide analogues with unnatural amino acid substituents designed for selected pharmacological activities, and/or improved therapeutic ratio (efficacy:safety) can only be produced via chemical synthesis.

Since each of the steps involved in the chemical synthesis of a polypeptide will not be complete (14,15) (i.e. $< 100\%$ yield) the resulting product will not likely be "pure." A major concern for synthetic peptides for use as therapeutics is, therefore, the possible significance of the impurities associated with the product. Impurities may be comprised of truncated or partial sequences, of peptides with deletions, subsitutions or modified (e.g., residually blocked) functional groups, of enantiomers, or molecules with altered conformation (14-17). The amount and the complexity of such peptide impurities vary from batch to batch and, in general, increase with the length of the polypeptide chain. Polypeptides synthesized by a route not involving stepwise assembly of purified fragments will, most likely, be less pure.

Sophisticated multi-dimensional analytical methods available today, such as high performance liquid chromatography, isoelectric focusing, peptide mapping, ELISA, peptide microsequencing and fast atom bombardment mass spectrometry, are powerful but may not identify, quantitate, or detect all such impurities associated with synthetic peptides. Evidence for the purity of a synthetic peptide must necessarily depend on the careful evaluation of biological activities as well as the use of a variety of analytical systems based on differing physicochemical principles. The ability to detect peptide impurities does not, by itself, imply that practical methodology could be developed to remove them during purification of the product.

For the reasons given above, products consisting of synthesized polypeptides should not be treated *a priori* as a simple chemical whose identity, purity, and/or safety can be shown by chemical and physical methods alone. As with rDNA-derived products, specific preclinical toxicity evaluation, and long-term animal testing for of synthetic peptide products are best addressed on a case-by-case basis.

Literature Cited

1. Talmadge, J.; Kaufman, J.; Gilbert, W. *Proc. Natl. Acad. Sci.* 1980, *77*, 3988.
2. Pollitt, S.; Zalkin, H. *J. Bact*. 1983, *153*, 27.
3. Hitzeman, R. A.; Leung, D. W.; Perry, L. J.; Kohr, W. J.; Levine, H. L.; Goeddel, D. V. *Science* 1983, *219*, 620.

4. Clamp, J. R. In The Metabolism and Function of Glycoproteins; Biochem. Soc. Symposium No. 40; 1981; p 3.
5. Hsieh, P.; Rosner, M. S.; Robbins, P. W. J. Biol. Chem. 1983, 258, 2548.
6. Anfinsen, C. B.; 27th Symposium of the Society for Developmental Biology; Developmental Biology Supplement 2, 1968, p 1-20.
7. Johnson, I. S.; Science 1983, 219, 632.
8. Liu, T.-Y. In The Proteins; Neurath, H.; Hill, R. L., Eds.; Academic: New York, 1978; Vol. 3.
9. Noble, G. R. In In Vitro; Monograph 6, 1985; p 173.
10. Van Metre, T. E. In In Vitro; Monograph 6, 1985; p 172.
11. Dinarello, C. A. In Methods for Studying Mononuclear Phagocytes; Adams, D.; Edelson, P.; Koren, H., Eds.; 1974; p 629.
12. Osborn, J. E. In In Vitro; Monograph 6, 1985; p 174.
13. Synthetic Peptides: Toxicity Tests and Control; British National Institute for Biological Standards and Control, Division of Hormones; Document V9 24; 1984.
14. Finn, F. M.; Hofmann, K. In The Proteins; Neurath, H.; Hill, R. L., Eds.; Academic: New York, 1976; Vol. 2.
15. Barany, G.; Merrifield, R. B. In The Peptides; Gross, A. E.; Meienhofer, J., Eds.; Academic: New York, 1979; Vol. 2.
16. Wunsch, E. Biopolymers 1983, 33, 493.
17. Moser, R.; Klauser, S.; Leist, T.; Langen, H.; Epprecht, T.; Gutte, B. Angewandte Chemie 1985, 24, 719.

RECEIVED October 30, 1987

ANALYTICAL

Chapter 13

Analytical Challenges in Biotechnology

John B. Landis

Upjohn Company, Kalamazoo, MI 49001

The development of r-DNA and hybridoma technology has revolutionized the application of biotechnology for the production of peptides and proteins in the laboratory and on a commercial scale. Using this new technology, it is now possible to produce biopolymers in large quantities with high purity. In contrast, proteins and peptides have historically been produced through isolation from natural sources as heterogeneous mixtures of low purity.

There are important differences between the small organic molecules typically encountered by the analytical chemist and the proteins and peptides produced from biotechnology. While most typical organic molecules have molecular weights less than 1,000 Daltons, the products of biotechnology will typically have molecular weights between 1,000 - 1,000,000 Daltons. Unlike their small molecule counterparts, the three dimensional structure of proteins and peptides dictates function and activity and is not entirely controlled by the chemical structure of the molecule. The larger molecules are likely to be isolated from a complex matrix in which they are a minor constituent among many species of similar size and composition. To meet these challenges, new analytical strategies and methods are under development for the characterization and analysis of biopolymers.

The routine application of cloning and expression of a large number of proteins will challenge the analytical chemist to rapidly characterize and analyze these molecules. Important parameters to characterize and control are identity (structure), purity, and impurity.

An understanding of protein structure is important to guide the design of new proteins, substrates, and inhibitors; to enhance protein stability and handling characteristics; and to modify biological function. A knowledge of protein structure is an essential ingredient in the development of a scheme for the batch-to-batch control of protein identity. Protein identity is determined by measuring composition, sequence, and conformation. The development

0097-6156/88/0362-0174$06.00/0

of microchemical methods for the determination of amino acid composition and sequence has greatly improved the speed, accuracy, and precision. Additional improvements are needed in sensitivity, reliability, and data analysis. Advances in spectroscopy have increased our detailed knowledge of protein conformation.

Spectacular advances have been made in the development of high resolution techniques such as x-ray, NMR, and Raman spectroscopy. X-ray crystallography has provided the most detailed and comprehensive picture of three-dimensional protein structure in the solid state. The advent of high intensity sources and new detectors promises to reduce the time required to obtain a structure and to improve the accuracy of the structure. Recent progress in NMR is extending the determination of detailed structure on small proteins in solution. Many structural problems are accessible only through NMR, notably where crystalline materials cannot be obtained. The advent of the tunable dye laser, high powered pulsed lasers, and the extension of Raman to the UV has greatly increased the sensitivity and selectivity of this technique for examining the peptide backbone and sulfhydryl groups.

Major advances in Mass Spectroscopy has extended this technique to the sequence of peptides, DNA, and carbohydrates. Optical methods such as absorbance, fluorescence, and circular dichroism are useful techniques for following structural changes in solution. Recent work has indicated that monoclonal antibodies raised against conformationally important epitopes may also serve as a conformational identity test.

The determination of protein purity and impurity are important for the establishment of product safety and effectiveness. Since r-DNA produced proteins are often isolated as a minor component from a fermentation process, potential impurities might include host cell proteins, residual DNA, product-related impurities including degradation products, process reagents and biological contaminants such as viruses, endotoxin, pyrogens, and bacteria. The presence of many of these impurities in trace amounts may have a negative impact on product quality. Protein contaminants, for example, may cause an immune response at the ppm level when used as a drug. The detection and accurate quantitation of contaminating proteins differing by as little as one amino acid in a complex matrix is an analytical challenge that necessitates the development of new approaches to analysis. Currently, immunochemical methods are the only techniques that approach this sensitivity and selectivity. The use of monoclonal antibodies will make these methods more reliable. Contaminants that may also arise from alterations in molecular structure during the production process include heterogeneity in amino acid sequence, chemical and physical alteration via isolations and purification, and post translational modifications via mechanisms such as glycosylation.

Analytical methods and instrumentation are needed to meet the challenge for the commercialization of biotechnology. Proteins can now be produced much faster than they can be characterized. New analytical schemes are needed to characterize proteins as chemicals. The added complexity of proteins make it impossible to characterize them as fully as the small molecules typically encountered by the analytical chemist. Many of the methods currently being used for protein characterization and analysis were developed for proteins as biologicals and not for proteins as chemicals. A new generation of methods and strategies are needed to increase sensitivity, selectivity, precision, and accuracy.

RECEIVED September 3, 1987

Chapter 14

Applications of Optical Spectroscopy to Protein Conformational Transitions

Henry A. Havel

Control Research and Development, Upjohn Company, Kalamazoo, MI 49001

Solution state protein structure investigations using optical spectroscopy (UV absorption, circular dichroism and fluorescence) are reviewed and are applied to studies of the forces which stabilize the native conformations of proteins. Structural rearrangements which occur as a protein folds or unfolds can be identified by using chemical denaturants (guanidine HCl or urea), heat or pH adjustment to unfold the protein and employing optical spectroscopic probes to monitor the unfolding transition. Recent results of unfolding studies with bovine growth hormone, a small (MW = 22,000 daltons) polypeptide hormone, are summarized. In particular, optical spectroscopy of the single tryptophan residue in bGH has contributed to the determination that a self-associated form of a partially unfolded intermediate is populated during equilibrium unfolding; in addition, it is shown that the tryptophan residues of self-associated bGH molecules are likely to be held rigidly in a polar environment which is near the interface between self-associating molecules.

The study of protein structure in the solution state has been, and continues to be, an important area for chemical research. In recent years there has been increased activity in this area as the revolution in biotechnology has allowed the production of large quantities of relatively pure proteins at modest cost via recombinant DNA techniques. This paper will discuss the utility of optical spectroscopic techniques, defined here as UV absorption, fluorescence and circular dichroism spectroscopy, as important tools in protein structure investigations. The information derived from these techniques is unique and complements that obtained by higher resolution techniques such

0097-6156/88/0362-0177$06.00/0

as NMR and vibrational spectroscopy in the solution state (as well as solid state X-ray structure studies) without the added complexities which these other methods possess in instrumentation, data analysis and/or data interpretation. This review will not provide an in-depth treatment of any of the optical spectroscopic techniques, but instead will highlight their more important features and illustrate how optical spectroscopic techniques can serve as important tools in the study of protein conformational transitions. The applications presented will be from equilibrium folding studies done at The Upjohn Company on bovine growth hormone (bGH, bovine somatotropin, bSt), a 22,000 dalton protein which is of interest because of its lactogenic and growth-promotant activities. It is an appropriate example for these studies because it has been shown to undergo equilibrium unfolding through a process involving at least one stable intermediate.

Protein Structure

Proteins are fundamental molecules for all living organisms (1) as they catalyze reactions, carry messages, defend against foreign agents and support the organism's structure. It is widely held that the molecular structure of a globular protein is determined by its amino acid sequence (2) and that the molecular structure determines its biological function. Hence, if structure-function relationships can be established it will be possible to engineer new biological activities into proteins through modification of amino acid sequences. For protein pharmaceutical products, desirable new activities include improved biological half-life, elimination of deleterious side effects, reduction in aggregation properties, etc.

Protein structure is described conventionally (3) using a hierarchical scheme. The first type of structure is termed primary structure and consists of the covalent bond structure of a protein; i.e., its amino acid sequence and the locations of any disulfide bonds. The next level in structure describes how local regions in a protein are arranged into organized assemblies and is termed secondary structure. Common secondary structure elements of globular proteins are the α-helix, β-pleated sheet, β-turn and disordered structures; a recent addition to this list is the ordered "Ω-loop" structure (4). Thirdly, there is the tertiary structure, which involves the orientation of, and non-bonded contacts between, the various secondary structure elements. The last level of structure is the quaternary structure which relates the tertiary structures of several protein chains to one another; e.g., it gives the relative positions of subunits in a multi-subunit protein.

The complete three-dimensional structure of a protein is the result of a myriad of chemical interactions (hydrogen-bonding, electrostatic, hydrophobic, etc.) between the twenty different amino acids which compose a protein chain and interactions of the amino acid residues with solvent (water). Given the number of interactions involved and the large number of conformational degrees of freedom, it is not surprising that predictions of protein structure from an amino acid sequence are difficult at best. The importance of this problem, however, has prompted numerous investigations into the comparison of theoretical and experimental results, with the most progress occurring in the prediction of secondary, rather than tertiary, structure. The most common prediction methods for protein secondary structure are those of Lim (5), Chou and Fasman (6) and Robson and co-workers (7). Although there has been criticism of the accuracy of these methods (8,9), they are useful when X-ray structures are not available and if the results are not extrapolated to uses for which they were not intended; combinations of prediction methods seem to work reliably (10,11). Recent work by Kuntz and co-workers (12,13) has shown some promise of improvement in predictability by applying an accurate prediction of the location of turns using a pattern-matching approach.

Conformational Transitions

Information derived from the study of conformational transitions of proteins, i.e., studies of the protein folding process, can provide contributions to the understanding of stability of protein structure (14). By examining the structural rearrangements which occur as a protein folds or unfolds, it should be possible to determine the critical attributes which confer stability. In particular, the molecular structures of intermediates which are populated during the folding process should provide clues to the forces which stabilize the native protein structure and the determination of the relationship between structure and stability. Equilibrium denaturation curves can be used to quantify stability and test predictions of the effects of amino acid substitutions on protein stability.

A practical goal of research into protein folding mechanisms is to provide a fundamental framework which can be used to develop and optimize manufacturing processes for proteins in their biologically active, folded state using a suitable host organism in large scale fermentation. Often the foreign protein is not excreted from the host but is sequestered into "inclusion bodies" where the protein is usually unfolded and has its disulfide bonds reduced. Efficient methods for isolation, solubilization, purification and folding of the protein-containing granules are important for commercial production of recombinant proteins in large quantities.

Several methods are available to alter protein conformation under equilibrium conditions in the laboratory (15-17); they may also be used in the preparation of folded proteins from inclusion bodies. The most convenient of these involve the addition of chemical agents such as urea or guanidine hydrochloride (Gdn HCl) which have been shown to unfold proteins reversibly and, at high enough concentration (12 M urea or 6 M Gdn HCl), yield protein chains that are random coils. In contrast, most surfactants bind irreversibly to protein molecules. Adjustment of solution pH and heating can also be used but often are accompanied by protein degradation which is not reversible and these methods are not assured of producing a completely unfolded protein chain. The comparison of results from the use of different means of unfolding a protein can provide further information about the nature of the unfolding pathway, for example, the addition of Gdn HCl increases solution ionic strength whereas urea does not.

Optical Spectroscopy

The use of optical spectroscopy to study protein structure is dependent on the sensitivity of electronic energy levels to protein structure changes. Structural information can be obtained by observing changes in spectral properties (intensities, wavelengths, band shapes, etc.) and correlating these data with results from model compounds. The principal absorbing components in proteins are peptide bonds and aromatic amino acids (tryptophan, tyrosine and phenylalanine) which all have absorption maxima in the UV (λ < 300 nm) (18). (The addition of prosthetic groups; e.g., hemes, flavins, or pyridoxal phosphate, can dramatically change the UV absorption spectrum, but proteins containing these groups will not be considered here.) Due to the large number of peptide bonds present in a protein, spectroscopic studies which probe the electronic energy levels of peptide bonds will necessarily provide "average" structural information, while spectroscopy of aromatic amino acids, which are represented less frequently in the protein chain, can often provide detailed structural information.

UV Absorption Spectroscopy. The absorption of radiation by peptide bonds occurs in the far-UV part of the spectrum due to a weak $n \rightarrow \pi^*$ transition (at about 215 nm, $\varepsilon \sim 100$) and a strong $\pi \rightarrow \pi^*$ transition (at about 190 nm, $\varepsilon \sim 7000$). Aromatic amino acids (tryptophan, tyrosine and phenylalanine) absorb energy in both the near- and far-UV due to strong $\pi \rightarrow \pi^*$ transitions ($\varepsilon \sim 5000$). The typical UV absorption spectra for peptide bonds (19) and aromatic amino acids (20) in aqueous solution have been recorded. The determination of protein concentration can be done conveniently using the near-UV absorption maximum of proteins due to the absorption of tyrosine and tryptophan residues. A compilation of molar absorptivity values for several hundred proteins and protein derivatives has been undertaken in a series of papers by Kirschenbaum (21). The secondary structure of the protein chain can alter the absorption maximum and intensity of peptide bond transitions (19), but the overlap of transitions makes the determination of secondary structure difficult from these data. It will be shown later that the circular dichroism spectrum in this same wavelength range provides a better determination of secondary structure.

The absorption spectra of aromatic amino acids are sensitive to the polarity (dielectric constant) of their environment due to the effect of solvent on electronic energy levels. Since the aromatic amino acids are non-polar, they tend to reside in non-polar environments in native protein structures but become exposed to polar solvent (water) upon unfolding. This change in environment is reflected in the UV absorption spectrum as shown in Figure 1 for bGH; there is a blue-shift of the absorption maximum of about 5 nm upon unfolding due to a change in solvation with the maximum change occurring near 290 nm. The unfolded bGH spectrum in Figure 1 is virtually the same as that for a mixture of the aromatic amino acids contained in bGH (1 tryptophan, 6 tyrosines and 13 phenylalanines). These data illustrate how the UV absorption spectrum can be used to monitor tertiary structure changes.

Figure 1 illustrates the poor resolution which is found in a typical protein UV absorption spectrum due to the large number of overlapping electronic transitions. This situation can be improved and peak positions located more accurately if the derivative of absorption with respect to wavelength is calculated (22). Usually the second-derivative spectrum is used (23-26) and maxima (peaks) in the zero-order spectrum become minima (troughs) in the second-derivative spectrum, but fourth-derivative spectra have been shown to have certain advantages (27,28) among which is that zero-order peaks are also peaks in the fourth-derivative spectrum. Some of the uses of these data include the determination of the tyrosine/ tryptophan ratio in unknown proteins (23), the determination of the number of tyrosine residues exposed to solvent in a protein (24) and conformational comparisons between native and unfolded proteins (25) or between proteins with similar amino acid composition (26). Using model compounds for tyrosine (N-acetyl-tyrosine ethyl ester) and tryptophan (N-acetyl-tryptophan ethyl ester), it has been demonstrated (25) that a change in solvent dielectric constant from non-polar to polar has the effect of shifting the second-derivative bands of tyrosine without changes in band intensities, while for tryptophan the effect is one of changing band intensities without shifting in the second-derivative bands. It will be shown later how these data can be used to interpret second-derivative spectra of the tryptophan in bGH.

Fluorescence Spectroscopy. The intrinsic fluorescence properties of proteins provide unique spectroscopic tools for protein structure investigations (29) using both steady-state and time-resolved techniques; by labeling with extrinsic probes the applications can be expanded even further. The utility of intrinsic fluorescence data is due to several factors, some of which are: (1) emission spectra are sensitive to fluorophore environment dielectric constant and the presence of substrates; (2) fluorescence provides a method to study protein dynamics as many protein motions can occur during an excited state lifetime (typically in the nanosecond regime); (3) the polarization properties of fluorescence emission can be exploited; (4) it is possible to use extrinsic quenching agents to probe fluorophore accessibility; (5) fluorescence energy transfer studies can be performed to determine distances between residues in proteins; (6) fluorescence emission is usually restricted to a small number of amino acids in a protein, providing a good degree of specificity.

Steady-State Fluorescence. The fluorescence emission spectra of the three aromatic amino acids in water yield an emission maximum for tryptophan at 348 nm, for tyrosine at 303 nm and for phenylalanine at 282 nm (30). The emission maximum for both tryptophan and tyrosine is affected by the dielectric constant of the environment with non-polar surroundings producing blue-shifted emission peaks. The most intense fluorescence emission in a protein is due to tryptophan residues with less due to tyrosine and phenylalanine residues. If tyrosine and tryptophan residues are both present (as in most proteins), it is not possible to observe separately the emission of tyrosine residues due to efficient energy transfer from tyrosine to tryptophan and quenching of tyrosine fluorescence by other functions groups of the protein (29). It is possible to isolate the fluorescence of tryptophan residues by exciting at the red edge of the tryptophan absorption spectrum, typically at 295 to 300 nm, and monitoring fluorescence emission at long wavelengths ($\lambda > 350$ nm). Under these conditions the emission from tyrosine is virtually zero. Fluorescence emission from phenylalanine residues is extremely weak.

The effect of protein unfolding on the fluorescence emission spectrum of bGH is substantial for excitation of the tryptophan residue alone (Figure 2) and excitation of both tryptophan and tyrosine residues (excitation at 280 nm, data not shown). In both cases a shift in the emission maximum is observed, reflecting a change in environment from non-polar to polar, and an increased intensity is seen in the unfolded state, probably due to intramolecular fluorescence quenching in the native state. There are two maxima observed when unfolded bGH is excited at 280 nm which are the resolved emission maxima of tyrosine at 305 nm and tryptophan at 350 nm; energy transfer from tyrosine to tryptophan in the native structure prevents the observation of most tyrosine fluorescence emission and only one emission peak is observed.

Time-Resolved Fluorescence. The study of protein structure with time-resolved fluorescence techniques has been reviewed recently (31,32) as have methods for the measurement of fluorescence intensity and anisotropy decay using time-correlated single photon counting (33); measurements using multifrequency phase techniques are discussed in descriptions of state-of-the-art instrumentation (34,35). Fluorescence lifetime measurements of proteins are complicated by the fact that even single tryptophan proteins such as phospholipase A2 (36), parvalbumin (37) and ribonuclease T1 (38) or single tyrosine proteins such as histone

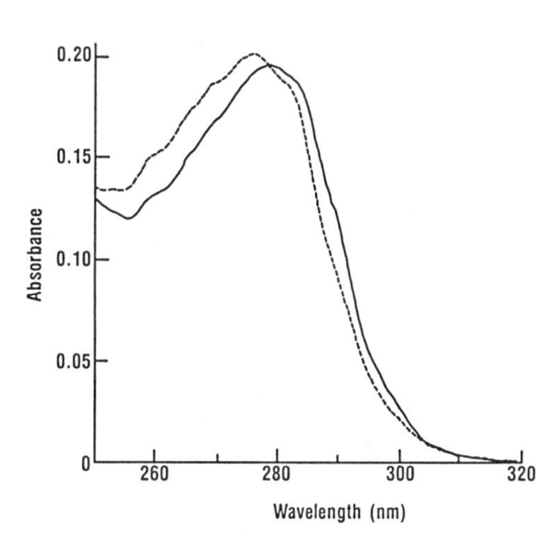

Figure 1. Ultraviolet absorption spectra of native (____) and unfolded (----) bGH. Solvent for the native spectrum was 0.05 M ammonium bicarbonate (pH 8.5) and for the unfolded spectrum was the same buffer plus 6 M Gdn HCl.

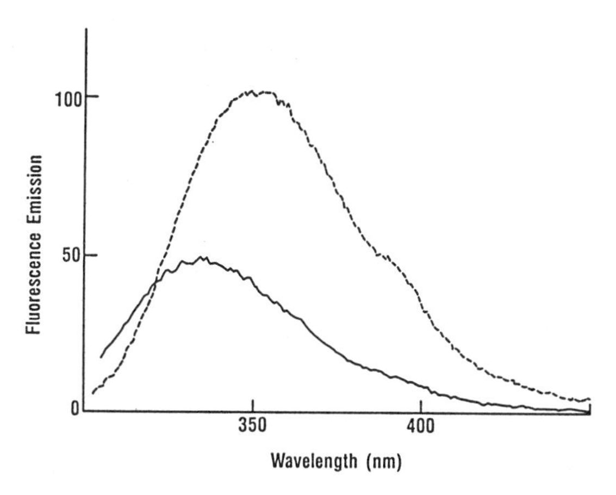

Figure 2. Fluorescence emission spectra of native (____) and unfolded (----) bGH with excitation at 295 nm (to excite only the tryptophan residue). Solvent for the native spectrum was 0.05 M ammonium bicarbonate (pH 8.5) and for the unfolded spectrum was the same buffer plus 6 M Gdn HCl.

H1 (39) are observed to have non-exponential decay kinetics. The fluorescence decay for free tryptophan in solution is also non-exponential (40,41) as is the decay for free tyrosine (42) with the origin in both cases ascribed to different rotamers of the side chain in the ground state; for proteins the causes of non-exponential decay are not understood entirely, but it is known that there is a variability in tryptophan decay kinetics among proteins with different amino acid sequences when in their native state (43). This variability disappears when proteins are denatured, as all proteins containing tryptophan exhibit dual exponential decay of emission with lifetimes of 1.5 and 4 nanoseconds. These results suggest that if the complex photophysics of a protein can be unraveled the potential exists to obtain detailed molecular information from fluorescence lifetime determinations.

Fluorescence anisotropy decay studies can be used to determine the rotational correlation time of a protein as well as providing a measurement of other dynamic processes which depolarize fluorescence emission (44,45). Recent work by Brand and co-workers (46) has demonstrated how it is possible to resolve the contributions of different components to total anisotropy decay in a rotationally heterogeneous system using anisotropy decay associated fluorescence spectra. The fluorescence emission spectra of the different components may then be used to make structural interpretations. The application of these techniques to the study of protein unfolding processes should allow the motions of different tryptophan residues to be separated and quantified, provided their emission spectra are different. In this manner, intermediates in the unfolding process may be characterized according to the range of motions available to its tryptophan residues.

Fluorescence Quenching. Fluorescence quenching studies of proteins provide information about the penetration of a quencher molecule into the protein matrix (47-49). With such studies it is possible to probe the accessibility of fluorophores; e.g., tyrosine or tryptophan residues, to the quencher as modulated by steric and dielectric constant effects. The use of different chemical species, such as iodide (50), oxygen (51), cesium ion, acrylamide (52) and trichloroethanol (53), can serve as a way of characterizing the environment around a fluorophore depending on the quenching efficiency of the different quenchers. The interpretation of these data must take into account the possibility that the quenching mechanism is dynamic (complexation of quencher with an excited state fluorophore) or static (complexation of quencher with a ground state fluorophore) or both. Further complications can arise if the quencher exhibits partitioning and/or binding to the protein "phase" (54).

Circular Dichroism Spectroscopy. Circular dichroism (CD) spectroscopy provides a measurement of molecular optical activity through measurement of the difference in extinction coefficient between left- and right-circularly polarized light (55,56). CD studies of proteins are informative because the amino acids of which they are composed are optically active individually. When the amino acids are combined in a protein they can produce large optically active structures (such as α-helices, β-sheets, etc.) whose structure can be probed with CD spectroscopy. This discussion will be limited to the study of protein electronic transitions with CD although recent investigations of vibrational CD of amino acids (57), peptides and polypeptides (58) have demonstrated that considerable structural information can also be gleaned from the VCD spectrum.

Near-UV CD. The electronic transitions of proteins in the near-UV are due to absorptions of the aromatic amino acids and were discussed previously. The significant overlap of transitions in the absorption spectrum often leads to poor resolution of spectral components, a problem which is less pronounced in a CD spectrum as transitions can have different signs (59). A CD spectrum, therefore, has higher inherent resolution than an absorption spectrum. This property has been exploited to assign the vibrational fine structure in absorption bands of phenylalanine (60), tryptophan (61) and tyrosine (62) in model compounds and proteins at low temperature (77 K). This work has been extended by Puett and co-workers (63) to bGH and other growth hormones in an assignment of their near-UV electronic transitions from CD spectra. These studies followed earlier work (64,65) which demonstrated the sensitivity of the near-UV CD spectrum of bGH (Figure 3) to solution conditions and provided evidence that the the near-UV CD spectrum is a reliable indicator of protein tertiary structure.

Far-UV CD. Proteins absorb energy in the far-UV due to electronic transitions of peptide bonds. The application of the CD spectrum for the determination of secondary structure has been standard practice for several years ever since the pioneering work by Fasman and co-workers (66). There has been considerable effort expended to improve the reliability of these determinations (67-70) and to go beyond the determination of α-helix, β-sheet and remainder percentages to separate parallel and anti-parallel β-sheet contributions and to determine β-turn percentages (71-75). Most of the methods rely on correlating the observed CD spectrum for reference proteins in the solution state with the secondary structure determination from X-ray crystal structures of the same proteins. The accuracy of the methods varies but the most reliable parameter to determine with all of them is the percent α-helix, due largely to its spectrum being isolated from the spectra of the other secondary structure elements (73). The effects of several experimental errors on secondary structure determinations from CD spectra can be significant and are to be avoided in order to achieve good agreement between experiment and theory (76). It should also be emphasized that spectra with high signal-to-noise ratios are required for accurate results.

The far-UV CD spectrum of bGH is characteristic for a protein containing large amounts of α-helix structure (minima at 222 and 208 nm, maximum at 193 nm) and has been used to estimate that bGH has about 50% α-helix and 10% β-sheet structure (63,65). The spectrum of unfolded bGH (in 6 M Gdn HCl) is consistent with a random coil spectrum; in particular, there is virtually no CD signal at 222 nm, a wavelength that can be used to monitor secondary structure during unfolding studies.

Optical Spectroscopy of the Single Tryptophan in bGH

The unfolding of bGH has been studied by several workers (63,77,78) and found to be a multistate process with at least one stable equilibrium intermediate:

$$N \rightleftharpoons I \rightleftharpoons U$$

where N represents the native state, I an intermediate and U the unfolded state of bGH. It has also been shown (79) that a self-associated form (or forms) of bGH is (are) populated under partially denaturing conditions (3.7 M Gdn HCl or 8.5 M urea) so that the complete unfolding process under equilibrium conditions can be represented as:

$$\begin{array}{c} N \rightleftharpoons I \rightleftharpoons U \\ \downarrow\uparrow \\ I_n \end{array}$$

where I_n represents the self-associated intermediate specie(s) and n is between 3 and 5. The equilibrium unfolding of bGH in Gdn HCl and urea are summarized in Table I.

Table I. Summary of Equilibrium Unfolding Results for bGH

	Transition Mid-point (M)	
Method	Gdn HCl	urea
Absorption (290 nm)	3.1	7.8
Second-derivative absorption (tyr)	3.2	
Second-derivative absorption (trp)	3.5	
Fluorescence (trp)	3.6	8.1
Size-exclusion HPLC	3.8	
Circular Dichroism (222 nm)	3.9	9.2

In the discussion below, optical spectroscopic methods will be described which characterize the molecular structure of each of these species as determined by probing of the lone tryptophan residue in bGH.

Second-Derivative Absorption Spectroscopy. It was demonstrated above that the second-derivative absorption spectrum of tryptophan can be used to probe the dielectric constant of the tryptophan environment in a protein. The spectra for bGH (79) in different conformational states show differences in polarity that are indicated by the different intensities of the peak at 295 nm and the trough at 291 nm. When the intensities are measured for the unfolding transition at high (0.3 mg/mL) and low (0.05 mg/mL) protein concentrations (79), it is concluded that the dielectric constant (ε) of the tryptophan environment for the self-associated intermediate state is the largest of all the conformational states with the others in the following order:

$$\varepsilon_{I_n} > \varepsilon_U > \varepsilon_I > \varepsilon_N$$

These data have been analyzed in more detail elsewhere (79).

Fluorescence Spectroscopy. When tryptophan fluorescence is used to monitor the unfolding of bGH with Gdn HCl at low protein concentration a transition curve is observed that has a peak near 4 M Gdn HCl (Figure 4); this is clear evidence that a simple two-state process cannot explain the unfolding of bGH: at least one equilibrium intermediate (I) is populated. Figure 4 also demonstrates that the fluorescence emission of the tryptophan in the native state is quenched intramolecularly when compared to the I and the U states. When the protein concentration is increased at 3.8 M Gdn HCl, the fluorescence intensity is observed to decrease monotonically and level off, indicating that the formation of I_n is accompanied by increased fluorescence quenching. The fluorescence quantum yields (ϕ) of tryptophan in the various forms of bSt are:

$$\phi_I > \phi_{I_n} > \phi_U >> \phi_N$$

The effects of the fluorescence quenching agents acrylamide, iodide and trichloroethanol (TCE) on the fluorescence emission of the single tryptophan of bGH in its various conformational states have been determined (Kauffman, E.W., The Upjohn Company, unpublished data) with the following results: (1) as expected, for all states TCE is the most effective quencher followed by acrylamide and iodide, the unfolded state is the most accessible of all the states and the native state is the most protected from TCE and acrylamide; (2) the self-associated intermediate state of bGH is the most protected from iodide of all the states. The latter result can be interpreted as indicating that there are negative charges near the tryptophan in the I_n state which prevent iodide from penetrating the protein matrix.

Circular Dichroism Spectroscopy. The near-UV CD spectrum of bGH (Figure 3) provides a convenient probe for the self-associated

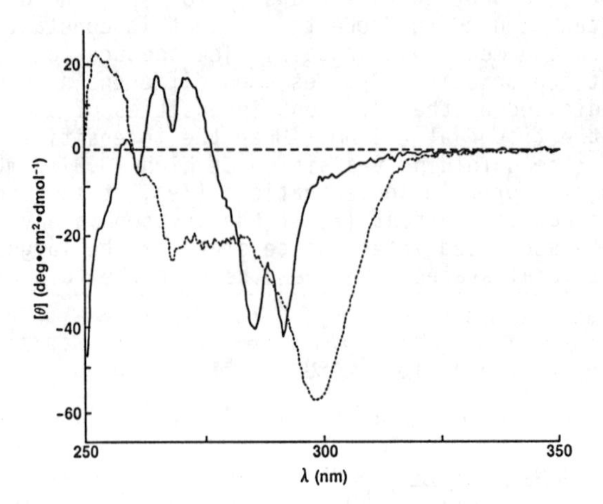

Figure 3. Near-UV CD spectra of native (____) and self-associated intermediate (----) bGH. Solvent for the native spectrum was 0.05M ammonium bicarbonate (pH 8.5) and for the self-associated intermediate was the same buffer plus 3.7 M Gdn HCl. (Reproduced from Ref. 79. Copyright 1986 American Chemical Society).

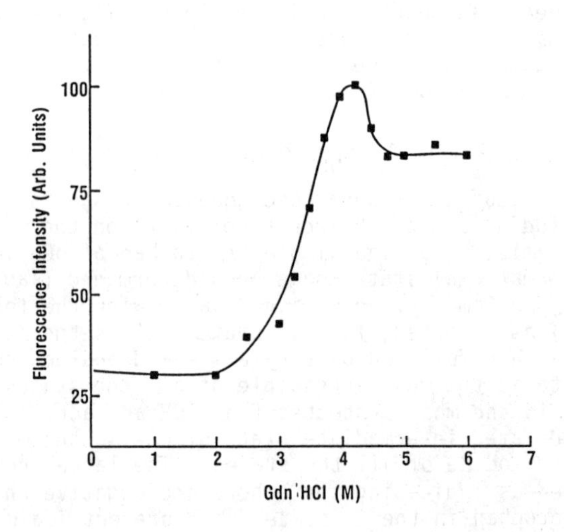

Figure 4. Unfolding transition for 0.01 mg/mL bGH in Gdn HCl as monitored by tryptophan fluorescence intensity.

intermediate form of bGH (79,80). A band is observed at 300 nm, due to the tryptophan residue, that grows in intensity in the I_n form. The situation is similar to that observed previously for insulin (81) where an increase in the CD of tyrosine residues upon self-association was ascribed to dipole-dipole coupling of residues on different insulin molecules. These data can be interpreted as an indication that the tryptophan residues of I_n are held rigidly near the interface between self-associating molecules.

Acknowledgments

I am grateful for the important contributions to this work by my collaborators at The Upjohn Company: Dr. D.N. Brems has been a great help in many areas of protein chemistry, E.W. Kauffman and S.M. Plaisted have contributed valuable technical assistance and Dr. R.D. White has been a continual source of support and advice.

I also acknowledge B.S. Hanna for conducting several fluorescence studies of bGH while completing a Kalamazoo College SIP internship at Upjohn and K. Hendricks for typing the manuscript.

Literature Cited

1. Doolittle, R.F. Sci. Am. 1985, 252 (4), 88-91,94-99.
2. Anfinsen, C.B.; Haber, E.; Sea, M.; White, F.H. Proc. Natl. Acad. Sci. U.S.A. 1961, 47, 1309-1314.
3. Lehninger, A.L. Biochemistry; Worth: New York, 1975; Chapter 5, 6.
4. Leszczynski, J.F.; Rose, G.D. Science (Washington, D.C.) 1986, 234, 849-855.
5. Lim, V.I. J. Mol. Biol. 1974, 88, 873-894.
6. Chou, P.Y ; Fasman, G.D. Ann. Rev. Biochem. 1978, 47, 251-276.
7. Garnier, J.; Osguthorpe, D.J.; Robson, B. J. Mol. Biol. 1978, 120, 97-120.
8. Nichikawa, K. Biochem. Biophys. Acta 1983, 748, 285-299.
9. Kabasch, W.; Sander, C. Proc. Natl. Acad. Sci. U.S.A. 1984, 81, 1075-1078.
10. Sternberg, M.J.E.; Cohen, F.E. Int. J. Biol. Marcomol. 1982, 4, 137-144.
11. Sawyer, L.; Fothergill-Gilmore, L.A.; Russell, G.A. Biochem. J. 1986, 236, 127-130.
12. Cohen, F.E.; Abarbanel, R.M.; Kuntz, I.D.; Fletterick, R.J. Biochemistry 1983, 22, 4894-4904.
13. Cohen, F.E.; Abarbanel, R.M.; Kuntz, I.D.; Fletterick, R.J. Biochemistry 1986, 25, 266-275.
14. Creighton, T.E. J. Phys. Chem. 1985, 89, 2452-2459.
15. Tanford, C. Adv. Prot. Chem. 1968, 23, 121-282.
16. Pace, C.N. CRC Crit. Rev. Biochem. 1975, 3, 1-43.

17. Lapanje, S. Physicochemical Aspects of Protein Denaturation; J. Wiley: New York, 1978.
18. Cantor, C.R.; Schimmel, P.R. Biophysical Chemistry; W.H. Freeman: San Francisco, 1980; Chapter 7.
19. Roseñheck, K.; Doty, P. Proc. Natl. Acad. Sci. U.S.A. 1961, 47, 1775-1785.
20. Wetlaufer, D.B. Adv. Prot. Chem. 1962, 17, 303-390.
21. Kirschenbaum, D.M. Appl. Biochem. Biotechnol. 1985, 11, 287-316.
22. Butler, W.L. Methods Enzymol. 1979, 56, 510-515.
23. Servillo, L.; Colonna, G.; Balestrieri, C.; Ragone, R.; Irace, G. Anal. Biochem. 1982, 126, 251-257.
24. Ragone, R.; Colonna, G.; Balestrieri, C.; Servillo, L.; Irace, G. Biochemistry 1984, 23, 1871-1875.
25. Terada, H.; Inoue, Y.; Ichikawa, T. Chem. Pharm. Bull. 1984, 32, 585-590.
26. Bewley, T.A.; Li, C.H. Arch. Biochem. Biophys. 1984, 233, 219-227.
27. Padros, E.; Dunach, M.; Morros, A.; Sabes, M.; Manosa, J. Trends Biochem. Sci. 1984, 9, 508-510.
28. Metzler, D.E.; Metzler, C.M.; Mitra, J. Trends Biochem. Sci. 1986, 11, 157-159.
29. Lakowicz, J.R. Principles of Fluorescence Spectroscopy; Plenum: New York, 1983; Chapter 11.
30. Teale, F.W.J.; Weber, G. Biochem. J. 1957, 65, 476-482.
31. Ghiggino, K.P.; Roberts, A.J.; Phillips, D. Adv. Polymer Sci. 1981, 40,
32. Beechem, J.M.; Brand, L. Ann. Rev. Biochem. 1985, 54, 43-71.
33. O'Connor, D.V.; Phillips, D. Time-correlated Single Photon Counting; Academic: New York, 1984.
34. Alcala, J.R.; Gratton, E.; Jameson, D.M. Anal. Instrum. 1985, 14, 225-250.
35. Lakowicz, J.R.; Laczko, G.; Gryczynski, I. Rev. Sci. Instrum. 1986, 57, 2499-2506.
36. Ludescher, R.D.; Volwerk, J.J.; de Haas, G.H.; Hudson, B.S. Biochemistry 1985, 24, 7240-7249.
37. Permyakov, E.A.; Ostrovsky, A.V.; Burstein, E.A.; Pleshanov, P.G.; Gerday, C.H. Arch. Biochem. Biophys. 1985, 240, 781-791.
38. James, D.R.; Demmer, D.R.; Steer, R.P.; Verrall, R.E. Biochemistry 1985, 24, 5517-5526.
39. Libertini, L.J.; Small, E.W. Biophys. J. 1985, 47, 765-772.
40. Creed, D. Photochem. Photobiol. 1984, 39, 537-562.
41. Engh, R.A.; Chen, L.X.-Q.; Fleming, G.R. Chem. Phys. Lett. 1986, 126, 365-372.
42. Laws, W.R.; Ross, J.B.A.; Wyssbrod, H.R.; Beechem, J.M.; Brand, L.; Sutherland, J.C. Biochemistry 1986, 25, 599-607.
43. Grinvald, A.; Steinberg, I.Z. Biochem. Biophys. Acta 1976, 427, 663-678.
44. Tran, C.D.; Beddard, G.S. Eur. Biophys. J. 1985, 13, 59-64.
45. Lakowicz, J.R.; Laczko, G., Gryczynski, I.; Cherek, H. J. Biol. Chem. 1986, 261, 2240-2245.

46. Knutson, J.R.; Davenport, L.; Brand, L. Biochemistry 1986, 25, 1805-1810.
47. Eftink, M.R.; Ghiron, C.A. Anal. Biochem. 1981, 114, 199-227.
48. Lakowicz, J.R. Principles of Fluorescence Spectroscopy; Plenum: New York, 1983; Chapter 9.
49. Ho, C.N.; Patonay, G.; Warner, I.M. Trends Anal. Chem. 1986, 5, 37-43.
50. Lehrer, S.S. Biochemistry 1971, 17, 3254-3263.
51. Lakowicz, J.R.; Weber, G. Biochemistry 1973, 12, 4161-4170.
52. Eftink, M.R.; Ghiron, C.A. J. Phys. Chem. 1976, 80, 486-493.
53. Eftink, M.R.; Zajicek, J.L.; Ghiron, C.A. Biochem. Biophys. Acta 1977, 491, 473-481.
54. Blatt, E.; Chatelier, R.C.; Sawyer, W.H. Biophys. J. 1986, 50, 349-356.
55. Charney, E. The Molecular Basis of Optical Activity; J. Wiley: New York, 1979.
56. Cantor, C.R.; Schimmel, P.R. Biophysical Chemistry; W.H. Freeman: San Francisco, 1980; Chapter 8.
57. Oboodi, M.R.; Lal, B.B.; Young, D.A.; Freedman, T.B.; Nafie, L.A. J. Am. Chem. Soc. 1985, 107, 1547-1556.
58. Narayanan, U.; Keiderling, T.A.; Bonora, G.M.; Toniola, C. J. Am. Chem. Soc. 1986, 108, 2431-2437.
59. Kahn, P.C. Meth. Enzymol. 1979, 61, 339-378.
60. Horwitz, J.; Strickland, E.H.; Billups, C. J. Am. Chem. Soc. 1969, 91, 184-190.
61. Strickland, E.H.; Horwitz, J.; Billups, C. Biochemistry 1969, 8, 3205-3213.
62. Horwitz, J.; Strickland, E.H.; Billups, C. J. Am. Chem. Soc. 1970, 92, 2119-2129.
63. Holladay, L.A.; Hammonds, R.G. Jr.; Puett, D. Biochemistry 1974, 13, 1653-1661.
64. Edelhoch, H.; Lippoldt, R.E. J. Biol. Chem. 1970, 245, 4199-4203.
65. Sonenberg, M.; Beychok, S. Biochem. Biophys. Acta 1971, 229, 88-101.
66. Greenfield, N.; Fasman, G.D. Biochemistry 1969, 8, 4108-4116.
67. Chen, Y.-H.; Yang, J.T.; Martinez, H.M. Biochemistry 1972, 11, 4120-4131.
68. Chen, Y.-H.; Yang, J.T.; Chau, K.H. Biochemistry 1974, 13, 3350-3359.
69. Baker, C.C.; Isenberg, I. Biochemistry 1976, 15, 629-634.
70. Provencher, S.W.; Glockner, J. Biochemistry 1981, 20, 33-37.
71. Chang, C.T.; Wu, C.-S.; Yang, J.T. Anal. Biochem. 1978, 91, 13-31.
72. Bolotina, I.A.; Chekhov, V.O.; Lugauskas, V.Y. Int. J. Quant. Chem. 1979, 16, 819-824.
73. Brahms, S.; Brahms, J. J. Mol. Biol. 1980, 138, 149-178.
74. Hennessey, J.P. Jr.; Johnson, W.C. Biochemistry 1981, 20, 1085-1094.

75. Compton, L.A.; Johnson, W.C. Jr. Anal. Biochem. 1986, 155, 155-167.
76. Hennessey, J.P. Jr.; Johnson, W.C. Jr. Anal. Biochem. 1982, 125, 177-188.
77. Burger, H.G.; Edelhoch, H.; Condliffe, P.G. J. Biol. Chem. 1966, 241, 449-457.
78. Brems, D.N.; Plaisted, S.M.; Havel, H.A.; Kauffman, E.W.; Stodola, J.D.; Eaton, L.C.; White, R.D. Biochemistry 1985, 24, 7662-7668.
79. Havel, H.A.; Kauffman, E.W.; Plaisted, S.M.; Brems, D.N. Biochemistry 1986, 25, 6533-6538.
80. Brems, D.N.; Plaisted, S.M.; Kauffman, E.W.; Havel, H.A. Biochemistry 1986, 25, 6539-6543.
81. Strickland, E.H.; Mercola, D. Biochemistry 1976, 15, 3875-3884.

RECEIVED July 8, 1987

Chapter 15

Sensitive Detection and Quantitation of Protein Contaminants in rDNA Products

Andrew J. S. Jones

Medicinal and Analytical Chemistry, Genentech, Inc., 460 Point San Bruno Boulevard, South San Francisco, CA 94080

Analytical methods are available for many of the contaminants possibly associated with pharmaceutically useful proteins available through rDNA technology. Such contaminants include process reagents, antibiotics, pyrogens, bacteria, viruses, etc. Two new major concerns about the safety of microorganisms and cell culture as sources of polypeptides are contaminating proteins and residual cellular DNA. While international consensus suggests that 10 pg DNA per dose is an acceptable and achievable level, no such consensus exists for protein contaminants. An acceptable level of contamination will be a function of the dose, frequency of treatment, seriousness of the disease etc. and will necessarily be evaluated on a case by case basis. The detection and accurate quantitation of contaminating proteins at the ppm level (i.e., ug of contaminant per gram of product) are analytical challenges that can only be met by immunological methods. Such multiple antigen immunoassays must be shown to be capable of detecting all the potential contaminants and to yield accurate values for their concentrations. The former constraint means that such assays will be "process-specific" while the latter requires that each sample be assayed as a dilution series to demonstrate antibody excess for each contaminant present. Detailed criteria for, and examples of, the validation of such assays are discussed.

The purification of polypeptide pharmaceuticals, whether from rDNA based systems or from natural sources, is achieved by the successive removal of undesirable components from the protein of interest until they are at a level acceptable for human use. Dr. Liu (1) has presented an overview of the major issues seen by the

0097-6156/88/0362-0193$06.00/0

Food and Drug Administration (FDA) to be associated with the new technology and this paper will be addressing one of these in detail: the measurement of protein impurities in products of rDNA technology. The FDA is taking a cautious attitude (1), which in this context relates to the lack of knowledge concerning the properties and desirability of proteins from *E. coli*, yeast or mammalian cells. Protein contaminants may arise from the host cell itself but also from medium components (serum additions and growth factors) and in certain cases from the process operations (e.g., monoclonal antibody columns). While there may be biological effects of some contaminants, the primary consequence of concern is the immunological responses to the contaminants. These range from relatively rapid allergic responses to the possibility of adjuvant-like effects on the immunogenicity of the product, the latter being potentially either a positive or negative feature depending on the product and the type of antibody response. The development of antibodies to contaminants is only a concern if the antibodies could lead to anaphylactic type reactions or if they are cross reactive to any of the patient's proteins. This latter possibility is clearly more likely in production systems which involve mammalian cells and serum than in bacteria and fungi.

In the past it has been a combination of accumulated clinical experience and the human origin of the starting material that has generated comfort with the safety record of protein pharmaceuticals developed without the need for purity levels measured as parts per million of contaminants. However, potential contamination of supplies by infectious agents (e.g., hepatitis, AIDS and Creutzfeld-Jacob agents) has caused a re-examination of many manufacturing processes with the same degree of concern currently being applied to rDNA based processes. In the case of the new technology, however, it is the unknown nature of the potential risks that prompts the cautious approach and consequently a close examination of the components.

At a recent symposium (2) it was generally agreed that DNA levels of picograms per dose represented an acceptable risk if transformed cells were used as the host cell. This was due largely to the fact that only one specific kind of DNA, i.e. that with transforming potential, was considered significant. Further, experimental data are available to evaluate the risk in terms of DNA half-life, route of injection, ability to cross physical barriers etc. In the case of proteins, however, no such consensus was reached. It is worth pointing out the complexities of evaluating purity requirements and to attempt to bring thinking in this area to a "per dose" basis rather than "percent". The parameters that need to be considered include 1) Dose - clearly the major parameter 2) Frequency - this clearly modulates the purity requirements, but especially when considered together with 3) Route of administration and half life. These three, together with the properties of the individual components, will determine the magnitude of the immunological or biological response to the contaminants. This response must then be balanced with the benefit of treatment in the conventional risk/benefit ratio. With examples ranging from a vaccine at two doses per lifetime of 20 ug and where antibodies are the desired consequence, to a hormone given several times a day or week for many years, or to a thrombolytic agent at 100 mg once, it is clear how complex the task would be to obtain a useful general purity requirement. Thus the case by case approach is the only way to evaluate the issue.

APPROACHES

Many reports dealing with purity of protein preparations use a combination of several analytical methods such as SDS-PAGE, IEF, HPLC, etc., and allow claims of "greater than", for example "99% pure". For most purposes, these are more than adequate for the work reported. It must be pointed out, however, that this represents 10,000 ppm or 1% contamination (on a weight basis) and secondly that all such protein separation methods suffer the criticism of co-migration or coelution of potential contaminants and have sensitivity limited to around 100 ppm for any individual component. Even by using several methods of purity analysis, the confidence of purity claims is not greatly improved, since the methods are usually analytical forms of the process steps used to purify the protein of interest.

In those cases where high purity seems a prudent "design criterion" for a long term therapeutic, such as growth hormone or insulin for treatment of dwarfism and diabetes respectively, levels of tens of ppm would seem to reduce the level of concern to that of pg DNA per dose. In order to achieve such levels of sensitivity, it is necessary to use immunological methods. There are many methods to choose from but we have concentrated on immunoassays with a double antibody sandwich format. This is because competitive methods cannot work and single antibody methods do not have sufficient sensitivity. We have routinely been able to measure down to 1 ppm of *E. coli* proteins (ECPs) in a variety of our rDNA products (3).

SANDWICH ASSAY DEVELOPMENT

The assay format is a conventional sandwich in which a capture antibody mixture is immobilized (in our case on microtiter plates) and exposed to the sample. The complex is then washed and incubated with the same antibody, but which has been suitably labeled, either by a radioactive or an enzyme tag. The amount of second antibody bound is then measured and related to the amount of contaminant bound in the first complex. The new challenges in this type of assay development are broadly the following: to make the assay 1) sensitive to all the potential contaminants, 2) insensitive to the product of interest, 3) sensitive to sub-ppm levels and, 4) to demonstrate that each contaminant is detected and properly quantitated. This obviously requires careful selection and preparation of the antigens and antibodies used, but also requires a complex validation effort to demonstrate that the values obtained are meaningful. Description and illustration of these will take up the rest of the paper.

ANTIGEN SELECTION

With rDNA technology it is possible to perform a "blank run" in which the operations of the complete manufacturing process from bacterium to vial are carried out with the gene for the product deleted (4). Thus one can prepare "pure" contaminants, i.e. immunologically pure in that they contain no product. It is desirable to be able to measure a sufficient number of components to show that both the process and the product are reproducible.

If the assay is designed to detect all possible components it will be extremely difficult to develop and validate. If it is developed to detect only the last few, the assay will be insensitive to minor upstream process failures. Thus we have usually selected our antigens at a point somewhere in between, e.g. where the product would be 95-99% "pure". It is critical to realize, however, that once the antigens have been selected, the upstream process is "fixed" and therefore any changes that might affect the contaminants will probably require a new, or at least modified, assay. We have experienced both situations during our assay development on various *E. coli* products (3). Conversely, the fixing of the assay allows its use to "fine tune" process operations downstream to improve product purity levels. Figure 1 shows data from two groups of final products, one of which is significantly purer than the other. This increase in purity (i.e. the decrease in contaminants by approximately 10 fold) was achieved by fine tuning some of the downstream processing operations using the ECP values as the basis for optimization. It is also important to note that no generic assay will be available and that only the manufacturer (and the FDA) can measure the purity level of a particular product unless the host cells and manufacturing processes are IDENTICAL.

The "blank run" is essential in order to obtain "pure" contaminants but it also must meet certain criteria. It must be run at full manufacturing scale, since we have observed scale-dependent differences in the behavior of antigens obtained this way. The absence of the product should not affect the chromatographic and physical behavior of the contaminants. The lack of product expression should also have little or no effect on the levels of the other proteins in the host cell. Both of these constraints are experimetally verifiable to an adequate level of assurance. During the blank run it is also wise to take broader pools from elution profiles etc. to assure the greatest ability of the assay to detect minor contaminants in the process.

An alternative to the blank run methodology to obtain suitable antigens is available for systems where high purity is not essential. This is immunoaffinity-based removal of the product from some suitable intermediate step during manufacture. The success of such an approach would depend on the efficiency of the affinity support to remove *all* the product and *no* contaminants. The demonstration of *both* conditions would not *be* trivial!

ANTIBODY GENERATION

Given the broad range of antigens, we have found it useful to immunize several animals to ensure responses to all the proteins. With an aggressive immunization schedule and a solid phase ELISA screening assay we have usually been able to begin pooling serum from the third or fourth month. We are also investigating methods for improving the responses to minor or weakly immunogenic proteins as suggested by Thalhamer and Freund (5,6).

There are two major reasons to purify the antibodies for this type of assay by immunoaffinity methods using immobilized antigens. Firstly it yields a high purity Ig preparation resulting in high signal to noise and therefore good sensitivity. However, probably more important for the validity of the multi-antigen immunoassay, it allows one to define the range of antibody specificities present in the assay reagent. This is because this purification step "normalizes" the antibody population so that it reflects the antigen population on the affinity column (4). This assures the detection of minor components and minimizes the possible domination caused by large responses to the more abundant or immunogenic contaminants.

ASSAY VALIDATION

The importance of affinity purification of the antibodies becomes apparent when one considers the importance of demonstrating antibody excess. In a useful and valid assay, the value should not be dependent on anything but the analyte. In single antigen immunoassays this condition is easy to demonstrate. In multiple antigen assays, however, the requisite condition of antibody excess for EACH contaminant is more complex to demonstrate. For routine purposes, a serial dilution of the sample is assayed and the parts per million contaminant value is calculated. If this value is independent of dilution over a significant portion of the assay standard curve, then we have the greatest assurance that the signal is directly proportional to the analytes being detected and that antibody is indeed in excess. In other words, the sample must dilute linearly.

In the case of the ECP's (E. coli proteins) obtained from the "blank run" of the process used for the manufacture of methionyl-human growth hormone (met-hGH) and used to construct the standard curve, this linearity is expected (Fig. 1). For several final product lots the condition of antibody excess is demonstrated by the linearity of plots of signal (ng/mL) versus relative concentration. By plotting ppm vs. dilution factor it is possible to evaluate what is happening in the assay of a particular sample (see Fig. 2). In this plot the signal should be independent of dilution if the antibodies are in excess (Curve 1, Fig. 2). If this is not true for concentrated samples, it may become true as the analyte(s) are diluted and the plot becomes a plateau (Curve 2, Fig. 2). In some cases the plateau may not appear before the sensitivity of the assay is reached (Curve 3, Fig. 2). While demonstration of antibody excess is required for

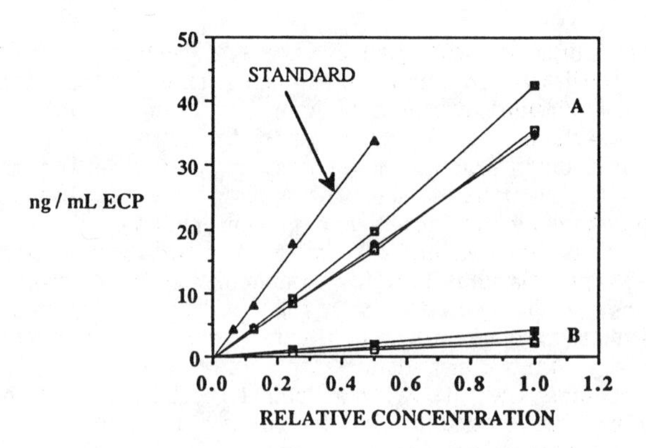

Figure 1. ECP values for samples of met-hGH.

Results are shown for the ECPs used to construct the standard curve and two groups of met-hGH final product batches. Batches in group A have typically 15-20 ppm while those from group B have typically 2 ppm as a result of process optimization using the ECP assay (see Text).

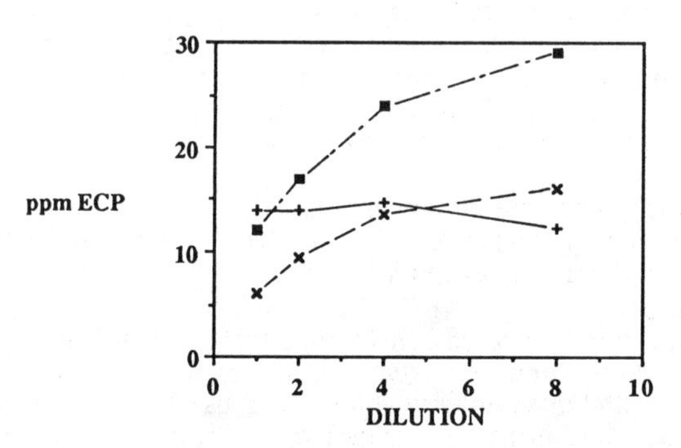

Figure 2. Effect of dilution on ECP values.

Three lots of met-hGH are shown. The ordinate is in parts per million and the dependence of this on dilution is illustrated. Curve 1 (+) is a lot manufactured by the process for which the assay was developed. Curves 2 (✕) and 3 (■), were manufactured by a similar, but not identical, process.

each sample and necessitates the analysis of every sample as a dilution series, the demonstration that all the potential contaminants would be detected in the assay is only needed once for each batch of antibodies. It is therefore advisable to prepare large batches of antibodies since their validation is not trivial. We have developed a relatively simple method for showing which antigens can be bound by a particular batch of antibodies. Immunoblotting methods (7,8) are very difficult to quantitate and detect only those epitopes which have survived the electrophoresis and transfer processes. By immobilizing the antibodies on a column it is possible to assess which proteins are bound to the antibodies in an undenatured state. This is conveniently achieved by saturating such a column with the antigen preparation, extensively washing and then eluting the bound proteins. The load and eluate can then be examined by one of the most sensitive non-immunological analytical methods: SDS-PAGE followed by silver stain (9). Side by side comparison allows the best evaluation and does not require precise quantitation (Fig. 3). Control studies with unrelated immobilized antibodies can be used to demonstrate the specificity of this binding. Additional specificity and interference information can be obtained by testing other products in the assay. Thus the ECPs which accompany gamma interferon are not detected by the ECP assay for met-hGH and vice versa.

Since competitive assays cannot work, the sandwich format is necessary. This requires that each contaminant be able to present two eiptopes simultaneously in order to be detected. We have evaluated the response in the assay of standard ECP's from hGH as a function of size.

Fig. 4 shows a gel filtration profile of the ECP's measured as protein concentration (by absorbance at 280 nm) and as ECP's (ug/mL in the ECP assay). It can be seen that there is good general agreement and that an extinction coefficient of 1 mL/mg.cm at 280 nm is indeed a good approximation for the majority of the proteins. There thus appears to be little dependence of detectability on size under nondenaturing assay conditions, and it appears that most proteins are detected by at least two epitopes.

CONCLUSIONS

1. Purity requirements for pharmaceutical proteins should be expressed on a per dose basis and evaluated on a case by case basis, in conjunction with safety data from clinical trials.
2. For high purity products, suitable assays are required to demonstrate low levels (parts per million) of impurities.
3. Such assays must be immunologically based and be shown to be capable of detecting and properly quantitating the likely impurities.
4. Multiantigen assays are only valid if a condition of antibody excess exists to the majority of the contaminants.
5. Such assays are not only manufacturer specific, they are process-specific.

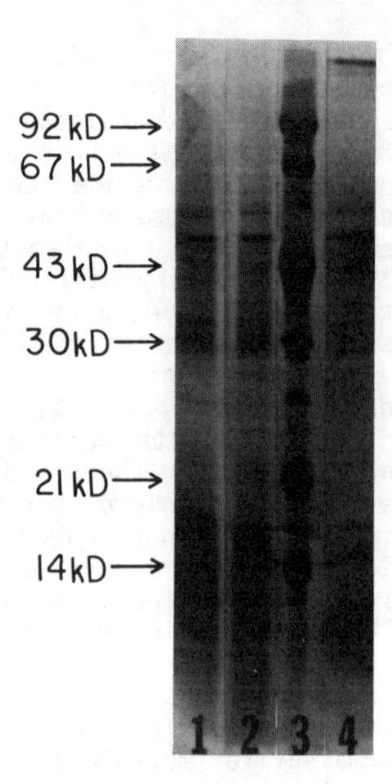

Figure 3. Evaluation of the reactivity of the antibodies used in the ECP assay. Antibodies were immobilized on a solid support (Affigel 10, Bio Rad Labs),and a small column was prepared. Lane 1 shows the ECPs that were loaded, Lane 2 shows the flowthrough after the column was saturated, Lane 3 shows the M.W. standards, and Lane 4 shows the ECPs that were specifically bound by the column. No bands were observed for control columns of other antibodies (not shown). (Reproduced with permission from Ref. 3. Copyright 1986 Elsevier.)

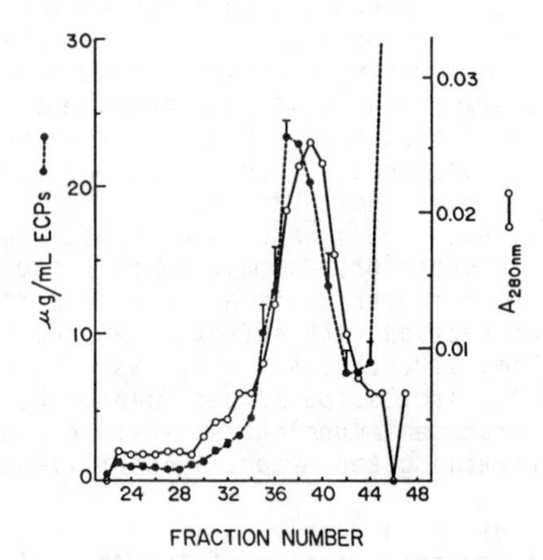

Figure 4. Comparison of assay values with protein content as a function of molecular weight. The ECP mixture was fractionated on a TSK 2000 size-exclusion column, and protein concentration was monitored as $A_{280\ nm}$. ECP values were obtained in the ECP assay. (Reproduced with permission from Ref. 3. Copyright 1986 Elsevier.)

ACKNOWLEDGMENTS

The author would like to thank Vince Anicetti and the authors of reference 3 and in particular Dr. Mike Ross of Genentech, for his encouragement and support. Thanks go also to Bernadette Jehl for preparation of the manuscript and to Dr. Darrell Liu for a preprint of his paper.

REFERENCES

1. Liu, D., Goldman, N. and Gates, F., II. This volume.
2. Noble, G. R., Biological Risk of Residual Cellular Nucleic Acid, in "Abnormal Cells, New Products and Risk", In vitro cellular and developmental biology, Monograph #6, Tissue Culture Association, Gaithersburg, MD. 1986 p. 173
3. Anicetti, V. R., Fehskens, E. F., Reed, B. R., Chen, A. B., Moore, P., Geier, M. D., and Jones, A. J. S., (1986) Immunoassay for the detection of E. coli proteins in recombinant DNA derived human growth hormone. J. Immunol. Methods 91: 213-224.
4. Jones, A. J. S. and O'Connor, J. V., Control of recombinant DNA produced pharmaceuticals by a combination of process validation and final product specifications, in: Developments in Biological Standardization: Symposium on Standardization and Control of Biologicals Produced by Recombinant DNA Technology, Vol. S9, (S. Kerger, ed.), International Association of Biological Standardization, Basel, Switzerland, 1985, pp. 175-180.
5. Thalhamer, J. and Freund, J., (1984) Cascade immunization: a method of obtaining polyspecific antisera against crude fractions of antigens. J. Immunol. Methods 66, 245-251.
6. Thalhamer, J. and Freund, J., (1985) Passive immunization: a method of enhancing the immune response against antigen mixtures. J. Immunol. Methods 80, 7-13
7. Towbin, H., Staehelin, T., and Gordon, J., (1979) Electrophoretic transfer of proteins from polyacrylamide gels to nitrocellulose sheets: Procedure and some applications. Proc. Nat'l. Acad. Sci. USA 76, 4350-4354.
8. Burnette, W. N., (1981) Western blotting electrophoretic transfer of protein from sodium dodecyl sulfate polyacryl-amide gels to unmodified nitrocellulose and radiographic detection of antibody and radioiodinated protein. Anal. Biochem. 112, 195.
9. Oakley, B. R., Kirsch, D. R. and Morris, N. R., (1980) A simplified ultrasensitive silver stain for detecting proteins in polyacrylamide gels. Anal. Biochem. 105, 361-363.

RECEIVED July 8, 1987

AGROCHEMICAL

Chapter 16

Agrochemistry: An Introduction

James N. Seiber

Department of Environmental Toxicology, University of California—Davis, Davis, CA 95616

Biotechnology provides many new opportunities to advance the field of agricultural pest control by augmenting chemical-based technology as well as affording alternative approaches to pest control. Examples include the development of herbicide-tolerant crops, insecticidal seed coatings and insect-resistant crops, bioengineering approaches to producing pest control agents, immunoassays for their analysis, and bacteria and other microorganisms for decontaminating waste products.

There is a growing realization that agriculture can benefit immensely from increased inputs of biotechnology, to complement and perhaps replace a dependence on chemical technology which has characterized much of the industry during the past several decades. To be sure, chemical technology has provided impressive gains in fertilization, growth regulation, pest control, animal health, and diagnostic techniques which have directly improved yields and quality of many agricultural products. But these advances have been accompanied by undesirable side-effects -- residue contamination, resistance, ecosystem impairments, and recalcitrant waste generation. Some have even ascribed the economic problems of agriculture of recent years to an over-reliance on chemical technology -- a simplistic view in many respects, but one which nonetheless has stimulated a renewed interest in "organic", "sustainable" and "low-input" agriculture. Whether profitability and environmental quality can be improved in modern farming by decreasing reliance on chemical technology will depend in large part on successful development of biotechnology-based alternatives. Certainly, many people believe that the potential benefits justify diversion of research and development capital into the biotechnology field, a trend which is already quite apparent in the state agricultural experiment stations, USDA's Agricultural Research Service, venture-capital firms, the food industry, and in fact in the agrochemical industry itself.

0097-6156/88/0362-0204$06.00/0

Biotechnology in the broad sense has long been practiced in agricultural science in general, and pest control in particular. Breeding programs have produced disease- and insect-tolerant crops; natural product chemists and biochemists have identified a host of secondary chemicals which have proved useful as pest and disease control agents, and as growth regulators; microorganisms and their enzymes have been applied to waste decontamination; fermentation provides indispensable intermediates and end products of a variety of agricultural uses; and biological control agents (predators, bacteria, and viruses) have proven their worth as components of pest control arsenals. The new biotechnology will build upon these leads, and carry them to new levels of efficacy and utility as organisms are genetically modified to optimize the desired end-use (1).

Much attention is being devoted to development of herbicide-resistant crop plants where classic genetic selection with whole plants is a slow, tedious, and expensive path. Using cell culture selection techniques, resistant mutants can be identified rather quickly, their chloroplasts combined with the nucleus of related species by cell fusion, and commercially viable resistant hybrids produced as the end result. Crops tolerant of triazines, sulfonyl ureas, and glyphosate represent examples of this technology. It is not far-fetched to envision that environmentally-friendly broad spectrum herbicides will be emphasized in the future, with selectivity achieved by genetically engineering resistance into crops (2).

Another example of the application of biotechnology to pest control lies in the transfer of the gene regulating protein toxin production in *Bacillus thuringiensis* -- a biocontrol agent in widespread commercial use -- to bacteria or plants. Toxin producing bacteria could be used in seed coatings to protect germinating seedlings from attack by soil insects. Direct transfer of the B.T. gene to crop plants represents another promising avenue for crop protection, particularly when the toxin can be confined to plant pests or life stages which are isolated from the harvested commodity of commercial value.

There is concerted interest in pest control agents which work at low dosage levels, thus minimizing the quantity of material potentially available for residue contamination of the commodity and its environment. Some such agents represent complex natural products whose synthesis by conventional chemistry is too difficult or expensive to be practical. Biotechnology affords an approach wherein the organism which produces the active agent can be modified for use in large-scale bioreactors, providing good yields, purity, and stereospecificity in the end product. An example lies in the production of ivermectins by the actinomycete, *Streptomyces avermitilis*. Modification of the fermentation medium and of the organism, by mutation, resulted in a 50-fold increase in the yield of avermectins from broth, opening the door to large-scale production of this promising family of remarkably potent animal parasite control agents (3). Further genetic manipulation of this and other organisms promises major advances in the field of bioengineering applied to the production of hormones, pheromones, and pest control agents.

In the area of analysis, operational benefits can be derived from the use of immunoassays to replace some classical physicochemical analytical methods in terms of speed, specificity, and lower costs. Immunoassays represent a here-and-now biotechnology with many examples of application to pesticide detection (4). The use of hybridoma cell technology offers the potential for providing a stable supply of antibodies for a given assay, of dependable efficacy. This technology can be used as the basis for field assay kits which might provide on-the-spot residue levels so that growers, pest control advisors, and regulators can assess residue levels in terms of harvest intervals, reentry intervals, water holding times, and the like.

Adapted microorganisms can also be used to destroy waste pesticides resulting from spills, spray equipment rinsates, and disposal operations -- applications orginally suggested by the prolific biodegradation capability of microorganisms in sewage sludge, eutrophic waters and sediments, and agricultural field soils (5). Once again, genetic engineering provides a potential for tailoring such organisms to better carry out what is natural for them, perhaps at higher rates and in unfamiliar environments.

Fulfilling the promise of biotechnology in these and other applications will require careful planning to address health and safety issues, particularly when genetically engineered organisms are to be released to the environment. Fortunately, these are familiar concerns to pest control researchers, and many of the safety testing protocols required now for chemical agents can be adapted to the products of biotechnology with appropriate changes in monitoring tools. Nevertheless, to convince a wary public that we know what we are doing, and that it will be beneficial to them, may require that we go to extra lengths in safety evaluation -- at least in the early going. The potential benefits to be gained justify this investment of time, energy, and capital.

Literature Cited

1. Clarke, N.P. In 1986 Yearbook of Agriculture: Research For Tomorrow; Crowley, J.J., Ed; USDA, Washington, DC, 1986; pp 37-41.
2. Schneiderman, H.A. "Overview of Innovation in Agriculture". Paper presented at the conference on Technology and Agricultural Policy, National Academy of Sciences, Washington, D.C., Dec 11-13, 1986.
3. Campbell, W.C.; Fisher, M.H.; Stapley, E.O.; Albers-Schönberg, G.; Jacob, T.A. Science 1983, 221, 823-828.
4. Hammock, B.D.; Mumma, R.O. In Recent Advances in Pesticide Analytical Methodology; Harvey, Jr., J.; Zweig, G., Eds; American Chemical Society Symposium Series, Washington, DC, 1980; pp 321-352.
5. Alexander, M. Science 1981, 211, 132-138.

RECEIVED August 6, 1987

Chapter 17

Bacillus thuringiensis Biological Insecticide and Biotechnology

T. R. Shieh

Sandoz Crop Protection Corporation, P.O. Box 220, Wasco, CA 93280

Bacillus thuringiensis, commonly known as BT, is a gram-positive flagellated rod shaped microorganism. It is an aerobic spore-forming bacilli characterized by the formation of para-sporal protein crystals in the course of sporulation.

These crystals consist of a protoxin containing active fragments, known as delta-endotoxin that are toxic to the larvae of a number of Lepidopterous species (1,2) as well as Dipterous (3,4) and Coleopterous species (5,6) of economic importance.

Since the first discovery of Bacillus sotto in 1901 (7) and isolation of variety Berliner in 1911 (8), over 20 varieties of BT strains have been isolated from nature (Table I). They are classified based on serotype of flagellar antigens. They differ in their immunological and in the insecticidal properties of the para-sporal crystals produced.

These BT strains of various serotypes typically harbor complex and variable arrays of extra chromosomal DNA or plasmids (9). Some strains contain more than 10 plasmids in a single subspecies.

Plasmids of a larger MW has been shown to be associated with the coding of protein crystal formation in B. thuringiensis strains (10,11).

It has been more than 20 years, since B. thuringiensis var. thuringiensis Berliner was first used as a microbial insecticide for control of agricultural pests. However, it is only recently, that the world wide usage of BT products has increased substantially for pest control in agriculture, forestry and public health sectors as well.

This increased usage of BT products around the world is due to the following reasons.

1. BT is active only to target insects and does not effect non-target species, wildlife or humans and it is biodegradable and environmentally safe.
2. More recently, a number of new strains of BT are being discovered to control not only Lepidoptera spp. but also other family of pests such as mosquitoes, blackflies of public health importance. Control of a beetle belonging to Coleoptera species also becomes possible with a new BT strain.
3. BT products are effective against pests that have developed resistance to many chemical insecticides.

0097-6156/88/0362-0207$06.00/0

TABLE I. SEROTYPE AND EXTRA CHROMOSOMAL DNA OF B. THURINGIENSIS SUB SPP.

SUB-SPP	SEROTYPE	MW X 10^6 D <200-50	<50-9	≤9	TOTAL NO.	SUB-SPP	SEROTYPE	MW X 10^6 D <200-50	<50-9	≤9	TOTAL NO.
THURINGIENSIS	1	2	3	6	11	DARMSTADIENSIS	10	2	0	0	2
FINITIMUS	2					TOUMANOFFI	11A11B	5	3	2	10
ALESTI	3A	1	3	7	11	KYUSHUENSIS	11A11C	3	5	6	14
KURSTAKI	3A3B	3	3	8	14	THOMPSONI	12	1	0	0	1
SOTTO	4A4B	1	0	0	1	PARKISTANI	13	4	0	4	5
DENDROLIMUS	4A4B	1	2	0	3	ISRAELENSIS	14	4	1	3	8
KENYAE	4A4C	2	0	6	8	DAKOTA	15	1	2	4	7
GALLERIAE	5A5B	1	2	2	5	INDIANA	16	1	1	1	3
CANADIENSIS	5A5C	1	1	4	6	TOHOKUENSIS	17				
SUBTOXICUS	6	1	1	0	2	KUMAMOTOENSIS	18				
ENTOMOCIDUS	6	1	1	0	2	TOCHIGIENSIS	19				
AIZAWA	7	4	4	5	13	YUNANENSIS	20				
MORRISONI	8A8B	3	3	0	6	WUHANENSIS	-	1	2	2	5
TENEBRIONIS	8A8B	1	3		4	COLMERI	21				
OSTRINIAE	8A8C	2	0	0	2	SHANDONGIENSIS	22				
TOLWORTHI	9	3	4	1	8	JAPONENSIS	23				

SOURCE: Reproduced with permission from Ref. 9.

4. BT products are economical.

The major commercial BT products being produced (Table II) since the early 70's in the U.S. are variety Kurstaki, 3a3b of HD-1 strain or its variants. They are used for suppression and control of Lepidopterous species in agriculture and forestry. Dipel is produced by Abbott and Thuricide produced by Sandoz and have been used widely for more than 15 years. In late 70's, a variety israelensis, serotype 14 was isolated from diseased mosquito larvae. It led to the registration of products for control of Diptera of public health importance such as mosquitoes and blackflies. The Vectobac which contains spores is being produced by Abbott and Teknar HP-D of asporogenic strain of israelensis is being produced by Sandoz. At the same time, a product called Certan utilizing Aizawai, serotype 7, was developed by Sandoz for control of wax moth on bee hives to which the direct spray of chemical insecticides is prohibited. Recently, we have developed and registered a new product called Javelin, which is prepared from a variant of Kurstaki strain. Javelin formulation contains high insecticidal activities against various Lepidoptera species than existing products, such as Dipel and Thuricide and is particularly effective against hard to control Spodoptera species. There are increasing evidences to support that more variants will be found within this subspecies for control of specific insects such as Heliothis.

The most exiciting discovery was made recently by Dr. Krieg, of Institute of Biological Pest Control, Darmstadt, West Germany, who isolated a variety, Tenebrionis serotype 8a8b which controls the Colorado Potato beetle, belonging to a Coleopteran family (5).

A patent application has been filed on this new strain of BT in the United States since 1983. Under a licensing agreement with Boehringer Mannheim of West Germany, Sandoz Corporation is developing this new strain of BT under the code name of SAN 418 for control of Colorado Potato beetle, which is increasingly becoming resistant to the variety of chemical insecticides in the north-east parts of the United States.

This discovery of utilizing BT strain for control of a beetle, along with the successful commercial utilization of israelensis for control of Diptera species and the development of other varieties of BT for control of many Lepidoptera species indicates a significantly increased potential for commercial role of BT for control of wide variety of insects. Since the insecticidal activities of various products are best measured by bioassay against their respective target insects, we have so designated the various potency units for each of new products. The potency of Dipel and Thuricide that are assayed against *T.ni* larvae are defined to contain so many International Units per mg of product. Javelin which is assayed against Spodoptera is defined in terms of Spodoptera units per mg. Similarly, *Aedes aegypti* units is used for Teknar for control of mosquitoes and *Galleria mellonella* units is defined for Certan for control of wax moths We propose to use Heliothis units for a product that controls Heliothis species such as Cotton bollworms and Tenebrionis units for measurement of toxic units produced by Tenebrionis strain.

Morphology and characteristics of the protein crystals of subspecies used for commercial products vary considerably (Table III). Crystals of variety Kurstaki and Aizawai strains that are active against

TABLE II. INDUSTRIAL STRAINS OF BACILLUS THURINGIENSIS PRODUCTS

PRODUCT TRADE NAME	VARIETY	SEROTYPE	STRAIN	POTENCY UNITS	USAGE	COMPANY
DIPEL	KURSTAKI	3A3B	HD-1 AND VARIANT	IU	LEPIDOPTERA AGRICULTURE AND FORESTRY	ABBOTT
THURICIDE	KURSTAKI 0	3A3B	HD-1 AND VARIANT	IUO	LEPIDOPTERA AGRICULTURE AND FORESTRY	SANDOZ
JAVELIN	KURSTAKI	3A3B	NRD-12 AND VARIANT SA-11	SU	LEPIDOPTERA HIGHLY ACTIVE AGAINST SPODOPTERA SPP	SANDOZ
VECTOBAC	ISRAELENSIS	14	HD-567 AND VARIANT	ITU	DIPTERA MOSQUITOES, BLACKFLY	ABBOTT
TEKNAR HP-D	ISRAELENSIS	14	HD-567 ASPOROGENIC	AAU	DIPTERA MOSQUITOES, BLACKFLY	SANDOZ
CERTAN	AIZAWA	7	VARIANT SA-2	GMU	LEPIDOPTERA WAX MOTH ON BEE HIVES	SANDOZ
SAN 414	KURSTAKI	3A3B	VARIANT SA-6	HZU	LEPIDOPTERA HIGHLY ACTIVE AGAINST HELIOTHIS	SANDOZ
SAN 418	TENEBRIONIS	8A8B	DARMSTADT BT256-82	TU	COLEOPTERA COLORADO POTATO BEETLES	SANDOZ

TABLE III. MORPHOLOGY OF CRYSTALLINE INCLUSIONS OF VARIOUS <u>BT</u> SUBSPECIES

SUBSPECIES	CRYSTAL MORPHOLOGY	MEAN LENGTH (UM)	PROTOXIN MW (KDA)	ACTIVE SUB UNIT (KDA)	INSECTICIDAL ACTIVITY
KURSTAKI, 3A3B	BIPYRAMIDAL CUBOID	0.3 - 1.5	135 65	62 60	LEPIDOPTERA
AIZAWA, 7	BIPYRAMIDAL CUBOID	0.3 - 1.12	-	-	LEPIDOPTERA
ISRAELENSIS, 14	CUBOIDAL AMORPHOUS	0.1 - 0.5	65	28	DIPTERA
TENEBRIONIS, 8A8B	RECTANGULAR	0.2 - 1.5	140	64-74	COLEOPTERA

Lepidoptera spp. show a typical bipyramidal and a cuboid shape with protoxin molecular weight of 135K and 65K respectively for these two types of crystal forms (12,13). The crystal of israelensis which is active against Diptera is much smaller in size with mostly cuboid, ovoid or amorphous forms and the molecular weight of active subunit of israelensis is about one half that of other strains (14). The crystal morphology of the newly discovered tenebrionis which is active against Colorado Potato beetle is distinctively different from that of all other subspecies. The crystals produced by Bacillus thuringiensis var tenebrionis exhibits a flat rectangular crystal shape with the protoxin molecular weight of about 65K (5, Schnetter, W. University Heidelberg, personal communication, 1985).
As mentioned earlier, our recent studies on the variants of Kurstaki subspecies demonstrated that the insecticidal activities can vary considerably within the subspecies.

The electron micrograph of crystals of variants, HD-1 and SA-11 shows both strains to contain bipyramidal and cuboid shape crystals in a similar proportion (Figure 1). Bioassay of these strains however demosntrated that the insecticidal activity of a Kurstaki variant SA-11 is 10 to 15 times more active than HD-1 strain against Spodoptera and is 8 times more active than HD-1 against Heliothis even though the activity against T.ni remains the same (Table IV).

Table IV. Insecticidal Activities of BT var kustaki, 3a3b variants Against Various Lepidoptera Species

VARIANTS	POTENCY TRICHOPLUSIA NI UNIT/QT	 SPODOPTERA EXIGUA LC_{50} (PPM)	 HELIOTHIS ZEA LC_{50} (PPM)
HD-1	2-3	8900	7000
SA-11	2-3	500-800	831

Comparison of plasmid distribution pattern on SDS-PAGE electrophoresis of SA-11 and HD-1 strains of commercial product also shows they are very similar (Figure 2).
To understand the natural specificity observed for various strains of BT, against various insects, the mode of action of BT delta-endotoxin in the mid gut is illustrated in Figure 3.

After ingestion of crystal and spore by the larvae, the crystal protoxin of molecular weight 135K is first solubilized by mid gut juice of high pH, followed by proteolytic digestion by gut enzymes to form active toxin of molecular weight 65K (13, 15, 16).

The active toxin molecule is then presumably bind specifically to a receptor site of the epithelial cell membrane, causing the disruption of ionic balance across the hemolymph membrane, and disintegration of the epithelial cells brush border membrane.

These events lead to feeding inhibition within a few minutes and eventual disintegration of gut walls.

At the same time, the germinated cells penetrate through the disintegrated gut walls entering hemolymph, and causing septicemia and eventual larval death (17).

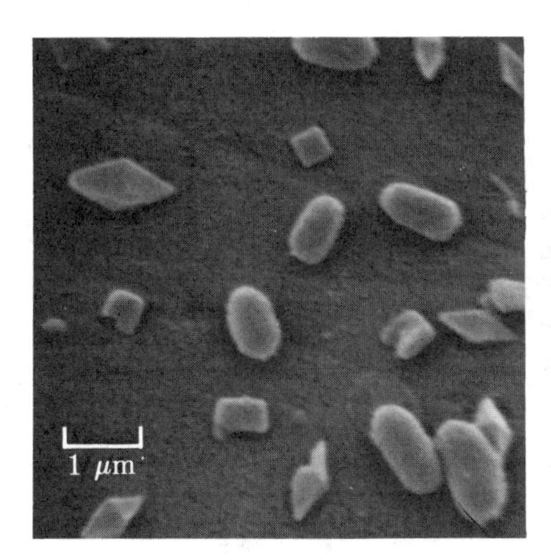

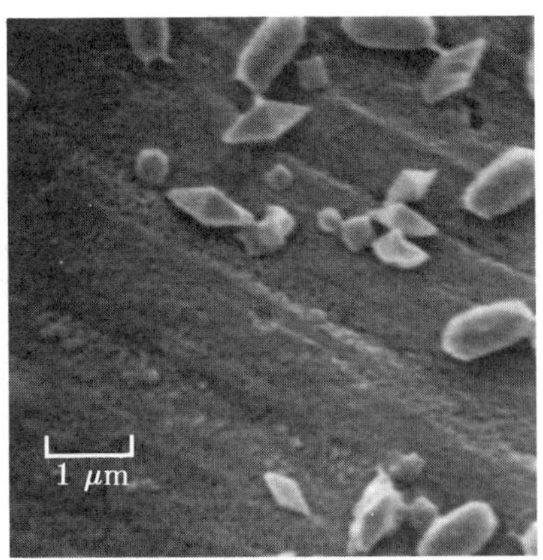

HD-1 SA-11

FIGURE 1. CRYSTAL MORPHOLOGY OF BACILLUS THURINGIENSIS VAR KURSTAKI 3A3B VARIANTS HD-1 AND SA-11

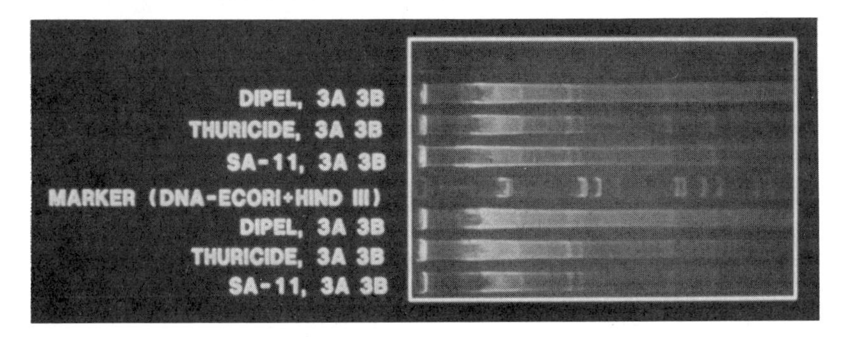

FIGURE 2. PLASMID DNA PATTERN OF BACILLUS THURINGIENSIS VAR KURSTAKI, 3A3B VARIANTS ON AGAROSE GELS SUBJECTED TO ELECTROPHORESIS

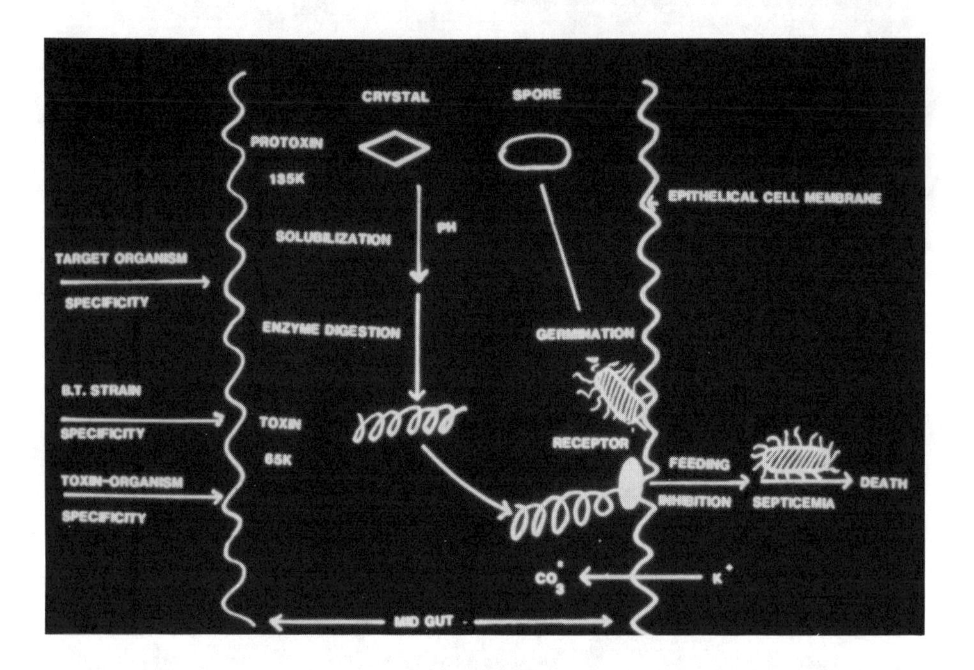

FIGURE 3. BT DELTA-ENDOTOXIN MODE OF ACTION

The model explains the specificity characteristics and non-toxic nature of BT crystals. Since the active toxin is formed only in the mid-gut of insects, crystal itself is not toxic to non-target organism and humans. The solubilization and enzyme digestion processes for conversion of protoxin to toxin in the mid-gut involve the target insect specificity. Tests indicate that human and non-target organisms do not have the similar system. The toxin molecule structure itself must be specific to BT strains and varieties. The toxin-receptor reaction appears to be specific to both toxin and organism with respect to their affinities.

We believe further understanding of the mode of action such as elucidation of the mechanims for conversion of protoxin to a specific toxin in mid-gut of host insects and understanding of toxin-receptor structure will contribute significantly to the biotechnology for development of better BT products for control of insects.

BT biotechnology involves not only the fermentation for production of toxin, but also involves the innovative approaches to the recovery of toxin and formulation of active ingredients for actual application in the field.

Fermentation technology and culture improvement can be achieved by the classic approaches of culture selection from natural as well as mutated populations.

Recent advances in genetic engineering techniques could allow addition to or exchange of plasmids between various strains to increase the spectrum of insecticidal activity of a strain.

In vitro recombination of different genes or mutagenesis by chemicals or gene splicing technique could potentially improve not only the insecticidal spectrum, but also the potency per unit of toxin produced.

It could also be feasible and important to improve the host cells characteristics, with respect to space, energy requirements and possibly the regulation of toxin formation.

Recent reports (18,19) on the amino acid sequence analysis of four subspecies which exhibit different insecticidal activity shows similar amino acid sequence in N-terminal section of the protoxin. The N-terminal and C-terminal portions of active fragments of these subspecies also show similar amino acid sequences with the exception of the absence of Threonine in N-terminal and C-terminal segment of Dendrolimus and Leucine in the C-terminal fragment of HD-73 variant of Kurstaki subspecies. The amino acids at the 148 and 302 positions, however, are different in each of these active fragments, suggesting that the difference in insecticidal activity might be due to just a few amino acids. This observation sharply suggests that improvements in insecticide spectrum and potency levels may be achieved by utilizing genetic manipulation techniques.

BT Biotechnology also involves innovation in the process development for recovery of the toxin. Efficient centrifugation, ultrafiltration or diafiltration, evaporation and drying processes may be utilized to accomplish the desired separation and concencentration steps.

Finally, emphasis must be placed on the formulation technology of BT products. Protein toxins are generally fairly easily denatured during the storage or after the release into the environment if not formulated properly to maximize the deposition of active BT toxin onto the target plants.

Innovative approaches to all of these areas of biotechnology are required in order to develop a significantly improved microbial insecticide.

Literature Cited

1. Hannay, C.L. Nature 1953, 172, 1004.
2. Angus, T.A. Nature 1954, 174, 545.
3. Goldberg, L.J. and Margalit, J. Mosquito News 1977, 37, 355-358.
4. Guillet, P. and Barjac, H. de. C.R. Acad. Sci. Paris 1979 289 D: 549-552.
5. Krieg, A.; Hurger, A.M.; Langenbrook, G.A. and Schnetter, W. Z. Aug. Ent. 1983, 96, 500-508.
6. Hernstadt, C.; Soars, G.G.; Wilcox, E.R. and Edwards D.L. Bio-technology 1986, 4, 305-308.
7. Ishiwata, S. Dainihon Sanshi Kaiko 1901, 2, (No. 114) 1-5.
8. Berliner, E. Z. Angew, Entomol. 1915, 2, 29.
9. Iizuka, T.; Faust, R. and Travers, R.S. J. of Sericul. Sci of Japan, 1981, 50, (2) 120-133.
10. Somerville, H.J. and James, C.R. In Microbial Toxins. Ann N.Y. Acad. Sci. 1973, 217, 93-108.
11. Gonzales, J.M.; Dulmadge, H.T. and Carlton, B.C. Plasmid 1981, 5, 351-365.
12 Iizuka, T.; Ishino, M. and Nakashima, T. Memoirs of the Faculty of Agriculture, Hokkaido University, 1982, 13, (3), 423-429.
13. Yammamoto, T. and Iizuka, T. Arch. Biochem. Biophy., 1983, 227, (1) 233-241.
14 Tyrell, D.J.; Davidson, L.I.; Bulla, L.A. and Ramska, W.A. Appl Environ. Microbial, 1979, 38 656-658.
15. Angus, T.A. Can J. Microbial. 1956, 2, 416-426.
16 Faust, R.M.; Adams, J.R. and Heimpel, A.M. J. Invertebrate Pathology 1967, 9, 488-499.
17. Luthy, P. and Ebersold, R. In Patholgenesis of Invertebrate Microbial Diseases; Davidson, E. Ed., Allanheld, OSMUR Publishers, 1981; 260.
18. Nagamatsu, Y.; Itai, Y.; Hatanaka, C.; Funatsu, G. and Hayashi, K. Ag. Biol. Chem. 1984, 48, (3) 611-619.
19. Wong, H.C.; Schnepf, H.E. and Whitely, J. Biol. Chem. 1983, 258, 1960.

RECEIVED August 17, 1987

Chapter 18

Pesticide Immunoassay as a Biotechnology

P. Y. K. Cheung[1], S. J. Gee, and B. D. Hammock

Departments of Entomology and Environmental Toxicology, University of California—Davis, Davis, CA 95616

Immunoassays are generally applicable to environmental chemistry. They are highly sensitive, specific, and precise, which leads to rapid, cost-effective assays. In many cases, immunochemical assays are most applicable for compounds that are difficult to analyze by classical methods. For instance, GLC is best suited for compounds that are volatile, heat stable, and which have functionalities easily detected by available selective detectors. HPLC is less restrictive, but still depends on specific functionalities for detection. However, these criteria that are essential for chromatographic methodologies are not necessary for a successful immunoassay. Thus, immunoassays expand our ability to monitor compounds in a cost effective manner that are difficult to detect using the GC/HPLC technology. With trends toward more complex classical pesticides including benzoylphenyl ureas such as diflubenzuron and sulfonyl ureas such as Gleen, immunoassay may prove to be the most suitable method of analysis. Furthermore, with biological insecticides, based on classical as well as genetically engineered products such as the toxins of *Bacillus thuringiensis*, immunoassay may prove to be the only suitable physical method of analysis of expressed protein.

In the past several years, immunoassay has begun to be recognized as a useful analytical technique in pesticide residue analysis. In 1980 when Hammock and Mumma (1) pointed out the potential for this application, only a handful of laboratories were utilizing the technique. Now a number of industrial and governmental laboratories are exploring the use of this versatile technology.

[1]Current address: Antec, Rockville, MD 20850

0097–6156/88/0362–0217$06.00/0

Since the award of the 1977 Nobel Prize in Medicine to Rosalyn Yallow for the development of the radioimmunoassay, immunoassay technology has had a major impact on many fields of science. A very sophisticated set of technologies has developed for the analysis of molecules which owe their great sensitivity and specificity to the biological systems that produce the reagent antibodies. All these assays are based on the high affinity, but reversible interaction of an antigen and antibody as determined by the law of mass action. However, only recently has some of this technology been applied to the environmental area. Environmental chemists slowly are beginning the realize that immunoassay is simply another analytical tool.

Advantages of Immunoassays

Any analytical technique has advantages as well as limitations. In some cases the very factors which are considered advantageous in one instance will present problems in another instance. In order to properly apply immunochemistry to analytical problems it is critical that one is aware of both these limitations and the advantages. We will try to describe here the basic advantages and how they might apply to pesticide chemistry as well as point out some of the pitfalls.

Speed of Analysis. There are a number of ways to describe the speed of analysis. One is to evaluate the assays that can be performed per man day, another is to evaluate the time required from receipt of sample to analytical result. Immunoassays score very high, regardless of which criterion of speed is applied. This tremendous speed of analysis is dependent, in turn, on the specificity and sensitivity of the immunoassay. It is important to realize that with immunoassay, as with any other analytical technique, there is a trade off among speed, sensitivity and cost. There are many immunoassay formats (2). Some require several hours to perform, others only minutes.

The actual time required to run a single sample by immunoassay is usually longer than the time required to run a single GLC analysis. However, numerous samples are usually run simultaneously with little extra time required. In most residue procedures, the major expenditure of time is in sample work up. The great advantage of immunoassay, is that in many cases, work up steps can be greatly reduced or eliminated. For example, the herbicide molinate (Figure 1) is used extensively in rice fields, and the ability to monitor levels of the pesticide in the field will assist in downstream water management. GLC workup requires extraction with an organic solvent and then analysis. We are currently developing an immunoassay which will analyze molinate in the water sample directly, with sensitivity to about 10 ppb. Similarly, the electron capture method for analysis of diflubenzuron (Figure 1) in milk has many clean up and derivatization steps. The corresponding immunoassay can quantitate diflubenzuron in milk without prior work up and with greater sensitivity (3).

The typical enzyme linked immunosorbent assay (ELISA) method requires six hours for the total analysis of a single bank of

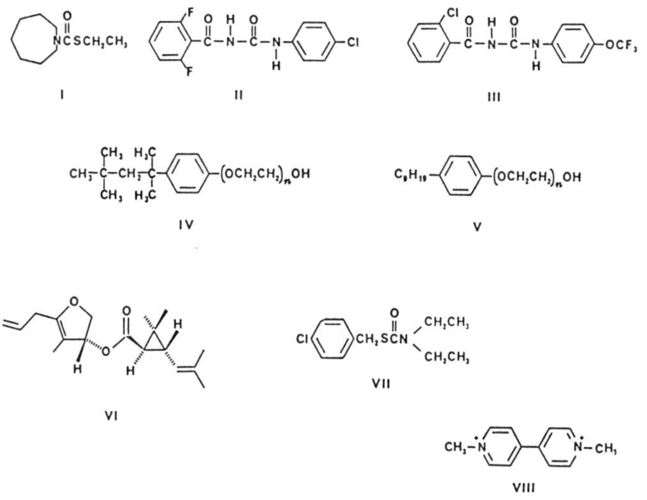

Figure 1. Structure of molinate (I), diflubenzuron (II), Bay Sir 8514 (III), Triton X (IV), Triton N (V), S-bioallethrin (VI), thiobencarb (VII) and paraquat (VIII).

samples, but this does not represent man hours required because most of the time is spent waiting; rather, it represents analysis time from start to finish. If the average retention time for a compound on GLC is three minutes, in six hours, one could optimally run 120 analyses. If each of these was a single sample at a single dilution, than 120 samples could be run. Comparably, a single operator, without automation could easily run 6-10 times this sample number by immunoassay, or, more information could be obtained about each of these 120 samples, by running more dilutions or replicates. Even if GLC is used for confirmation, immunoassays are useful for rapidly eliminating negatives and ranking samples in terms of concentration for subsequent analysis.

Ease of Automation. Solid phase assays (such as ELISA) are especially amenable to semiautomation with little investment in sophisticated equipment. Our laboratory's first ELISA reader was a system consisting of a small Gilford spectrophotometer interfaced with an Atari 400. Each cuvette holder held 50 cuvettes. Once reagents were prepared, 20-100 assays could be readily performed. Currently, we utilize a Flow Titertek Multiskan interfaced to an IBM-PC for data collection and management. With the aid of multichannel pipettors, 100-1000 assays can be performed per day in a 96 well plate format. The addition of plate washers and diluters to the scheme can further increase both speed and precision. The immunoassay format is especially amenable to robotic procedures for the handling of very large numbers of samples such as those generated by the monitoring of pesticide residues in food.

Specificity. The specificity of immunoassays can be very high. Immunoassays can readily distinguish the same functional enzyme from different but related species, or even from different organs in the same species. This may be translated as a single amino acid change in a primary sequence of over one hundred residues. For small molecules, we have demonstrated the ability to distinguish very closely related structures. For example, Triton X and Triton N (Figure 1), nonionic surfactants, require extensive sample workup and yield multiple peaks on HPLC during analysis. An immunoassay developed in this laboratory could easily distinguish between these structurally related compounds (4). We have also shown with S-bioallethrin (Figure 1) that even geometrical and optical isomers can be distinguished by immunoassay (5).

One drawback is that cross reactivity can occur to structurally related compounds. If one requires an assay specific for a parent compound, this would be a disadvantage. On the other hand, if one's goal is to detect a class of compounds or the parent and its metabolites, this cross reactivity would be advantageous. For instance, diflubenzuron is very closely related to another benzoylphenyl urea insecticide (Bay Sir 8514, Figure 1). A very high efficiency HPLC column is required to separate these compounds. We have developed several highly specific immunoassays which can distinguish between these very closely related compounds, while other assays can detect the benzoylphenyl ureas compounds as a class (3).

Sensitivity. ELISAs are not usually considered to be as sensitive as radioimmunoassays, however, sensitivities in the mid picogram/ml range have been obtained (5,6). Assays in the ng/ml range are sufficient for most analytical needs.

Sensitivity can also be considered in terms of ppb detectable in the presence of some matrix. In this case, sensitivity is a function of selectivity, and immunoassays are capable of detecting the target antigen at low concentrations even in the presence of large amounts of contaminating material. In this context, molinate is detectable at 10 ppb in water, 0.1 ppb or less in water extracts, but only 30-60 ppb in soil extracts without cleanup.

With chromatographic analysis, assay sensitivity is usually defined as a function of peak height relative to baseline noise. With immunoassays, detection limits can be defined as limit of detection or limit of quantitation. For instance, using a highly selective antibody, the limit of quantitation of thiobencarb (a thiocarbamate herbicide used in rice culture, Figure 1) is approximately 10 ng/ml when it is based upon the linear region of the standard curve (or the mid region of a sigmoidal curve, Figure 2). The actual limit of detection relative to background noise is less than 1 ng/ml when the lower portion of the linear region of the standard curve is used.

Cost Effectiveness. The most expensive part of an assay is the analyst's time. For diflubenzuron, over 100 immunoassays can be carried out in the same time that it takes to perform 0.5 GLC based assays. This apparent advantage is obviously greatest for those compounds that require many workup steps prior to immunoassay. Even with compounds that are relatively easy to assay by classical methods, immunoassay may prove cost effective if numerous samples must be handled. Solid phase assays are relatively easy to automate completely so that many more samples can be run per unit of analyst or machine time by immunoassay than by automated HPLC or GLC.

Finally, it must be emphasized that these characteristics of the assay depend upon one another. If one wants a highly sensitive assay, it may require more cleanup, and so one sacrifices speed. If one desires an assay to detect a series of compounds, one will have to compromise specificity and expect increases in cross-reactivity. Therefore, an inexpensive assay of moderate sensitivity that can be performed in the field with no specialized equipment or highly trained personnel can be developed, but one cannot expect these assays to be highly sensitive and precise as well.

Applicability. Immunochemical technology is applicable to most analytical problems, but it can be more easily applied to some analytical problems than others. Immunoassay sensitivity ultimately depends on the K_d of the antibody-antigen complex. Since this K_d, in turn, usually depends upon summation of weak molecular interactions, immunoassays for very small molecules are rarely sensitive. Because immunoassays are performed in aqueous solutions, they are not easy to apply directly to hydrolytically unstable materials. However, one can expect success in an attempt to develop an immunoassay to a molecule of moderate to large size.

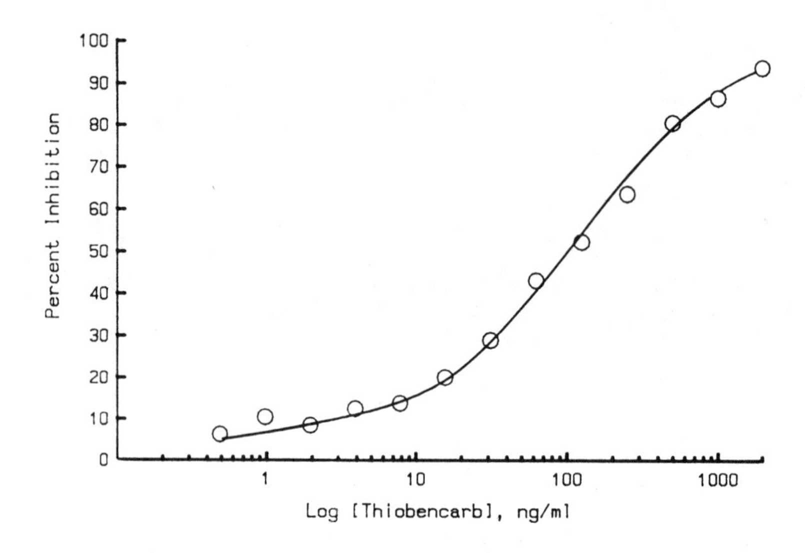

Figure 2. Typical standard curve for the thiocarbamate herbicide thiobencarb. Ninety-six well microtiter plates were coated with 0.2% glutaraldehyde for 20 min. After removing glutaraldehyde each well was coated with 0.2 ml of a 3 ug/ml solution of a hexanoic acid derivative of thiobencarb coupled to ovalbumin in 0.5M carbonate buffer pH 9.8 and allowed to incubate in the refrigerator overnight. The standard curve was prepared by incubating various concentrations of thiobencarb with a 1/3000 dilution of antiserum in phosphate buffered saline containing 0.05% Tween-20 (PBS-Tween) overnight at room temperature. The antiserum was raised in rabbits to a p-aminophenyl derivative of thiobencarb coupled to keyhole limpet hemocyanin. The following day the plate was washed three times with PBS-Tween and the standard curve preparations added. After a two hour incubation at room temperature, 0.2 ml of a solution of goat anti-rabbit IgG conjugated to alkaline phosphatase was added. Two hours later the plate was again washed three times with PBS-Tween and a 1 mg/ml solution of p-nitrophenylphosphate in 10% diethanolamine buffer pH 9.8 was added. After a 30 minute incubation at room temperature the optical density was read at 405 nm. This is a representative of a competitive sandwich ELISA. Any antibody which binds to thiobencarb in the standard curve preparations, cannot then bind to the coating on the plate. The goat anti-rabbit antisera and p-nitrophenylphosphate serve to detect the amount of antibody bound to the plate. Thus the more thiobencarb that was in the standard curve preparation, the less antibody is free to bind to the plate and thus the less color development occurs.

This is especially true if the molecule has several easily recognizable functionalities including potential sites for formation of ionic or hydrogen bonds and regions of alternating hydrophobicity and hydrophilicity.

Small, highly lipophilic molecules may be more easily analyzed by methods other than immunoassay for a variety of reasons. However, lipophilicity alone does not preclude the development of a sensitive immunoassay. At the part per billion or trillion level of sensitivity of good immunoassays, most compounds are soluble. Even if the compound exists as a micelle or on the surface of proteins, antibodies can pull the target materials into their binding site by mass action. Antibodies vary dramatically in their sensitivity to water miscible organic solvents. However, most antibodies will tolerate a few percent, and many 30 percent or more, of solvents such as methanol and tetrahydrofuran. Thus, for a skilled analyst, simple water solubility is seldom a problem in immunoassays of even highly lipophilic materials. However, the cost of separation of a lipophilic target compound from a lipid rich matrix may greatly reduce the attraction of immunoassay development for such materials.

Role of Monoclonal Antibodies

The development of hybridoma technology is one of the most exciting events in biology because the availability of monoclonal antibodies allows many very difficult problems to be approached for the first time. However, there may be a tendency to apply the technology to problems which can be solved as well, and in some cases better, with the use of polyclonal (classical) antibodies. In evaluating the advantages and disadvantages of mono- versus polyclonal antibody technology for pesticide analysis, it is important to understand that most assay formats are not influenced by the source of the antibody (poly- versus monoclonal).

Hammock and Mumma in 1980 (1) discussed some of the advantages and disadvantages of hybridoma technology with regard to pesticide residue analysis. In the intervening years, the hybridoma field has matured sufficiently to evaluate its potential more fairly. There are numerous scientific as well as popular articles (7,8) on this rapidly evolving technology including a number of articles in this text.

The antibodies which one obtains from the serum of an animal usually consist of a population of specific types of antibodies which recognize a variety of antigenic determinants with varying degrees of specificity and affinity. In contrast, monoclonal antibodies are obtained from a cell line ultimately traceable to a single cloned cell. If there are no problems with mixed chains, the cell line will produce a single type of antibody molecule. Hopefully, this antibody will recognize a single antigenic determinant with constant affinity and specificity.

The major advantage offered by monoclonal antibodies to pesticide analysis may be as much administrative as scientific. One unresolved example of this is the patent position that one may encounter with mono- versus polyclonal assays. In theory, a cell line producing a monoclonal antibody is immortal. Thus, one moves away from the fear of the "magic rabbit" that will ultimately die leaving a world lacking its precious serum. Realistically, the magic rabbit problem is quite rare and a reasonably good antiserum can be reproduced in another animal. However, it is rare that two antisera are totally identical in their specificity and affinity, even when they come from the same rabbit. The properties of each new bleed must be evaluated. With the expense of developing hybridoma antibodies decreasing, there may be a time when the cost of producing and maintaining a small library of monoclonal antibodies to a pesticide will be less than the cost of periodically raising and reevaluating a new antiserum, or of using the common procedure of combining many bleeds to obtain an average pool of serum. Certainly, a permanent supply of monoclonal antibodies will have an appeal to regulatory agencies, as well as to patent attorneys. However, it must be remembered that these cell lines may contain an unstable chromosome complement, and their immortality depends upon proper storage and maintenance. Realistically for an important molecule, one would work with pooled polyclonal sera stored in small aliquots or with a large amount of monoclonal antibody produced at one time and stored in small aliquots. Thus both systems usually rely on storing antibodies.

In theory, one could screen a large number of clones, and find those that are producing antibodies of exceptionally high affinity and specificity. In practice, the screening effort required to obtain a few positive clones is often so major that the further effort of finding those producing the optimum antibody is not undertaken. Designing a rapid screen for a producer of an optimum antibody is obviously much more difficult than simply finding producers. Some part of the screening effort may be reduced because of automation, yet even then, the monoclonal antibody found may not prove superior to a good serum. Murine antibodies are not well known for high affinity, and the optimum clone resulting from a major screening program may yield an antibody with a K_d which is not as low as the $K_d$ave of a rabbit serum. At this time, there appears to be an attempt to solve problems in immunochemistry using thousands of dollars of screening and hybridoma technique that could be solved by hundreds of dollars of planning and hapten synthesis. Even after extensive screening, poor selection of hapten structures or poor screening strategies can result in an unworkable monoclonal based assay.

The defined specificity of a monoclonal antibody is unique among biological reagents, yet an antiserum may offer greater specificity. A single interfering substance which binds to the monoclonal antibody could lead to complete cross reactivity in the resulting immunoassay. This same interference may only be a minor problem in a polyclonal system where it would bind to only a small subpopulation of total antibodies reacting with the molecule of interest. Obviously, the optimum situation would result from the employment of a library of well characterized monoclonal antibodies as either a mixture, (analyzing for several components in one assay) or a series of assays (analyzing one sample in several assays).

Monoclonal antibodies do offer some advantages now, and many advantages in the future. For an institution wishing to limit the sensitivity of a technique, the defined specificity and lower affinity offered by hybridoma technology could offer a promising approach. The reactions of a monoclonal antibody are easier to handle mathematically than those of a polyclonal mixture. This fact, coupled with a known kinetic k_a and k_d, makes monoclonal antibodies much easier to use in many assay formats, especially with some "biosensors". It is important to realize that hybridoma technology is advancing. With automated systems, it is possible to screen large numbers of clones and thus develop a library of antibodies of varying specificity and sensitivity. Such libraries can be invaluable, especially with complex antigens. Finally, techniques are being developed using multiple mouse strains to realize the potential of hybridoma technology to produce antibodies of extraordinary specificity and sensitivity.

Until immunoassay is more widely accepted in the field of agricultural chemicals, it seems questionable that the added expense of hybridoma technology is warranted scientifically for routine analytical problems. However, it seems certain that hybridoma technology will dominate the immunoassay field in the future.

Biological and Genetically Engineered Pesticides

Up to this point, the advantages of immunochemical assays in pesticide analyses have been discussed with an emphasis placed on classical pesticides. However, development of the new biotechnology of molecular genetics may have a dramatic impact on pesticide chemistry and place immunochemical assay in an indispensable position in pesticide analysis. Cognizant of the powerful tools of molecular biology, researchers and the pesticide industry are rapidly developing new classes of pesticides. These "new" pest control agents are usually of a biochemical and biological nature (eg. peptides, microbial toxins, and micro-organisms). For instance, fermentation technology has already given us avermectin and ivermectin. While a complex exotoxin from *Bacillus thuringiensis (BT)* is not allowed in current formulations, it has been considered for development as an insecticide itself. These materials are far more complex than the compounds normally analyzed by GLC and HPLC. Although their analysis can certainly be approached by classical means, immunoassays seem particularly advantageous with these molecules. One can anticipate the development of more such complex molecules, and the need for analytical technology to deal with them. Furthermore, given the general public's attitude toward "genetic engineered" microbes, it is certain that the vast majority of products from this type of research will instead be peptides and proteins. For such materials, immunochemical methods are unsurpassed.

Although proteins and peptides resulting from research in biotechnology followed by fermentation are unlikely to present environmental contamination problems, a whole new concept in pollution may arise from this area of research. Since nucleic acids have the potential to reproduce in the environment, we do have the potential to pollute some segments of a "wild" gene pool with man-made or contrived genes. In the last year, the protein toxin of *BT kurstaki* has been cloned into, and expressed in plants (e.g. tobacco, cotton), and Monsanto is still trying to test the same toxin in the field following expression in an alternate species of bacterium (personal communication). The crude mixture from a genetically engineered bacterium containing the *BT* toxin gene already has been field tested in California. In situations where foreign genes are released into the environment, immunochemical procedures are necessary to monitor the expression, as well as the products of expression in the environment. Immunoassays could be used directly on suspect biological material, or on the protein produced in a cell free translation system using the DNA from the suspect materials. Thus, immunoassays become complementary to the genetic probes, which will be necessary to monitor the actual gene.

Immunoassays not only represent a biotechnology themselves, but they may well be the best way to analyze for products of biotechnology research. In this regard, immunoassays for the whole crystal toxin of the *BT kurstaki* and *israelensis* strains have been developed (9-13). These assays all demonstrated good correlation with biological assays for the toxins, and they are routinely used for monitoring production and quality control in the fermentation process. Adaptation of these assays to actual residue procedures will be a logical alternative for monitoring these "new" biological pesticides.

Immunoassay vs. Bioassay in the Detection of Biological and Genetically Engineered Pesticide. It has been rather widely assumed that the biological assay is the only way to perform residue analysis on biological insecticides in the field. However, residue analysis either by biological or by immunochemical assay for a toxin is analogous in that one desires a selective detector. This is achieved in bioassay by using test animals that are susceptible to the toxic materials, and in immunoassay by an antibody which selectively recognizes antigenic determinants on the target molecule. Therefore both bioassays and immunoassays can be as specific as the design of the respective assay itself.

On the other hand, sensitivity is a different matter. In bioassays, the test animal acts as a detector and a concentrator simultaneously. For example, the mosquito larvae bioassays for *BT israelensis* δ-endotoxin is extremely sensitive because of the feeding behavior of the mosquito larvae. The animal selectively ingests particulate matter and thus accumulates the crystal toxin, even when exposed to very low toxin concentrations. An LC_{50} of 200pg/ml was reported for the purified crystal toxin (14). Sensitivity of immunochemical assays operates on a different principle. All enzyme immunoassays are basically amplification systems in which enzyme conjugated to the antibody or the antigen is used to drive an enzyme reaction which amplifies a weak signal (eg. low concentration). Therefore the sensitivity of an immunoassay is ultimately limited by the binding constant (10^8 to 10^{13}) between the antibody and the antigen. Although various well established techniques (eg. fluorescent immunoassay, Avidin/Biotin systems, radio-labelled immunoassays, etc.) can be used to further enhance the detection limit of an immunoassay, it still may not detect biologically significant levels of many peptides or proteins in environmental samples.

As with any other analytical residue procedure where the sensitivity of an assay is insufficient, further sample workup is required. However, microbial degradation, heat sensitivity, pH sensitivity, and solubility should all be taken into consideration when one is attempting to concentrate or extract biological materials during sample workup.

Dot-Blotting. In our experience, the *Dot blot technique* is most simple and versatile in sample workup for biological materials. The dot-blot assay takes advantage of the fact that pure nitrocellulose strongly adsorbs proteins, nucleic acids and cellular compounds. Antigens can be quickly immobilized onto a nitrocellulose membrane, and then be analyzed with any of the many EIA procedures. Concentrations of the antigen are revealed by the intensity of color development on the membrane. Such color development can be determined qualitatively, or quantitatively, if one chose, by instruments available commercially. An added advantage of the dot-blotting technique is that multiple sample application is possible. Repeated applications further concentrate the antigen on the membrane, and thus extends the detection limits for solutions of lower concentration (Figure 3). This, theoretically, can increase the detection limit indefinitely. Thus the dot-blotting technique essentially achieves what biological assays have to offer. It acts simultaneously as a concentrator and a detector.

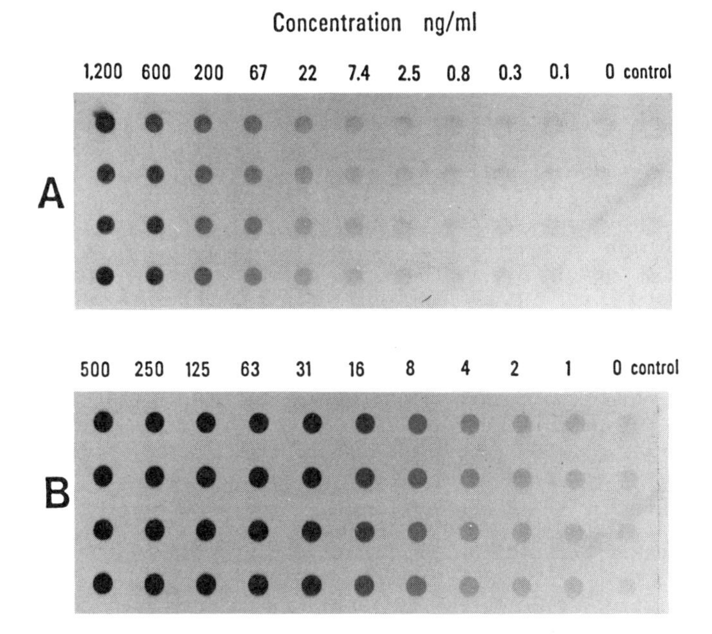

Figure 3. Dot-blotting of *BT israelensis* δ-endotoxin (A) one sample application vs. (B) five sample applications.

Applications to Environmental Health

Pest Management. The capability of immunochemical technology to provide reliable, low cost analytical data to individuals lacking intensive training and sophisticated equipment will probably be its greatest contribution. Due to the expense of current analytical methods, industry and government normally only ask those analytical questions for which an answer is required by regulatory agencies. Individuals who can make the best practical use of these analytical data to increase the agricultural productivity seldom have access to such information. The ability to determine deposit, coverage, residue, and if a pesticide even reached the microenvironment of the target host are obviously critical for the effective use of an agricultural chemical. In a more pragmatic sense, any chemical which fails to reach the target could be considered a pollutant. However, the analytical capability to answer such questions is far out of reach for most pest management scientists.

Rapid Field Assays. The high speed and simplicity of many immunoassay formats allow them to be performed in the field. Therefore it is likely that the immunochemical technology will do more than supplement existing technology. It will greatly extend our power to monitor environmental health. For example, farmers can insure that a herbicide is at a safe level before planting; applicators can be assured that drift is limited and that the chemical reached the target; farmworkers can be assured that a field is safe to enter, and wholesalers and regulatory agencies will know whether residue levels are acceptable.

Human Exposure. Immunoassays are especially useful in monitoring materials in human body fluids. When immunoassays are developed for key environmental pollutants or indicator compounds, the toxin itself can be used as a marker of human exposure by monitoring urine or blood samples. Paraquat (Figure 1) is a commonly used herbicide which is responsible for the majority of deaths attributed to pesticides in the U.S. An immunoassay for paraquat is being used for clinical diagnosis of poisoning, and we have found that crude blood, urine and lymph can be analyzed directly in an immunoassay for paraquat with greater sensitivity than other published techniques. This same immunoassay has been used in a worker exposure study. Here, the immunoassay and GLC data compared favorably (6), however, the immunoassay offered the advantage of directly analyzing human body fluids and the antibody could even be used to extract paraquat from filter patches, eliminating the acid extraction step. Such rapid assays make it possible to carry out the pharmacokinetic studies necessary to evaluate the significance of occupational exposure to pesticides.

Released Time. Pesticide residue analysis as practiced today is both equipment and labor intensive. In many laboratories, the experienced scientists who manage the laboratories are so involved with handling the day to day problems caused by an ever increasing work load that they lack the time to apply their skill and creativity to long term problems. A very important result of implementing immunochemical technology in a residue laboratory could be the release of equipment, monetary resources and especially human resources from the routine residue analysis to the many pressing problems in analytical and environmental chemistry.

Literature Cited

1. Hammock, B.D.; Mumma, R.O. ACS Symposium Series, 1980, 136, 321-52.
2. Ngo, T.T.; Lenhoff, H.M. (eds.) Enzyme-Mediated Immunoassay, 1985, 489 pp., Plenum Press, New York.
3. Wie, S.I.; Hammock, B.D. J. Agric. Food Chem. 1982, 30, 949-57.
4. Wie, S.I.; Hammock, B.D. Anal. Biochem. 1982, 125, 168-76.
5. Wing, K.D.; Hammock, B.D. Experientia 1979, 35, 1619-20.
6. Van Emon, J.; Hammock, B.D.; Seiber, J.N. Anal. Chem. 1986, 58, 1866-73.
7. Vinas, J. Pure & Appl. Chem. 1985, 57, 577-82.
8. Shimizu, S.Y.; Kabakoff, D.S.; Sevier, E.D. Enzyme-Mediated Immunoassay, 1985, 433-450, Plenum Press, New York.
9. Wie, S.I.; Andrews, R.E., Jr.; Hammock, B.D.; Faust, R.M.; Bulla, L.A., Jr.; Schaefer, C.H. App. Environ. Microbiol. 1982, 43, 891-94.
10. Wie, S.I.; Hammock, B.D.; Gill, S.S.; Grate, E.; Andrews, R.E., Jr.; Faust, R.M.; Bulla, L.A., Jr.; Schaefer, C.H. J. Appl. Bacteriol. 1984, 57, 447-54.
11. Cheung, P.Y.K.; Hammock, B.D. Appl. Environ. Microbiol. 50, 984-88.
12. Andrews, R.E., Jr.; Iandolo, J.J.; Campbell, B.S.; Davidson, L.I.; Bulla, L.A., Jr. Appl. Environ. Microbiol. 1980, 40, 987-90.
13. Smith, R.; Ulrich, T.T. Appl. Environ. Microbiol. 1983, 45, 586-90.
14. Tyrell, D.J.; Davidson, L.I.; Bulla, L.A., Jr.; Ramoska, W.A. Appl. Environ. Microbiol. 1979, 38, 656-58.

RECEIVED June 17, 1987

Chapter 19

Enzymatic Processes for Pheromone Synthesis

Philip E. Sonnet

Animal Biomaterials Unit, Eastern Regional Research Center, Agricultural Research Service, U.S. Department of Agriculture, 600 East Mermaid Lane, Wyndmoor, PA 19118

Some general features of enzyme technology applied to synthetic organic chemistry are discussed and recent uses of enzymes to obtain critical configurations in pheromone structures are described. Commercially available lipases were screened to assess their potential to resolve methyl *n*-alkyl carbinols. A lipase from the fungus *Mucor miehei* was employed to obtain configurational isomers of the carbinol carbon of 8-methyl-2-decanol, the parent alcohol of a sex pheromone of several economically important *Diabrotica* (rootworm) species.

The desire to develop alternative methods for controlling insect pests led to the development of research programs in fields such as insect pheromones. Of special concern has been the synthesis of stereoisomers of those pheromones having chirality. Although usually only one stereoisomer is active while the enantiomer has no biological activity, there are exceptions. Therefore, in addition to requiring pure stereoisomers to facilitate identification and to promote fundamental studies in insect behavior, there have been instances where correct stereochemistry was critical to insect attraction (gypsy moth, japanese beetle, several important rootworm species).

A potential resource for those interested in stereochemical syntheses are enzymes, proteins that have biological activity and are responsible for catalyzing large numbers of transformations of biologically important molecules. Judging from the literature about enzymes, one might expect to see them in more general use. A glance at chemical and biochemical catalogs reveals that indeed some enzymes can be purchased, although the majority must be searched out from lists of manufacturers or "home grown". Many have been characterized, but remain yet to be isolated and formulated in a usable fashion. Additionally, the costs may seem high for all but specialty products. Stability is another concern; many enzymes will suffer irreversible change in solution during use although others will not. The problems of cost and catalyst longevity are, however, not independent parameters, and would be discussed in concert with considerations of catalytic rate, degree of difficulty of the particular synthetic step, and techniques that stabilize the enzyme and allow for its recovery. Substrate selectivity may be perceived as a technical problem also, though evidence has been rapidly accumulating to indicate that many familiar enzymes accept other than their natural substrates and that catalytic rates, though usually less, may still be useful.

The problems of longevity and reusability have been addressed by an array of immobilization procedures (1) in which enzymes have been alternatively deposited, adsorbed, chemically bound, and included. Deposition can be produced by salting (ammonium sulfate) or adding acetone to a crude solution containing relatively impure enzyme and a powdered solid such as diatomaceous earth. The recovered powder then has enzymatic activity, though exposure to aqueous reaction mixtures will quickly remove the enzyme. Similarly, adsorption to ion exchange resins, though interactions are stronger, serves the purpose of carrying out reactions in non-aqueous media better. Inclusion into gels or capture by controlled pore glass have also been explored. Desorption, or leaching, of enzyme activity is reduced compared to celite deposits. The most permanent technique involves chemical bonding, frequently accomplished by employing the free amino groups from lysine residues that are expected to be present at the surface of globular proteins. Such methodology produces long lasting activity in aqueous reaction mixtures, but the yield of activity is often low. One can imagine that the random binding to a solid support could occasionally produce a structure that could no longer undergo the manifold of conformational changes necessary to catalyze the desired chemical reaction. Given a particular transformation that one wishes to perform, the scale of the reaction, and degree of concern about enzyme recovery, one can screen several common immobilization techniques using the desired catalyst. The following is a brief review of recent reports of asymmetric syntheses that were targeted for selected chiral insect pheromone structures. These can be viewed as exemplifying two broad categories of reaction. In one a prochiral center is reduced asymmetrically; in the other a racemic material is resolved by selective reaction of one enantiomer.

Asymmetric Induction

The reducing power of *Saccharomyces cerevisiae* (Baker's yeast) has been exploited frequently for the preparation of chiral alcohols. Ethyl acetoacetate was reduced by Mori (2) achieving 70-74% ee (enantiomeric excess) of ethyl (*S*)-3-hydroxybutyrate that was then converted to sulcatol, the aggregation pheromone of the Ambrosia beetle, *Gnathotricus sulcatus* (Figure 1). Optical yields in this reduction reported from different laboratories vary, and this may reflect differences in yeast strain and/or conditions of reaction.

In a subsequent study to produce (R)-4-hexanolide, a pheromone of the dermestid, Trogoderma glabrum (Figure 1), an enzymatic reduction of ethyl 3-oxopentanoate was sought. For this substrate Baker's yeast produces the (R)-enantiomer although the sterobias is much less. By enlarging the hydrophobic substituent (ethyl → phenylthioethyl) and screening several reducing yeasts, a suitable reductant, Pichia terricola, was found that produced the desired configuration in 94% ee (3). Subsequent chemical elaboration of the carbomethoxy group to permit generation of the lactone ring was followed by reductive removal of the phenylthio unit.

An additional important use for reducing-enzymes (Figure 2) is for the hydrogenation of double bonds that are conjugated to a carbonyl. For example, citronellol can be oxidized to an α,β-unsaturated aldehyde that undergoes Baker's yeast reduction at both carbonyl and carbon-carbon double bond generating an (S)-configuration on the new saturated methyl branched carbon atom (4). Using this synthetic sequence (R)-and-(S)-citronellols have been converted to (2S,6R) and (2S,6S)-2,6-dimethyl-1,8-octanediols. These diols can be viewed as useful chiral synthons for the many natural products that are 1,5-dimethylated acyclics such as pheromone structures of pine sawflies, red flour beetle, and tsetse fly species. The authors point out that one could use the (2S, 6R)-isomer to build the (R, R)-isomer of 17,21-dimethylheptatriacontane.

Kinetic Resolution

Kinetic resolution makes use of a preferential reaction of one of two enantiomers of a racemate. In contrast to the reductions just described that require cofactors (the cells of the cultures were being used rather than an enzyme preparation), hydrolyzing enzymes (proteases, esterases, lipases) generally do not require cofactors and powders derived from cultures can be employed. Whereas carbonyl reduction can in principle provide a complete conversion to one enantiomer, resolutions are usually performed with the goal of obtaining quantities of both enantiomers pure. Then, if one seeks just one enantiomer, the unwanted isomer must be transformed by stereospecific chemical processes to the desired one. In any case, the selectivity of the reaction must be quite high or much material will be lost before the remaining (slower reacting) enantiomer is pure enough for the purposes of the effort.

A strain of an Aspergillus species (Amano Co.) has been cultured for N-deacylase activity. The selectivity for (S)-acylated amino acids is nearly absolute. Racemic threo-2-amino-3-methylhexanoic acid was synthesized and converted to an acetamide (5) (Figure 3). Treatment with the deacylase produced the (2S,3R)-amino acid. Further treatment of recovered undeacylated material provided a sample of the enantiomer as its amide that was then hydrolyzed to the (2R,3S)-amino acid. These amino acids were employed initially to obtain the remaining diastereomers by inverting the 2-position, then to prepare the enantiomers of threo-4-methyl-3-heptanol, a pheromone component of the smaller European elm bark beetle, Scolytus multistriatus.

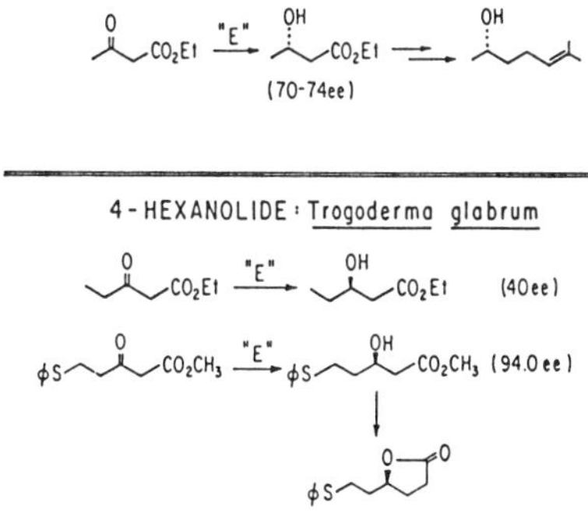

"E"
OH
CO_2Et
(70-74ee)
OH

4-HEXANOLIDE: Trogoderma glabrum

CO_2Et "E" OH CO_2Et (40ee)

ϕS CO_2CH_3 "E" ϕS OH CO_2CH_3 (94.0ee)

Figure 1. Sulcatol: Gnathotricus sulcatus.

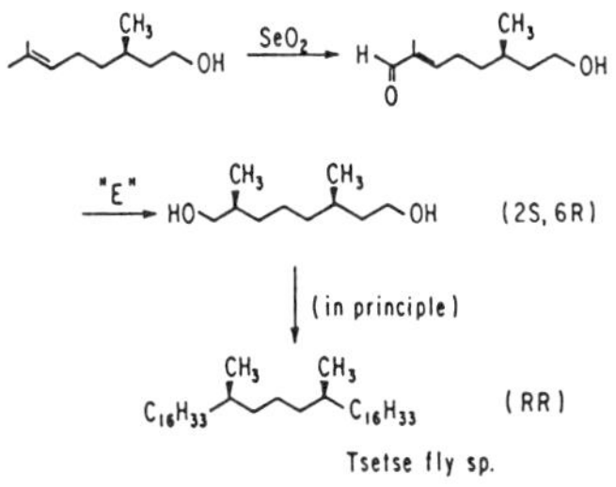

CH_3 OH SeO_2 H O CH_3 OH

"E" HO CH_3 CH_3 OH (2S, 6R)

(in principle)

$C_{16}H_{33}$ CH_3 CH_3 $C_{16}H_{33}$ (RR)

Tsetse fly sp.

Figure 2. A useful synthon: 2,6-Dimethyl-1,8-Octanediol

In similar fashion, racemic 2-aminodecanoic acid was resolved as its N-chloroacetyl derivative (6) (Figure 3). The amino-bearing carbon retained its configuration during deamination that replaces NH_2 with OH, and the carboxyl terminus was extended by conventional means that preserved the compound's stereochemistry. The ultimate products were the enantiomeric 4-dodecanolides that are produced in the pygidial glands of rove beetles. This lactone was also synthesized by microbial reduction by another group (7).

A study of hydrolysis of acetates of alkynyl alcohols using _Bacillus subtilis_ var Niger was studied (8) (Figure 4). The optical purities obtained were highly dependent on the alkyl group present and, for the 4-methyl-3-pentenyl unit (Figure 4) were only fair. Nevertheless such a structure is convertible to the pheromone of the Japanese beetle and an improved enzymatic process would probably be a worthwhile objective, since the current industrial preparation begins with the unnatural isomer of glutamic acid.

Mathematical equations that describe enzymatically driven resolutions have been developed (9). These allow calculation of the ratio of specificity constants (V_{max}/K_M) for a pair of enantiomers and give a constant value for this ratio with stipulated provisions. A convenient form of this expression that may be more useful generally for synthetically oriented chemists has also appeared (10) (Figure 5). Although the rate ratios are actually ratios of specificity constants with attendant limitations to their validity (9), the use of K_R and K_s was employed for simplicity and the sake of comparison (10). Since lipase catalyzed resolutions to be described always resulted in faster reaction of R-enantiomers, the rate ratios are depicted as k_R/k_s. Operationally, one need only determine the fraction of starting material converted (C) and the enantiomeric excess (ee) of the starting material. The product ee could, of course, be determined instead and transformed to starting material ee. A way to view this is as follows: if you had set the goal of 95% ee in your unreacted starting material, a rate ratio of 2.6 requires that the reaction proceed to 95% conversion. The product ee would only be 5. For an infinitely large rate ratio, one only needs to go to 50% conversion. A "useful" rate ratio is subjective, and can be maximized by screening sources of enzymes, strain selection processes, optimizing reaction conditions and altering substrate choice.

For a rate ratio of 10 and a 34% conversion, the product ee still exceeds that of starting material (78:40). But even with this relatively low rate ratio, 90% ee in the starting material is obtained after 63% conversion.

Screening Lipases to Kinetically Resolve Secondary Alcohols

Apart from the evident utility of configurationally pure alphatic secondary alcohols as chiral building blocks, a number of pheromone structures are esters of methyl alkylcarbinols. In particular, 8-methyl-2-decanol is the parent alcohol for esters that have been identified for several closely related species of

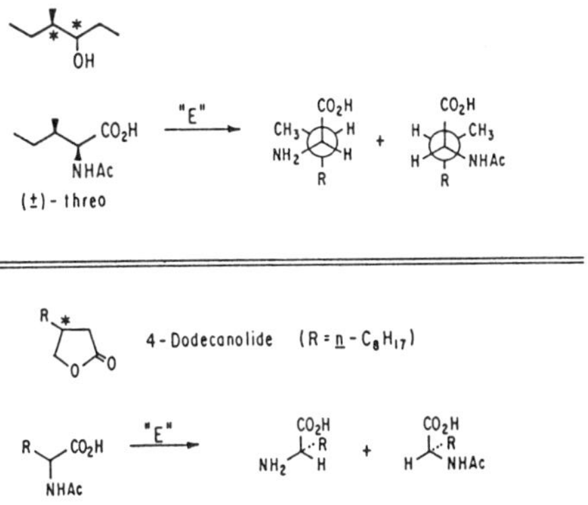

(±)-threo

"E"

4-Dodecanolide ($R = \underline{n}-C_8H_{17}$)

"E"

Figure 3. Smaller European Elm Bark Beetle.

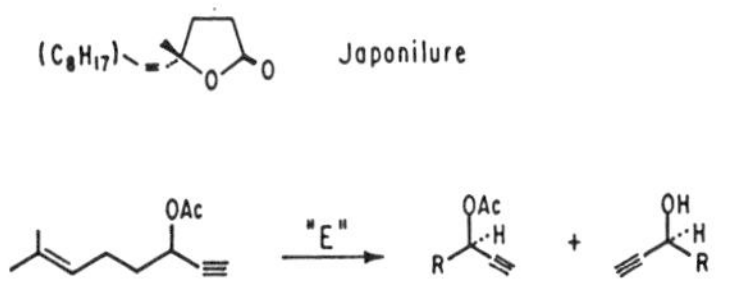

Japonilure

"E"

Figure 4. Kinetic resolution of alkynyl alcohols using Bacillus subtilis.

RATE RATIOS (K_R/K_S) WERE CALCULATED FROM THE EXPRESSION:

$$\frac{K_R}{K_S} = \frac{\mathrm{Ln}(1-C)(1-ee)}{\mathrm{Ln}(1-C)(1+ee)}$$

C = FRACTION OF RACEMIC STARTING MATERIAL CONVERTED.

ee=+ ENANTIOMERIC EXCESS OF RESIDUAL STARTING MATERIAL.

FOR HYDROLYSIS, THE "ee" WAS DETERMINED IN THE PRODUCT AND THEN RELATED TO THE STARTING MATERIAL:

$$ee_A = \frac{C\,(ee_B)}{1-C}$$

A REFERS TO STARTING MATERIAL; B TO PRODUCT.

Figure 5. Equations that characterize enzymatic resolution (9). Since R-alcohols and their esters react faster with the lipases studied than S-alcohols, rate ratios are expressed as K_R/K_S.

rootworm (genus Diabrotica). These compounds contain two asymmetric centers, one involving the secondary alcohol (carbon 2) and the other is designated as the hydrocarbon center (carbon 8). The responses of seven species to the stereoisomers in which the hydrocarbon center is R are tabulated in Figure 6. The insects apparently do not respond to isomers with an (S) hydrocarbon center - synthetics can be racemic at that site and, excepting for effects of dilution, elicit the same response as the corresponding (R)-8 stereoisomer. However, responses to the carbinol site are varied, and evidence of inhibition has been obtained for tests of mixtures that contain blends of (R)-2 and (S)-2 materials. Since the racemic alcohol can be readily synthesized, such as by a procedure developed for the USDA by Zoecon Corporation (12), an efficient enzymatic resolution of the racemic carbinol center could be a valuable synthetic process. The ability to monitor these species selectively by using synthetics as baits in traps may depend critically on the stereochemical constitution of those chemicals.

We initially examined reactions involving 2-octanol and its esters as models using commercially available lipase preparations. Lipases have indeed been employed to perform resolutions (13), but the alcohols (esters) involved were usually alicyclics. Enzymatic resolution of methyl n-alkylcarbinols is much more difficult and has only been occasionally reported (13). In addition, general screening of lipases for this purpose does not seem to have been done. The enzymes were calibrated for activity on olive oil (ca. 85% triolein) using an initial rate assay (Figure 7). Considerable differences exist in these lipase preparations, and this can be attributed to the degree of purification as well as to intrinsic differences between the enzymatically active proteins. At this point little is known of the structure or mechanism of action of triglyceride lipases. We opted therefore to treat these materials simply as undefined, but potentially useful catalysts for organic synthesis.

An examination of the esterification of 2-octanol using octanoic acid in hexane at 30°C with several lipases (Figure 8) gave a strong indication that the M-miehei lipase sold by NOVO Co. would be useful for our purposes (14). This listing of lipase is far from exhaustive, and merely represents a few readily available materials some of which are in commercial use now.

Intriguingly, the esterification's stereochemical consequences are dependent on the length of the fatty acid chain (Figure 9). Acetic acid did not become esterified in the presence of the M-miehei lipase, and stereoselection for (R)-2-octanol increased proceeding to hexanoic acid, diminished beyond nonanoic, and increased again to hexadecanoic. Substituting 2-hexanol as the alcohol to be esterified produced a similar profile that showed overall lowered stereoselection. For example, the rate ratio calculated for 2-octanol and hexanoic acid was >50:1. The corresponding rate ratio for 2-hexanol was 9.5:1. A similar profile was also obtained for 2-decanol; rate ratios were >50:1 for a wider range of fatty acids. The usual conceptualization of the mechanism of hydrolase activity invokes the intermediacy of an

INSECT	RESPONSE TO CARBINOL CENTER 2 (R)	2(S)	ESTER
D. VIRGIFERA VIRGIFERA			
WESTERN CORN ROOTWORM	++	+	PROPIONATE
D. VIRGIFERA ZEA	++	+	PROPIONATE
MEXICAN CORN ROOTWORM			
D. BARBERI	++	I	PROPIONATE
NORTHERN CORN ROOTWORM			
D. PORRACEA	-	++	PROPIONATE
D. LEMNISCATA	+	++	PROPIONATE
D. LONGICORNIS	I	++	PROPIONATE
D. CRISTATA	_	++	ACETATE

D = DIABROTICA; I = INHIBITORY

THOSE INSECTS WITH TRIVAL NAMES ARE ECONOMIC PESTS

Figure 6. Responses of Diabratica (rootworm) species to esters of 8-(R)- methyl-2(R or S)-decanol.

LIPASE	MANUFACTURER-CODE	UMOLMIN^{-1} MG^{-1}	P. SELECTIVITY[B]	F.A. SELECTIVITY[C]
ASPERGILLUS NIGER,1	AMANO-AP	0.154	1,3	18(Δ9)
ASPERGILLUS NIGER,2	AMANO-K	11.2	1,2,3	10,12
CANDIDA RUGOSA	ENZECO	9.82	1,2,3	18(Δ9)
	SIGMA	9.70	1,2,3	18(Δ9)
MUCOR MIEHEI	AMANO-MAP	2.67	1,3	<12
	GIST BROCADES-S	40.0	1,3	<12
	NOVO (POWDER)	5.33	1,3	<12
	NOVO 3A (RESIN)	0.095	1,3	<12
PORCINE PANCREATIC[E]	SIGMA	15.6	1,3	4
RHIZOPUS ARRHIZUS[F]	GIST BROCADES	33.4	1,3	8,10

Figure 7. INITIAL RATE ASSAYS OF COMMERCIAL LIPASES ON OLIVE OIL.[A]

[A]PH MAINTAINED AT 7.3 UNLESS OTHERWISE INDICATED. [B]POSITION SELECTIVITY WITH RESPECT TO TRIGLYCERIDE HYDROLYSIS. [C]FATTY ACID SELECTIVITY PREVIOUSLY ESTABLISHED; THE PREFERENCES ARE SLIGHT IN MOST CASES. [D]THE POWDER FORM IS NOT CURRENTLY AVAILABLE, THOUGH THE LIPASE IS NOW FORMULATED ON BOTH AN ION EXCHANGE RESIN SOLD AS "3A" AND AS A SOLUTION AS "225". [E]pH 8.0 WITH 0.2 ML OF 0.3 M $CaCl_2$ TO 5 ML EMULSION. [F]0.2 ML OF 0.3 $CaCl_2$ TO 5 ML EMULSION.

ESTERIFICATION OF (±)-2-OCTANOL WITH OCTANOIC ACID (HEXANE, 40 C)

LIPASE	K_R/K_S
ASPERGILLUS NIGER (AMANO "AP")	TOO SLOW
ASPERGILLUS NIGER (AMANO "K")	6.1
CANDIDA RUGOSA (ENZECO)	1.1
MUCOR MIEHEI (AMANO "MAP")	TOO SLOW
MUCOR MIEHEI (G.B "S")	2.8
MUCOR MIEHEI (NOVO)	100
RHIZOPUS ARRHIZUS (SIGMA)	5.5
PANCREATIC (SIGMA)	5.4

Figure 8. Rate ratios for esterification of racemic 2-octanol with octanoic acid using several commercial lipase preparations.

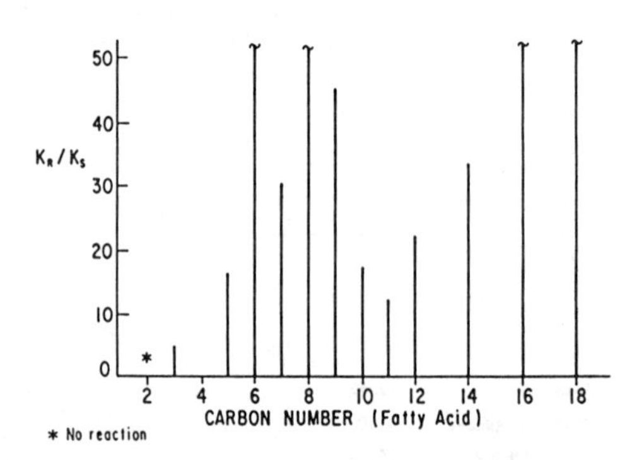

Figure 9. Enantiomeric ratio for esterification of racemic 2-octanol versus fatty acid chain length using *Mucor miehei* lipase.

acyl-enzyme structure. Evidently information involving fatty acid chain length can be transmitted back to the vicinity of the acyl-bearing portion of the complex resulting in altered selection for enantiomeric alcohols. This phenomenon may be general for ester hydrolases and does not appear to have been investigated.

We subsequently evaluated resolution *via* ester hydrolysis and transesterification using these lipases. Hydrolysis of several esters of racemic 2-octanol are shown in Figure 10. Again, the octanoate ester appeared to be the best substrate for resolution. Transesterification with selected triglycerides gave interesting results, but it was felt that workup procedures would be more complicated. The reason for evaluating triglycerides as sources of acid residues for the resolution was that naturally occurring triglycerides such as vegetable oils or animal fats might prove to be cheaper reagents than organic fatty acids.

The information that we had obtained in evaluating esterification and hydrolysis of racemic 2-octanol was then employed to resolve 8-methyl-2-decanol, the pheromone precursor. Throughout these studies resolutions were monitored by derivatizing the recovered alcohols from reactions with (*S*)-α-methylbenzylisocyanate. The resulting diastereomeric carbamates are easily separated by capillary GLC. Figure 11 shows the chromatogram of the carbamates formed from the racemic alcohol; the (*S*,*S*)-diastereomer eluted first. The accompanying chromatogram is of the carbamate derivative of the hydrolysis product (26% yield after distillation of 98.8% *R*). Typically, esterifications produced 93% (*S*)-alcohol, and gave ester that was saponified to 95% (*R*)-alcohol each in greater than 80% theoretical yield.

Prospects

It seems likely that the current interest in enzyme technology will spur studies resulting in more complete information on the enzymes that in many cases are currently being generated by recipes designed empirically to satisfy an industrial customer. Better reference material that will tabulate reactions conducted with homogeneous enzymatically active substances would be useful. New sources of enzymatic activity will lead to a greater range of choices of substrate structure. Enzymes that have been altered chemically may become available, or will be prepared by the user. For example, we are currently working with *C. rugosa* lipase that has been derivatized to add polyethylene glycol chains thereby rendering the protein soluble in organic solvents. Such a procedure allows reactions to occur homogeneously in benzene, methylene chloride, etc, and offers some interesting alternatives to conducting reactions with the native material (15). The potential for changing enzyme selectivity by such conditions has yet to be investigated. Finally, the methods of recombinant DNA would allow the evaluation of site-selective amino acid replacements that would lead to enzymes whose catalytic activity was better tailored for the desired substrate.

HYDROLYSIS OF (±)-2-OCTANOL ESTERS

(M. MIEHEI LIPASE = NOVO, 25°C, PH STAT 7.0)

FATTY ACID	K_R/K_S
C_3	31
C_8	110
C_{12}	24

TRANSESTERIFICATION TECHNIQUES (30°C)

ESTER (CONDITIONS)	K_R/K_S
TRIBUTYRIN (NEAT)	13
TRIBUTYRIN (HEXANE)	5.8
TRIBUTYRIN (~ H_2O)	100
TRIBUTYRIN (HEXANE, ~ H_2O)	100
TRIPROPIONIN (HEXANE, ~ H_2O)	28
TALLOW (HEXANE, ~ H_2O)	6

Figure 10. Rate ratios for the transfer of the indicated fatty acid residue to racemic 2-octanol using Mucor miehei lipase.

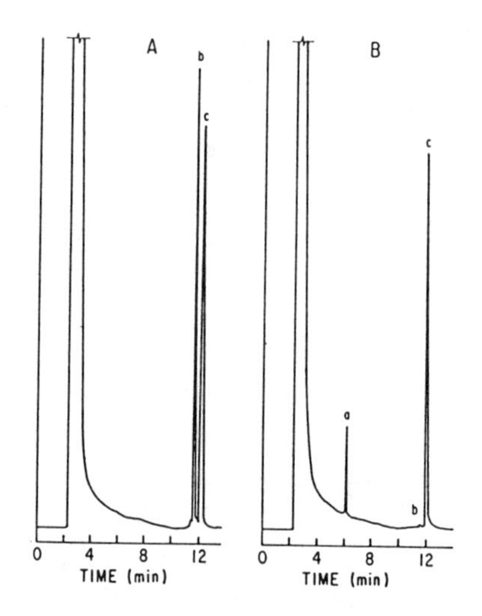

Figure 11. (A)-Chromatogram of diasteromeric carbamates formed from (S)- α -methylbenzyl isocyanate and racemic 8-methyl-2-decanol; (B)- product of lipolysis with Mucor miehei lipase: peak (a)=octanoate ester, (b) = carbamate of 2(S)-alcohol, and (c) = carbamate of 2(R) -alcohol.

Acknowledgment

I am very grateful to Dr. M. W. Baillargeon for helpful discussions during the course of this research.

Literature Cited

1. Rosevear, A. J. Chem. Tech. Biotechnol. 1984, 34B, 127-150.
2. Mori, K. Tetrehedron 1981, 37, 1341-1342.
3. Mori, K., Mori, H., Sugai, T. Tetrehedron 1985, 41, 919-925.
4. Gramatica, P., Mannito, P., Poli, L. J. Org. Chem. 1985, 50, 4625-4627.
5. Mori, K., Iwasowa, H. Tetrahedron 1980, 36, 2209-2213.
6. Sugai, T., Mori, K. Agric. Biol. Chem. 1984, 48, 2497-2500.
7. Muys, G. T., Van der Ven, B., de Jonge, A. P. Appl. Microbiol. 1963, 11, 389-393.
8. Mori, K., Akao, H. Tetrahedron 1980, 36, 91-96.
9. Chen, C.-S., Fujimoto, Y., Girdaukas, G., Sih, C. J. J. Am. Chem. Soc. 1982, 104, 7294-7298.
10. Martin, V. S., Woodard, S. S., Katsuki, T., Yamada, Y., Ikeda, M., Sharpless, K. B. J. Am. Chem. Soc. 1981, 103, 6237-6240.
11. Krysan, J. L., Wilkin, P. H., Tumlinson, J. H., Sonnet, P. E., Carney, R. L., Guss, P. L. Am. Entomol. Soc. Am. (in press).
12. Specific Cooperative Research Agreement No. 58-7B30-1-176, August 14, 1981.
13. Kirchner, G., Scollar, M. P., Klibanov, A. M. J. Am. Chem. Soc. 1985, 107, 7072-7076.
14. Sonnet, P. E., Baillargeon, M. W. J. Chem. Ecol. (in press).
15. Ajima, A., Takahashi, K., Matsushima, A., Saito, Y., Inada, Y. Biotechnol. Lett. 1986, 8, 547-552; and earlier references cited.

RECEIVED June 17, 1987

Chapter 20

Microbial Production of Avermectin

Prakash S. Masurekar

Merck and Company, Inc., Rahway, NJ 07065

The avermectins, a group of eight closely related compounds produced by Streptomyces avermitilis, are potent, broad spectrum antiparasitic agents, effective in animals and plants. These compounds are oleandrose disaccharide derivatives of a pentacyclic 16 membered macrolide ring and act as GABA agonists. It has been shown that a hydroxyl at C-5 and the disaccharide moiety are essential for good activity. Initial fermentation medium development in shake flasks resulted in the selection of dextrose, peptonized milk and autolysed yeast as medium ingredients. Process scale-up studies were done in highly instrumented fermentors. Kinetics of the fermentation showed that the product is formed after near exhaustion of glucose and completion of cell growth. Glucose was also found to repress substantially an extracellular neutral lipase. Postulated biosynthetic scheme involves the formation of the macrolide ring from acetate, propionate and isoleucine via polyketide pathway followed by methylation and glycosylation. Genetic evidence in support of this scheme was obtained by characterization of mutants which have altered compositions of avermectins. Further, avermectin B2 O-methyltransferase was shown to increase coordinately with the yield of the avermectins in productivity mutants.

Avermectins are very potent, broadspectrum antiparasitic agents. These were detected by Merck scientists in fermentation broth of Streptomyces avermitilis in a screen which involved the use of mice infected with Nematospiroides dubius. The microbial culture was originally isolated at the Kitasato Institute from a soil sample collected at Kawana Ito City in Japan (1,2). Subsequent high performance liquid chromatographic analysis showed that the broth contained eight closely related compounds. The four major components were designated as A1a, A2a, B1a and B2a and the four minor components were designated as A1b, A2b, B1b and B2b.

0097-6156/88/0362-0242$06.00/0

Ivermectin, which is sold commercially, is 22,23-dihydroavermectin B1. It contains at least 80% 22,23-dihydroavermectin B1a and not more than 20% 22,23-dihydroavermectin B1b.

CHEMISTRY

The structures of the eight components were determined by ^{1}H and ^{13}C NMR and mass spectroscopy as well as X-ray crystallography (3,4). The basic structure is shown in Figure 1. Avermectins contain a 16-membered macrolide ring, a spiroketal couched in two 6-membered rings, a cyclohexene diol fused to 5-membered cyclic ether and α-L-oleandrosyl-α-L-oleandrosyloxy disaccharide. The eight avermectins differ from each other in the nature of the substituents on C-5, C-23 and C-25. In those designated as A, R5 is methyl as compared to those designated as B where R5 is H. Avermectins of series 1 contain 22,23-olefin, whereas in those of series 2 this olefin is reduced. Furthermore they contain a hydroxyl group at C-23. A sec-butyl substituent on C-25 characterizes "a" group while those of "b" group contain an isopropyl substituent. Thus, all possible combinations of these three structural differences give rise to the eight components.

In order to understand the structure-activity relationship, a number of avermectin derivatives were prepared chemically. Some of the types of reactions carried out include acylation, (5) alkylation (6), hydrogenation (7) and oxidation (8). Comparisons of the antiparasitic activities of the eight compounds produced in the fermentation along with those of chemically modified avermectins provided information on the structure activity relationship. The disaccharide was essential for good activity (7). The activity of the aglycone against gastrointestinal helminths was reduced 30-fold as compared to that of the disaccharide. Similarly, those which contained a 5-hydroxy group were more potent than those containing a 5-methoxy, i.e, avermectins of "B" series were more potent than those of "A" series (5,7). Reduction of 22,23 double bond had only small effect on the activity (1). However, differences between the activities of avermectin "1" and "2" seemed to be dependent on the sensitivities of particular parasites and the mode of administration (2,9). Acylation of 4"-hydroxy group did not have a detrimental effect on the activity. The subtle differences between the activities of avermectins "1" and "2" indicated 22,23-dihydroavermectin B1 to be a more desirable product.

MODE OF ACTION OF AVERMECTIN

This was studied in nematodes (10), in lobster (11) and in mammalian brain (12,13). For example in the studies done with the nematode *Ascaris* the dorsal excitatory motoneuron was indirectly stimulated via its ventral nerve cord (10). Avermectin addition eliminated this response (Figure 2). The GABA antagonist picrotoxin restored the response. These finding suggested that avermectins act by interference with GABA mediated neural signal transmission. Studies with lobster walking leg stretcher muscle showed that avermectins interfere with GABA mediated neuromuscular transmission (11). These data suggested that avermectin B1 acted as a GABA agonist. This was confirmed by the studies with mammalian brain (12,13).

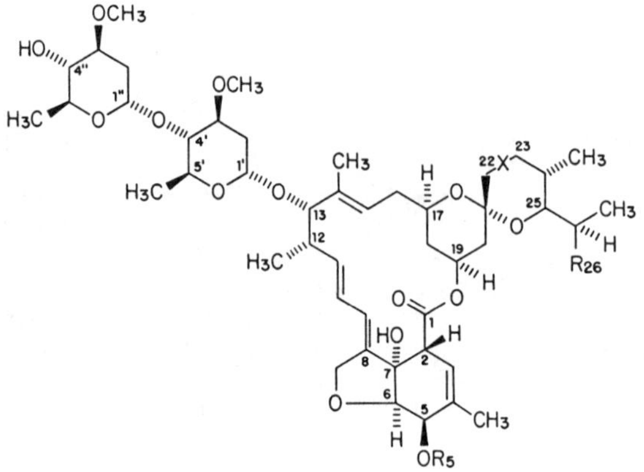

Figure 1. Structure of avermectin. See text for the definitions of R5 and R26. Reproduced with permission from Ref. 1. Copyright 1984 Academic Press Inc.

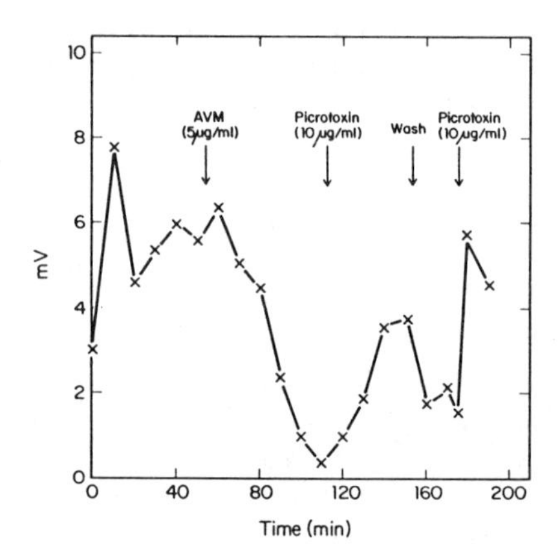

Figure 2. Effect of avermectin on the response of the dorsal excitatory motoneuron of nematode Ascaris. The arrows indicate the time of additions. Reproduced with permission from Ref. 1. Copyright 1984 Academic Press Inc.

BIOLOGICAL ACTIVITY

Ivermectin which is the marketed product, has been shown to be effective against a wide range of endoparasites in sheep, cattle and horses (14,15,16). A single application either orally or parenterally was adequate for the elimination of the parasites. It was also found to be effective against a number of ectoparasites belonging to all the major groups like flies, mites, lice, ticks and fleas (1,9). In man, activity against *Onchocerca volvulus*, which causes blindness, was shown (17). In agriculture, activity was shown against a number of pests and insects such as two spotted spider mites on roses and beans (18,19), leaf miner on ornamentals (20), bean aphids on beans (19) and citrus rust mites on oranges (19). Thus, avermectins are truly broad spectrum and highly effective antiparasitic agents.

FERMENTATION PROCESS DEVELOPMENT

MEDIUM STUDIES. Initial development of the process to produce avermectin was done in shake flasks. This involved optimization of the production medium and strain improvement.

For medium development studies the seed was grown in medium containing cerelose, starch, beef extract, Ardamine PH, N-Z amine and trace elements for 1-2 days (21). The composition of initial production medium is described in Table I.

Table I. Complex Medium for the Production of Avermectin

Dextrin	40 g
Distillers solubles	7 g
Autolyzed yeast	5 g
$CoCl_2.6H_2O$	0.05 g
Distilled water	1 liter
pH 7.3	

Reproduced with permission from Ref. 2. Copyright 1982 Japan Antibiotic Research Association.

This medium was best of a number of complex media screened. Yield in this medium was 9 µg/ml and it served as the basis for medium improvement studies (21). As is customary in these types of studies, various carbon and nitrogen sources along with vitamins and trace elements were tried (21). Results of three such experiments are shown in Table II.

In the first experiment the carbon source was varied. Dextrose was found to give the best titer. In the second experiment different nitrogen sources were tried. A combination of peptonized milk and polyglycol P-2000 resulted in increase in production to 53 µg/ml. Further improvement in the yield was obtained when Ardamine PH was used as a vitamin source. Optimization of dextrose, peptonized milk and Ardamine PH concentrations resulted in medium shown in Table III.

Table II. Effect of Various Carbon and Nitrogen Sources on Production

Experiment Number	Carbon Source	Nitrogen Source	Vitamin Source	Polyglycol	Avermectin (μg/ml)
1	Dextrin	Pharmamedia	Primary yeast	-	5
	Sucrose	Pharmamedia	Primary yeast	-	0
	Dextrose	Pharmamedia	Primary yeast	-	30
	Lactose	Pharmamedia	Primary yeast	-	0
2	Dextrose	Fermamine I	Yeast extract	-	13
	Dextrose	Sheften WP100	Yeast extract	-	15
	Dextrose	Primatone	Yeast extract	-	8
	Dextrose	Peptonized milk	Yeast extract	-	17
	Dextrose	Peptonized milk	Yeast extract	+	53
3	Dextrose	Peptonized milk	None	-	38
	Dextrose	Peptonized milk	Yeast extract	+	75
	Dextrose	Peptonized milk	Ardamine PH	+	83

Table III. Optimized Complex Medium for the Production of Avermectin

Cerelose	45 g
Peptonized milk	24 g
Autolyzed yeast	2.5 g
Polyglycol P-2000	2.5 ml
Distilled water	1 liter
pH 7.0	

Presence of P-2000 was critical to the success of this medium. Avermectin production by the original soil isolate in this medium was about 100 μg/ml. Thus, at this point, the medium improvement resulted in a 10-fold increase in the yield.

STRAIN IMPROVEMENT. Ultra-violet irradiation was used to mutate the original culture. The mutated cultures were randomly selected and tested for avermectin production. One of the mutants was found to have superior ability to produce avermectins. It was able to synthesize avermectin at a higher rate and for a longer time than its parent. This mutant designated ATCC 31271 produced 500 μg/ml as compared to 120 μg/ml by its parent (21).

KINETICS OF FERMENTATION. Four variables, dry cell weight, pH, glucose utilization and avermectin production are described in Figure 3. The growth was essentially complete in the first 1/3 of the cycle, subsequently, the dry cell weight remained constant. About 85% of glucose was utilized during the growth phase. The rest was used up in the production phase. The pH dropped to 5.9 during the growth and increased to 6.9 as the growth slowed down. The pH then slowly declined to 6.3 during avermectin production period and finally increased to 7.0 in the last 20% of the cycle. Thus, the pH changes were consistent with the pattern of glucose utilization. Avermectin synthesis began after growth was complete and continued linearly until the glucose was used up. The rate of synthesis declined after the exhaustion of glucose. These results indicate that pH can be used to monitor the progress of fermentation. Since we noted that a limited amount of glucose was used during the production phase, we studied the possibility of other catabolic enzymes like lipases might be active at that time.

KINETICS OF LIPASE PRODUCTION. For this purpose the culture was grown in a production medium similar to the one described. Samples were taken periodically and assayed for growth, pH, avermectin production, glucose utilization, neutral lipase production and phospholipase production. Extracellular neutral lipase was measured by determining release of ^{14}C-palmitic acid (22). The activity of extracellular phospholipase was assayed by measuring the inorganic phosphate released by the enzyme (23). Essentially no lipase was found as long as glucose was present (Figure 4). It was derepressed almost 100-fold when the glucose was exhausted. It rapidly reached the maximum and then declined. The reason for this might be that the synthesis of lipase was not continued in the last 10% of the cycle and the enzyme synthesized up to that time was degraded. Phospholipase did not show this sensitivity to glucose. Its activity seemed to be at the maximum during the early part of the

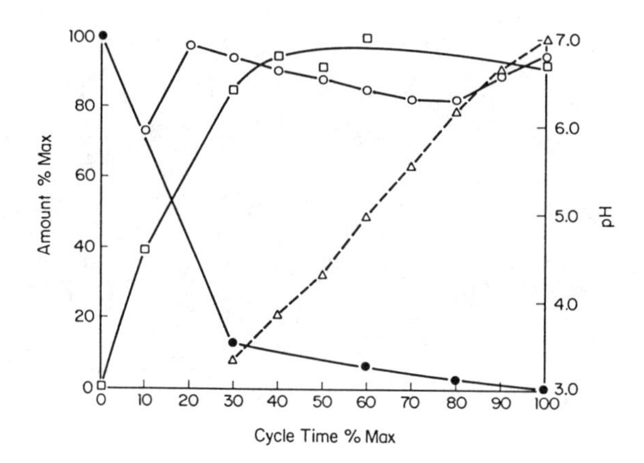

Figure 3. Kinetics of avermectin fermentation in shake flasks. Legend: pH (○), dry cell weight (□), avermectin (△) and glucose (●).

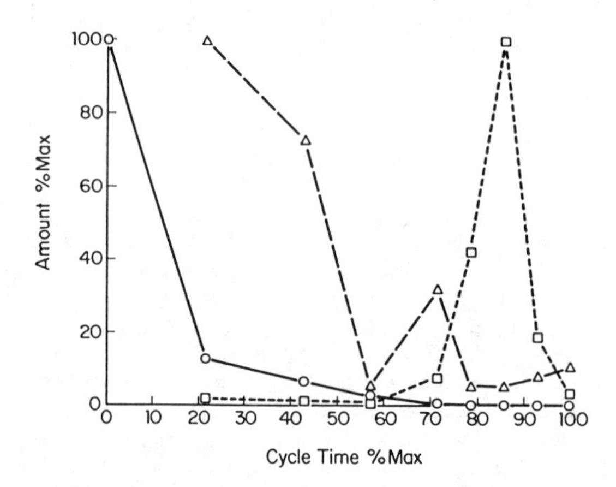

Figure 4. Kinetics of lipase production by Streptomyces avermitilis in shake flasks. Legend: glucose (○), lipase (□) and phospholipase (△).

fermentation and then reduced as the fermentation proceeded. The maximum phospholipase activity observed was 50-fold lower than that of neutral lipase activity. These results suggested that alternative carbon sources may be used during the production phase.

SCALE-UP STUDIES. Various size fermentors, 14-L to 800-L, were used to determine the critical variables for scale-up. These fermentors were instrumented to measure pH, temperature, dissolved oxygen, aeration and agitation rates, exhaust gas composition etc. A distributed digital control network based on a Honeywell TDC-2000 was used (24). Valuable process information was obtained from these instrumented fermentors, which not only improved its understanding but also made it possible to implement various control strategies. These studies resulted in a process which was used in the production fermentors to manufacture avermectin.

An example of this is taken from the work of Buckland *et al* (24). From the process control point of view, specific growth rate is an important variable. However, it is not possible to measure it on line in mycelial fermentations by conventional means. If the relationship between growth rate and oxygen uptake rate is considered, it can be seen that it is feasible to estimate the specific growth rate from data collected by exhaust gas analysis. Similarly, growth can be estimated from the total amount of oxygen used. For example, the growth rate can be expressed as a constant times the oxygen uptake rate and the dry cell weight to be the same constant times integral of oxygen uptake rate in appropriate limits. The ratio of these two equations gives an estimate of the specific growth rate. Figure 5 shows the data obtained in an 800-L fermentor. Total oxygen uptake and specific growth rate are shown as a function of time. The specific growth rate was determined from exhaust gas analysis. The dry cell weight is plotted on the second Y-axis for purpose of comparison. The growth rate was high in the first eight hours indicating possibly exponential growth. After that it declined and reached a very low value by 24 hrs. Avermectin production did not begin until after the specific growth rate had fallen virtually to zero, which was similar to the observation made in shake flasks. For the first 30 hrs, there was a good correlation between the dry cell weight and the total oxygen uptake. These results demonstrate that the specific growth rate and dry cell weight estimated from the exhaust gas analysis represent true values and hence can be used for feedback control of the process.

Very often in process development the problem is as much of scale-up as of scale-down. Production facilities cannot be modified without a large outlay of capital, and hence it is essential to model production fermentors in pilot plant fermentors. This is greatly facilitated by having instrumented fermentors, as demonstrated by the following example (24).

A new production medium was developed in shake flasks and was found to have a beneficial effect on the yield. However, before it could be recommended for production, it was necessary to determine if it could be used in the production fermentors. For this purpose new medium was run in an 800-L fermentor. The agitator speed was controlled to prevent the dissolved oxygen concentration from falling below 20% as it was known that dissolved oxygen concentration below that level was detrimental to avermectin production. Previous experience indicated that if the peak oxygen demand was below 35 mmoles/l.hr and the broth viscosity was the same as that obtained in the old medium then the new medium would work in the production fermentor. Oxygen uptake rates in the new and the old medium are plotted against time for the first 50 hrs of the fermentation in Figure 6. The peak oxygen demand for the new medium was lower than that for the old medium and it was less than 35 mmoles/l.hr. As a cross-check, power consumption was measured for the new medium in the 800-L fermentor, which too indicated that the power per unit volume requirements for the new medium would be met in the production fermentor. These results showed that the new medium could be recommended for production.

BIOSYNTHESIS OF AVERMECTINS

ISOTOPE INCORPORATION STUDIES. In initial studies, [1-^{13}C] acetate and [1-^{13}C]propionate were used as precursors for incorporation into avermectins. At the end of the fermentation, avermectins A1a, A2a, B1a and B2a were isolated and analysed by ^{13}C-NMR spectroscopy (25). The results showed identical carbon distribution for the macrolide ring as expected from the assumption of polyketide pathway for its synthesis. Seven acetate and five propionate units were incorporated as shown in Figure 7. This figure shows the position where the acetate and propionate precursors are incorporated. C-25 carbon and its sec-butyl substituent were not labeled by (^{13}C) acetate or propionate. These were established to be derived from L-isoleucine (1). The C-25 and its isopropyl substituent in avermectin "b" series are derived from L-valine (1). The disaccharide is derived from glucose (26). Further studies with [1-$^{18}O_2$,1-^{13}C]propionate and [1-$^{18}O_2$,1-^{13}C]acetate incorporation and NMR analysis of four avermectins showed that oxygens attached to C-7 and C-13 of all avermectins and C-23 of "2" series avermectins come from propionate; and those attached to C-1, C-5, C-17 and C-19 are from acetate (25).

BIOSYNTHETIC PATHWAY. Figure 8 shows the biosythetic scheme postulated from these data (27). Avermectin biosynthesis is initiated from the precursors isoleucine, acetate, propionate and glucose and proceeds through the intermediate "X". At this point, a dehydration reaction separates the biosynthesis of avermectin "1"s from the "2"s. We have not isolated any intermediates in this proposed pathway up to 6,8adeoxy-5-keto B1a algycone (28) and B2a aglycone. The reactions from either B1a aglycone and B2a aglycone to four avermectins are similar and involve glycosylation and methylation (27). One exception to this similarity is that B1a cannot be converted into A1a.

In addition to the labeling studies described earlier, evidence in support of this pathway has come from the isolation of mutants which produce altered composition of avermectins and from the enzymatic studies. I have listed some of the compositional mutants in Table IV.

The first class of mutants is defective in 5-O-methyl transferase activity and as a result it cannot convert avermectin "B"s into "A"s and thus, accumulates avermectin Bs (29). The second class

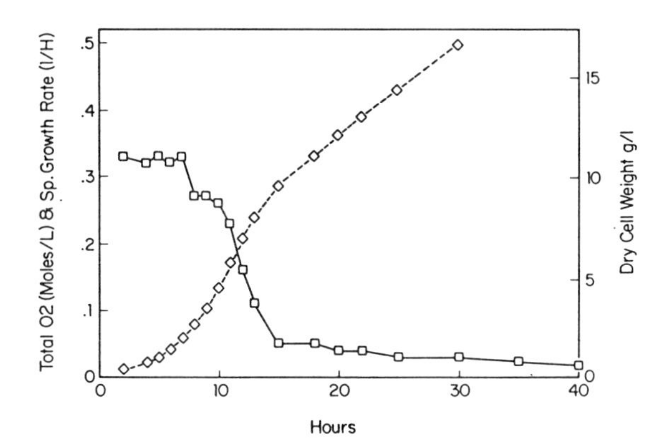

Figure 5. Estimation of the dry cell weight and the specific growth rate from the total oxygen consumed. Legend: total oxygen consumed and proportional dry cell weight (——) and specific growth rate (---). Reproduced with permission from Ref. 24. Copyright 1985 Nature Publishing Co.

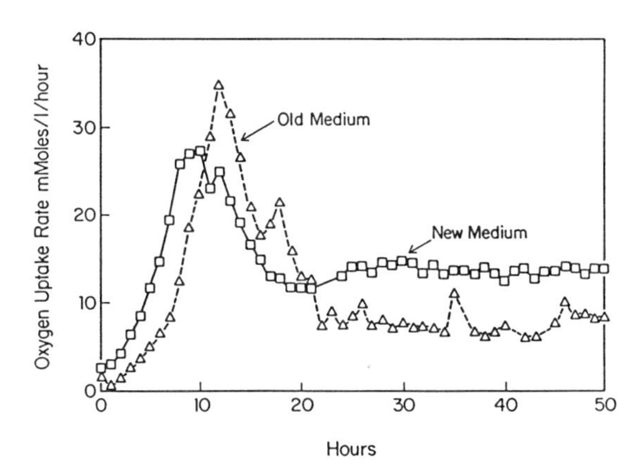

Figure 6. Comparison of oxygen uptake rates in two production media. Legend: old medium (△) and new medium (□). Reproduced with permission from Ref. 24. Copyright 1985 Nature Publishing Co.

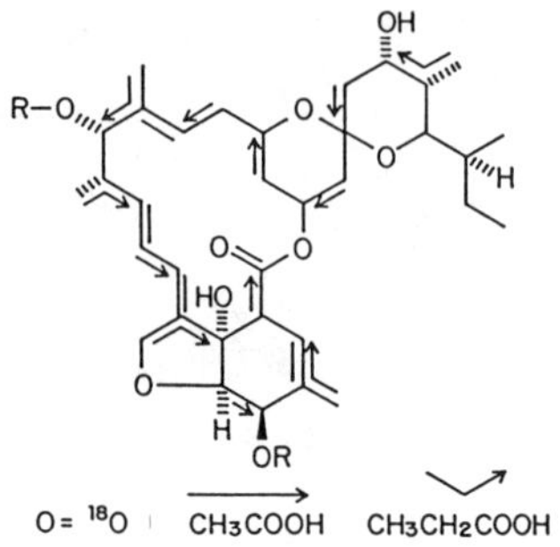

O = ^{18}O CH_3COOH CH_3CH_2COOH

Figure 7. Incorporation of acetate and propionate units into macrolide ring of avermectin. Reproduced with permission from Ref. 1. Copyright 1984 Academic Press Inc.

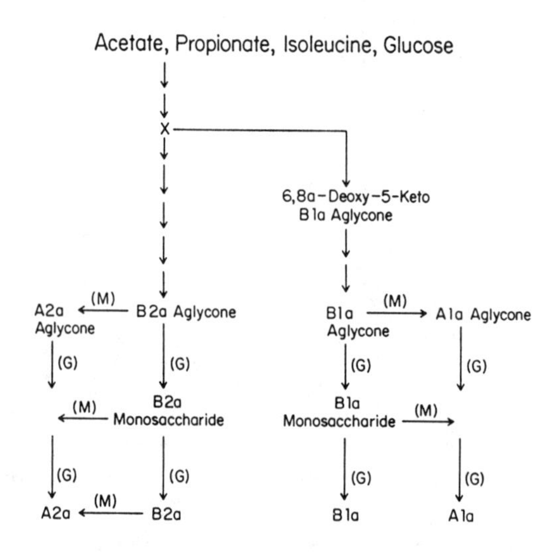

Figure 8. Proposed scheme for the biosynthesis of avermectins. G = glycosylation, M = methylation. Reproduced with permission of the authors from Ref. 27.

Table IV. Mutants Which Produce Different Compositions of Avermectin

Mutant No.	Defect	Product
1	5-O-methyltransferase	Bs
2	Glycosyl-O-methyltransferase	A & B desmethyls
3	Glycosyltransferase	A aglycones
4	Furan 'cyclase'	many

of mutants lacks glycosyl-O-methyltransferase and forms avermectin A and B desmethyls (29). The third class of mutants is impaired in the glycosylation reaction. They form mainly avermectin A aglycones. The fourth class of mutants is defective in furan "cyclase" and cannot form the 5-membered cyclic ether (28). These produce many different products. Thus, these classes of mutants provide genetic evidence in support of the proposed reactions.

Studies were done to isolate enzymes which catalyze the reaction in the proposed biosynthetic pathway. One which is catalysed by 5-O-methyltransferase is shown in Figure 9. Avermectin B component is converted into A component with S-adenosyl methionine as a methyl donor. The enzyme was detected in the cell-free extracts after removal of cell debris (30). The assay involved conversion of avermectin B2a aglycone into A2a aglycone, which was measured by determining the incorporation ^{14}C-labeled methyl from S-adenosyl methionine. The enzyme is specific for C-5 hydroxyl and does not catalyse the transfer of a methyl group to C-3' or C-3" in the disaccharide moiety. This was independently supported by the isolation of mutants of classes 1 and 2 described earlier.

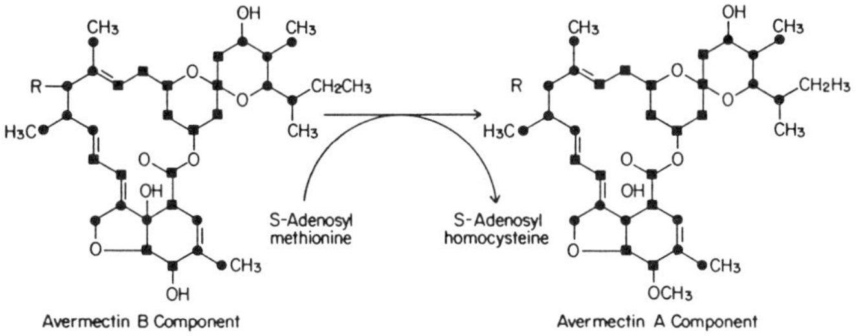

Figure 9. Reaction catalysed by avermectin B2 O-methyltransferase. Reproduced with permission from Ref. 30. Copyright 1986 American Society for Microbiology.

In order to further investigate its role in avermectin biosynthesis, its activity was determined in four strains of S. avermitilis (Table V).

These four strains are sequentially improved producers of avermectin obtained by UV and N-methyl-N-nitrosourethane mutagenesis (30). They were grown in the production medium described earlier for 6 days. Each day from 48 hrs on, avermectin production and the

Table V. Avermectin B O-methyltransferase Activity and Avermectin Production in Strains A, B, C, and D

Strain	Maximum Sp. Act. of O-met. Trans. (nmol/h per mg of Protein)	Total Avermectin at 144 h (Relative Units)	Fold increase in: Maximum Sp. Act. of O-met. Trans.	Fold increase in: Avermectin
A	0.37	0.85	1	1
B	0.86	2.10	2.3	2.5
C	1.06	2.4	2.9	2.8
D	1.35	3.0	3.6	3.5

activity of the O-methyltransferase was determined. Avermectin was measured by HPLC. The enzyme assay was as described. The maximum specific activity and the avermectin production is shown here. The specific activity of the enzyme increased coordinately with the increase in avermectin production. It is interesting that although the specific activity of the enzyme increased 3.5-fold, the ratio of B to A components remained unchanged. These results suggest that in any one of the strains this enzyme is not rate limiting for the conversion of avermectin B to A. Such information is crucial in the design of more rational approaches to improve yields.

In summary, avermectin is a potent, truly broad spectrum antiparasitic agent, produced by S. avermitilis. It was discovered through the use of a targeted screen. The fermentation process was developed first in shake flasks, where medium optimization and strain improvements gave a 50-fold increase in yield. The process was further refined and scaled up in highly instrumented fermentors with sophisticated process control. Biosynthetic studies with labeled precursors, isolation of altered compositional mutants and enzymological studies have given a fairly good picture of how avermectins are synthesized.

LITERATURE CITED

1. Fisher, M.H.; Mrozik, H. In Macrolide Antibiotics Chemistry, Biology and Practice; Omura, S., Ed.; Academic: New York, 1984; Chapter 14.
2. Stapley, E.O.; Woodruff, H.B. In Trends in Antibiotic Research; Umezawa, H., Demain, A. L., Mata, T. and Huchinson, C.R., Ed.; Japan Antibiotic Research Association, 1982; p. 154.
3. Albers-Schonberg, G.; Arison, B.H.; Chabala, J.C.; Douglas, A.W.; Eskola, P.; Fisher, M.H.; Lusi, A.; Mrozik, H., Smith, J.L.; Tolman, R.L. J. Am. Chem. Soc. 1981, 103, 4216-21.
4. Springer, J.P.; Arison, B.H.; Hirshfield, J.M.; Hoogsteen, K. J. Am. Chem. Soc. 1981, 103, 4221-24.
5. Mrozik, H.; Eskola, P.; Fisher, M.H.; Egerton, J.R.; Cifelli, S.; Ostlind, D.A. J. Med. Chem. 1982, 25, 658-63.
6. Fisher, M.H.; Lusi, A.; Tolman, R.L. U.S. Patent 4,200,581, 1980.

7. Chabala, J.C.; Mrozik, H.; Tolman, R.L.; Eskola, P.; Lusi, A.; Peterson, L.M.; Woods, M.E.; Fisher, M.H.; Campbell, W.C.; Egerton, J.R.; Ostlind, D.A. J. Med. Chem. 1980, 23, 1134-36.
8. Chabala, J.C.; Rosegay, A.; Walsh, M.A.R. J. Agric. Food Chem. 1981, 29, 881.
9. Campbell, W.C.; Fisher, M.H.; Stapley, E.O.; Albers-Schonberg, G.; Jacob, T.A. Science, 1983, 221, 823-28.
10. Kass, S.; Wang, C.C.; Walrond, J.P.; Stretton, A.O.W. Proc. Nat. Acad. Sci. 1980, 77, 6211-15.
11. Fritz, L.C.; Wang, C.C.; Gorio, A. Proc. Nat. Acad. Sci. 1979, 76, 2062-66.
12. Pong, S.S.; Wang, C.C. Neuropharmacology 1980, 19, 311-17.
13. Pong, S.S.; Wang, C.C. J. Neurochemistry 1982, 38, 375.
14. Egerton, J.R.; Birnbaum, J.; Blair, L.S.; Chabala, J.C; Conroy, J.; Fisher, M.H.; Mrozik, H.; Ostlind, D.A.; Wilkins, C.A.; Campbell, W.C. Bri. Vet. J. 1980, 136, 88-97.
15. Egerton, J.R.; Eary, C.H.; Suhayda, D. Vet. Parasit. 1981, 8, 59-70.
16. Egerton, J.R.; Brokken, E.S.; Suhayda, D.; Eary, C.H.; Wooden, J.W.; Kilgore, R.L. Vet. Parasit. 1981, 8, 83-88.
17. Aziz, M.A.; Diallo, S.; Diopp, I.M.; Lariviere, M.; Porta, M. Lancet 1982, 1982-II, 171-173.
18. Green, A. St.J.; Dybas, R.A. Proc. 1984 British Corp Protection Conf. 1984, 3, 11A-7.
19. Dybas, R.A. U.S. Patent 4,560,677, 1985.
20. Green, A. St.J.; Hejne, B.; Schreus, J.; Dybas, R.A. Med. Fac. Landbouww. Rijksuniv. Gent 1985, 50/2b, 603-22.
21. Burg, R.W.; Miller, B,M.; Baker, E.E.; Birnbaum, J.; Currie, S.A.; Hartman, R.; Kong, Y.L.; Monaghan, R.L.; Olson, G.; Putter, I.; Tunac, J.B.; Wallick, H.; Stapley, E.O.; Oiwa, R.; Omura, S. Antimicrob. Agents Chemother. 1979, 15, 361-67.
22. Khoo, J.C.; Steinberg, D. In Methods in Enzymology; Lowenstein, J.M. Ed.; Academic: New York, 1975, Vol. 35, p.181.
23. Krug, E.L.; Kent, C. In Methods in Enzymology; Lowenstein, J.M.; Academic: New York, 1981, Vol. 72, p. 347.
24. Buckland, B.; Brix, T.; Fastert, H.; Gbewonyo, K.; Hunt, G.; Jain, D. Biotechnology, 1985, 3, 982-88.
25. Cane, D.E.; Liang, T.-C.; Kaplan, L.; Nallin, M.K.; Schulman, M.D.; Hensens, O.D.; Douglas, A.W.; Albers-Schonberg, G. J. Am. Chem. Soc. 1983, 105, 4110-12.
26. Schulman, M.D.; Valentino, D.; Hensens, O. J. Antibiotics, 1986, 39, 541-49.
27. Chen, T.S.; Inamine, E. Twenty Seventh Annual Mtg. of Am. Soc. Pharmacognosy 1986. Abst #4.
28. Goegelman, R.T.; Gullo, V.P.; Kaplan, L. U.S. Patent 4,378,353, 1983.
29. Ruby, C.L.; Schulman, M.D.; Zink, D.L.; Streicher, S.L. Proc. 6th Int'l. Symp. on Actinomycetes Biol. 1985, Vol. A., p. 279.
30. Schulman, M.D.; Valentino, D., Nallin, M.; Kaplan, L. Antimocrob. Agents Chemotherap. 1986, 29, 620-24.

RECEIVED July 27, 1987

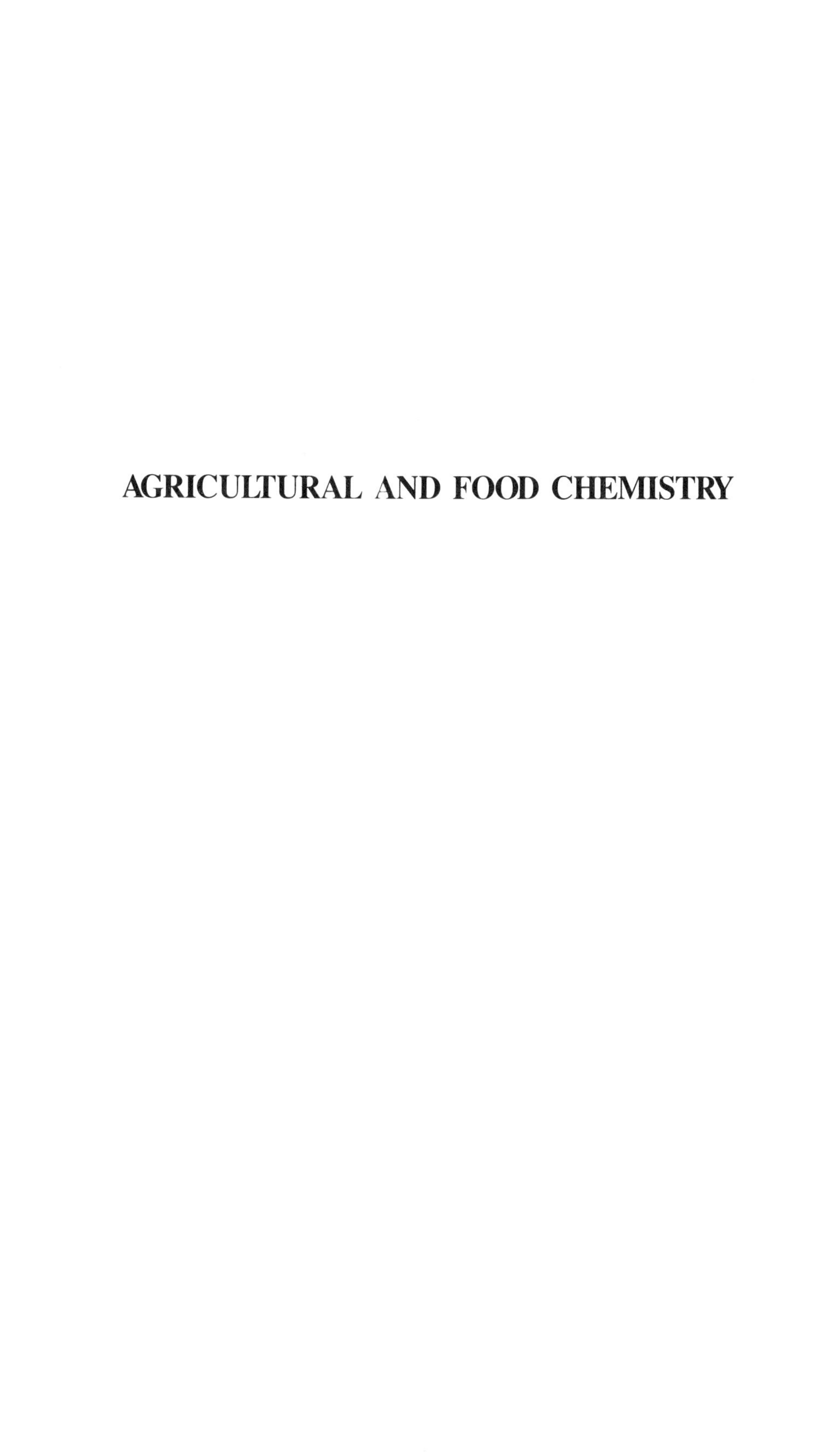

AGRICULTURAL AND FOOD CHEMISTRY

Chapter 21

The Role of Biotechnology in Agricultural and Food Chemistry

Donald W. De Jong[1] and Marshall Phillips[2]

[1]Tobacco Research Laboratory, U.S. Department of Agriculture, Agricultural Research Service, Oxford, NC 27656

[2]National Animal Disease Center, U.S. Department of Agriculture, Agricultural Research Service, Ames, IA 50010

The application of biotechnology to agriculture requires the interaction of many disciplines, and basic as well as applied aspects of science have been essential to advances in agricultural biotechnology. There is little doubt that the field has evolved as rapidly as it has because commercial interests have been eager to put new discoveries into the marketplace as quickly as possible. In this section on Ag & Fd chemistry we have sought to provide a balanced representation of the basic and the applied--the plant and the animal. This choice was made to demonstrate the broad scope of agricultural biotechnology. Furthermore, we decided to treat biotechnology in its broadest sense rather than the narrow sense of only recombinant DNA. We realize that biochemistry which originated from the fusion of enzyme chemistry with cellular metabolism, provides the framework in which molecular biology functions. Biotechnology uses the molecular processes and the manipulation of the chemical structures established in biochemistry and chemistry laboratories. This is the central impact of chemistry on biotechnology.

In this section we present two chapters that are fundamental in nature. They both deal with very complex biochemical processes in plants--*nitrogen fixation* and *photosynthesis*. In the first case, i.e., nitrogen fixation, we are dealing with a synergistic association between mutually beneficial organisms living symbiotically. The case described is a cooperative relationship between microorganisms--*Rhizobia*--and host plants--legumes. The chapter by Stacey discusses the salient features in what is presently known about the interaction between plant and bacterial genes in symbiotic nitrogen fixation. It is conceivable that this research could lead to enlarging the host range of the most efficient bacterial strains, including even monocots as well as nonleguminous dicots. At this point in time the overwhelming complexity of the system precludes genetic tinkering with the apparatus itself.

The second paper of a fundamental nature concerns photosynthesis wherein the cooperation occurs intracellularly between two organelles, the nucleus and chloroplasts. In response to light there occurs coordinated expression of genes that code for proteins that reside and function in the chloroplasts. The chapter by Timko, et. al. provides an excellent example of the utility of recombinant DNA methodology for resolving some of complex problems encountered in plant biochemistry. The long-term objective of this research, of course, is to find means for improving the efficiency of the photosynthetic process. A better understanding of the genetic control systems for enzymes which catalyzes CO_2 fixation could eventually lead to strategies that alter the function of the enzyme. Suggestions have been made that repression or elimination of its competing oxygenase activity would improve photosynthetic rates. Whichever strategy proves feasible, intimate knowledge of the genetic regulatory mechanism is essential and biotechnology research is providing this information.

It is important to note the impact of biotechnology on these two areas of research. From time to time in the advance of technology, limits to understanding are reached that hold back further progress. Data continues to accumulate but theoretical advances stagnate. Then a new technique or creative insight arrives on the scene and progress resumes. This scenario is true for both nitrogen fixation and photosynthesis. Ultimately, of course, hopes are high that by genetic manipulation, the natural systems can be improved upon to increase the efficiency or extend the range of each of these critical life-support mechanisms. Both papers admit that practical applications are not yet attainable; even so the research they describe is adding valuable information relative to genetic expression of key events regulating the two fundamental processes.

Three presentations address applied aspects of agricultural biotechnology. (Chapter by Carlson is included in the Chemical Marketing Section.) The general subject matter is plant tissue culture. Two very practical and powerful applications of this field have emerged: 1) Screening multitudes of tissue culture entries for evidence of somaclonal variation. With or without selection pressures this has lead to the generation of novel plant forms that in theory at least could have certain desirable features not normally found in extant cultivars. Successful results have been reported from a number of laboratories. The contribution by Orton and Reilly discusses the relevance of tissue culture to biotechnology directed toward the improvement of plant food products. Specific objectives include features that concern industry: processing quality; those that impact consumers: nutritional and aesthetic properties; and those that are agronomic in nature and therefore concern farmers: better yield, stress tolerance and pest resistance. The second major tissue culture enterprise of practical significance is: 2) Propagating vegetative clones of important crop plants to ensure disease-free and genetically stable plantlets ready for transplanting to the field. Crop Genetics International has been successful in introducing tissue culture techniques into the production of sugar cane plantlets on a commercial scale. Carlson cites the difficulties encountered when a perfectly logical and scientifically feasible project, i.e., developing a disease-free banana cultivar, is proposed for application in the Third World. In countries where a given product of biotechnology is apt to be most desperately needed, barriers to development may totally prevent research as well as implementation. These two papers focus on different aspects of a common theme: that is, impediments to adoption of biotechnology practices and products. To Orton with an eye toward the sophisticated market structure of developed economies the major constraint is regulatory, whereas Carlson stresses the strategic constraints that one faces in the Third World setting. Both issues must be recognized in realistic terms regardless of whatever technical hurdles need to be overcome. The issue of socio-political constraints to implementation of agricultural biotechnology also tempers the generally optimistic projections in Hardy's look into the future.

The paper by Reed describes the fortunes of a relatively new company that has concentrated its efforts on the development of products for the farm animal business. Reed describes both successes and failures in the company's efforts to introduce products of biotechnology into the marketplace. Most of the products he describes are vaccines or immunizing agents for diseases of livestock that have historically been difficult or expensive to control. Other products discussed provide diagnostic services to veterinary medicine. The products of Molecular Genetics have resulted primarily from recombinant DNA procedures including utilization of monoclonal antibody technology; but Reed suggests that other technologies besides rDNA might be more appropriate to a given need. He states that "...goals must be linked to ... improved livestock performance not to...a particular technology." According to Reed the fate of biotechnology products is more often affected by economic or regulatory considerations than by actual capability to manufacture the substance.

What can we expect with regard to the projected impact of biotechnology upon the value of farm or associated agricultural enterprises? It has been estimated that biotechnology could generate an additional $100 billion to the value of American agriculture--which represents a doubling in value due to biotechnology applications alone. This projection may not take into consideration the nontechnical hurdles--environmental and public resistance among others--that could impede adoption of biotechnology on a massive scale. Hardy deals with these and related issues in his projections into the future. Although, the prospects for advancement into new agricultural frontiers are mind-boggling and could affect almost every aspect of agriculture predicting the precise direction of development requires more than a crystal ball. Even today, at the threshold of new advances in molecular genetics a very sobering assessment by three prominent plant scientists concluded that "...our technical ability to isolate and transfer genes into plants has surpassed our level of biochemical understanding required for their rational manipulation" (1).

Our preferred definition for biotechnology is that it is the application of biological systems and organisms to technical and industrial processes. The target of biotechnology in agriculture is to help solve high-priority agricultural problems such as conserving the quality of natural resources, reducing farm production costs, improving crop protection and production efficiency, enhancing market value and quality of farm products, and promoting human health through better nutrition.

Literature Cited

1. Fraley, R. T.; Rogers, S. G.; Horsch, R. B. CRC Critical Reviews in Plant Sciences; 1986, 4, 1-46.

RECEIVED August 19, 1987

Chapter 22

Genetics of Symbiotic Nitrogen Fixation

Gary Stacey

Department of Microbiology and Graduate Program of Ecology, University of Tennessee, Walters Life Sciences Building, Room M–409, 1414 West Cumberland Avenue, Knoxville, TN 37996–0845

Rhizobium and *Bradyrhizobium* species are Gram-negative, aerobic, soil bacteria with the unique ability to infect plant roots and establish a nitrogen-fixing symbiosis. The establishment of this intimate plant-microbe interaction is complex and requires several plant and bacterial genes. Research into the genetic basis of this process is revealing important principles of how plants and microorganisms interact. For example, the regulation of those bacterial genes necessary for plant infection involves a form of plant-microbe "communication." The bacteria sense plant-produced secondary products thereby "recognizing" the plant and inducing the necessary gene functions. In a general sense, definition of communication networks between microorganisms and their hosts can theoretically lead to schemes for improving or terminating the interaction.

Bacteria of the genera *Rhizobium* and *Bradyrhizobium* possess the ability to infect leguminous plants and establish a nitrogen fixing symbiosis. This process is called nodulation. The morphological structure formed on the root in which the bacteria reside is termed a nodule. The formation of a nodule is a developmental process both from the standpoint of the bacteria and the plant. Each step in the process likely involves one or more bacterial and plant genes. The elucidation of the genetics of nodulation is uncovering significant information on bacterial and plant development, as well as serving as a model for the investigation of other plant-microbial interactions.

This review will exclusively cover the genetics of rhizobia. It should be remembered, however, that the genetic determinants of the plant host are essential for successful nodule formation. Host functions also play crucial roles in modulating the efficiency and utility of the nitrogen fixing symbiosis.

The general class of genes in rhizobia involved in nodule formation and function are sometimes referred to as *sym* (for

0097–6156/88/0362–0262$06.00/0

symbiotic) genes. The sym genes are divided into three broad categories: nif, fix, and nod. The distinction between these classes of genes is not always clear. In addition, it should be noted that the nomenclature to be applied to the sym genes has not been universally agreed upon. For the purpose of this review, nif genes will be those that are responsible for construction of the nitrogen fixing enzyme, nitrogenase (ie. nifKDH) and all other genes identified that are analogous to nif genes already identified in Klebsiella pneumoniae (1). The fix genes are those genes necessary for nitrogen fixation but not readily comparable to the nif genes of K. pneumoniae. The nod genes are those genes involved in the formation of the nodule.

Genetics of Rhizobia

Rhizobium and Bradyrhizobium species are Gram-negative, aerobic, soil bacteria that possess the unique ability to nodulate plants and fix nitrogen. The formal distinction between fast-growing Rhizobium and slow-growing Bradyrhizobium is a recent one (2). The distinctive differences between the two taxonomical divergent genera are listed in Table I. Studies of the molecular genetics of rhizobia have largely utilized fast-growing Rhizobium species primarily because most of the essential sym genes are conveniently located on plasmids in these bacteria (3). The sym genes are assumed to be chromosomally encoded in Bradyrhizobium species due to the fact that plasmids encoding symbiotic function have never been reported.

Not every plant species is infected by any given rhizobia. Most leguminous plants (4) can be nodulated and nodulation of one non-legume (Parasponia) has been reported (5). Any given plant species can only be nodulated by a relatively narrow group of strains of rhizobia; that is, the interaction of rhizobia and a respective host exhibits specificity. This led early to the formation of cross-inoculation groups (eg. B. japonicum-soybean, R. meliloti-alfalfa, R. phaseoli-bean, R. leguminosarum-pea, and R. trifolii-clover). It is now recognized, however, that a stringent species specificity does not exist with some cross-infection occuring usually with less efficiency. In other cases, a single strain of rhizobia may show surprising promiscuity. However, genetic and physiological studies have now unequivocally shown that rhizobia do possess functions which optimize interactions with only specific host plant species (reviewed in 6, 7). The genetic basis for the specific interaction between rhizobia and host plants is of obvious interest and may provide insights into the specificity of other plant-microbe interactions (eg. plant pathogens).

The Genetics of Nodulation. Nodule initiation and development is a complex process involving several bacterial genes. The bacteria present free-living in soil must colonize the root of a germinating seedling. Attachment of the compatible rhizobia to the host root surface is likely one of the first committed steps to nodulation. Once attached, initial invasion occurs through root hair cells that become tightly curled in response to the bacteria. Curling of the root hair may allow concentration of the bacteria within the curl so that the root hair wall is penetrated. The bacteria enter the root

Table I. General Phenotypic Differences between Fast-Growing Rhizobium and Slow-Growing Bradyrhizobium Species*

Phenotype	Rhizobium	Bradyrhizobium
Doubling Time	2-4 h	7-20 h
Mannitol Medium	Acid Production	Alkaline Production
Disaccharide utilization	(+)	(-)
Phosphogluconate dehydrogenase	(+)	(-)
Chemoautotrophy	(-)	(+)
Antibiotic sensitivity	5-50 ug/ml	100-1000 ug/ml
Free-living N_2 Fixation (acetylene reduction)	(-)	(+)
Flagellation	petrichous	subpolar
Guanine/Cytosine	50-63.1%	62.8-65.5%
Salt Tolerance	0.2-0.4 M	0.1 M

* Adapted from review by Prakash and Atherly (3). This table represents a generalization; specific strains may differ in one or more traits.

hair cell enclosed in a growing cellulose thread (ie. infection thread) that penetrates and ramifies into the cortex of the plant. In advance of the growing thread, plant cortical cells divide and the growing nodule emerges through the root epidermal layers. The bacteria are released from the infection thread into plant cells surrounded by a membrane, the peribacteroid membrane. The bacteria within the plant cells fix gaseous nitrogen and are morphologically and physiologically distinct from those cells that initiated the infection (Figure 1). The plant derives a steady supply of fixed nitrogen from the symbiosis while the bacteria persist and multiply within the soil environment due to the intimate association with the plant.

The Common Nodulation Genes. Long et al. (8) were the first to isolate genes essential for nodule formation. A gene bank prepared from wild type *R. meliloti* was used to complement a Tn5 induced Nod⁻ mutant. The advancement of the genetic understanding of nodulation has been rapid since this initial report. Work on fast-growing *R. meliloti*, *R. leguminosarum*, and *R. trifolii* provides the bulk of information in this area (reviewed in 6, 7, 9). However, research on the genetics of nodulation in *Bradyrhizobium* species is also developing rapidly (eg. 10-13). The nodulation genes identified in *Rhizobium* species are encoded on large plasmids ranging in size from approximately 150 kb (as in *R. leguminosarum*, 14-22) up to 1000 kb or larger in *R. meliloti* (ie. megaplasmids, 23-27). Some *R. meliloti* strains possess two megaplasmids (28). Although one of the plasmids appears to encode most of the symbiotic

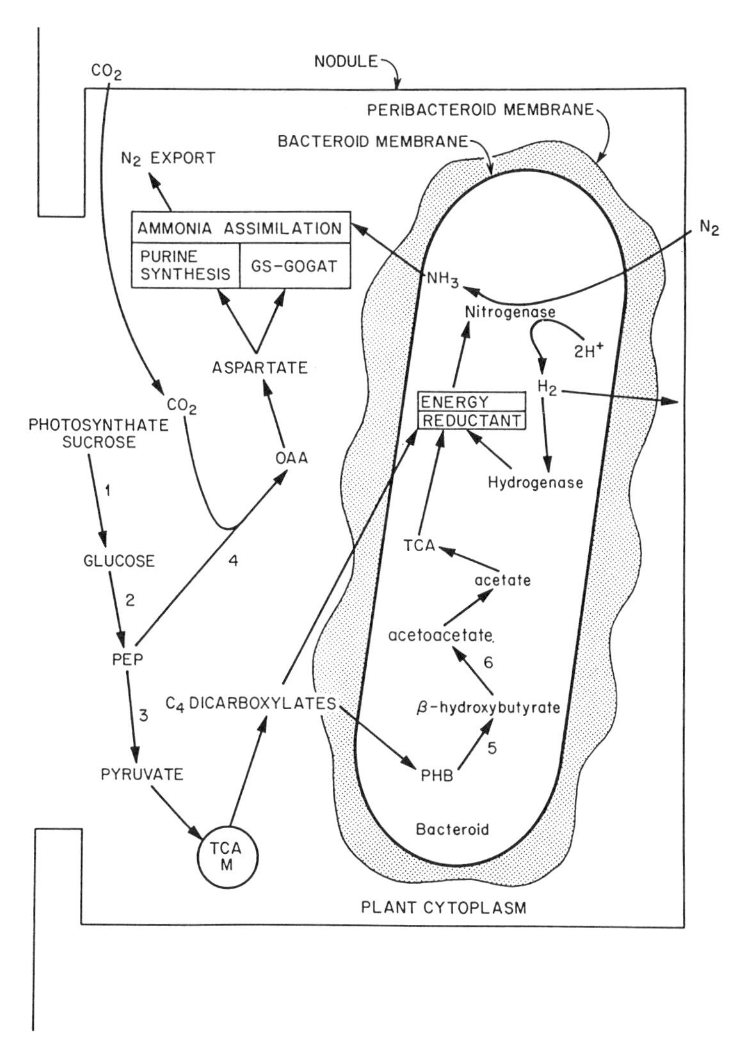

Figure 1. Simplified scheme showing a possible flow of carbon, nitrogen, and energy within a leguminous root nodule. The figure has been drawn to accentuate the mutual interaction of bacteria and host. 1, invertase; 2, glycolysis; 3, pyruvate kinase; 4, PEP carboxylase; 5, depolymerase; 6, β-hydroxybutyrate dehydrogenase; OAA, oxaloacetate; PEP, phosphoenolpyruvate; M, mitochondrion; TCA, tricarboxylic acid cycle; purine synthesis, involved in ureide synthesis for nitrogen transport in some legumes; GS-GOGAT, glutamine synthetase-glutamate synthase pathway for amide synthesis for nitrogen transport in some legumes. (Reprinted with permission from ref. 6. Copyright 1986 CRC Press Inc.)

functions, recent evidence indicates that the other megaplasmid is also essential for efficient nodulation (29, 30).

The first nodulation genes clearly identified were those found to encode root hair curling ability (Hac, ex. 8, 23, 31). The appearance of curled root hairs is one of the first microscopic indications of successful infection by rhizobia. In most of the Rhizobium and Bradyrhizobium species examined, these root hair curling genes are clustered (reviewed in 6, 7, 9); six genes have been identified in this region (nodABCDIJ, 31-35). Recently, a potential seventh gene (nodK) has been identified as an open-reading-frame upstream of nodA in B. parasponia (36) and an apparently unrelated gene (ORF1) identified upstream of nodA in B. japonicum (37). Hybridization and sequencing studies of the nodABCDIJ genes have demonstrated significant homology among different Rhizobium and Bradyrhizobium species (10-12; 31-38). For example, the nodABC genes of R. meliloti are 69-72% homologous to the nodABC genes of R. leguminosarum (32). Due to the broad conservation of these genes, this region is generally referred to as the "common" Nod region.

The function of the nodABC or IJ genes is presently unknown. The nodD gene likely encodes a positive regulator and will be discussed in greater detail below (39). The hydrophilicity plots of the DNA sequence of nodABC indicate that the encoded proteins contain hydrophobic regions (32, 34, 35) and, therefore, may be associated with membranes. Antibody to the nodC protein product from R. meliloti has been used to localize this protein in the membrane (33, 40). Antibody to nodC protein partially blocks nodulation by R. meliloti suggesting that the nodC protein is exposed on the bacterial surface (40). The nodA gene product has been localized in the bacterial cytosol (41). Comparison of the protein sequence of the nodI gene to a bank of E. coli protein sequences revealed some homology to the hisP protein (42). The hisP gene is involved in histidine transport. Greatest similarity between the nodI and hisP proteins was found to those regions of hisP known to be involved in ATP binding. Whether or not this sequence comparison is meaningful remains to be established.

The above studies give some suggestions as to the role the nodABCIJ genes may play in nodulation. Mutants defective in the nodIJ genes show, depending on the host, various levels of symbiotic capability (Wijffelman, personal communication). Therefore, these genes may not be absolutely essential for nodule formation. Mutants in the nodABC genes are invariably Nod$^-$. These genes are essential for nodule formation. One hypothesis for nodABC function is that these genes produce a bacterial signal compound that induces root hair curling and cortical cell division by the plant. For example, Van Brussel et al. (43) have shown that R. leguminosarum will induce thick-short roots (TSR) when inoculated onto plants of Vicia sativa subsp. nigra. R. leguminosarum lacking nodABC function do not cause this effect. R. leguminosarum possessing only the nodDABC genes, but lacking other Sym plasmid-encoded functions, can still produce the TSR phenotype. These data suggest that the nodABC genes are involved in synthesizing or controlling the synthesis of a factor which can affect root development.

Regulation of Common Nod Genes. The nodABC genes appear to be tightly regulated being induced only in the presence of the plant host. Mulligan and Long (39) showed that a nodC::lacZ fusion in R. meliloti was induced to form B-galactosidase only in the presence of alfalfa seed and root extracts. A functional nodD gene was essential for induction, suggesting that nodD is a positive regulatory gene. The nodD gene transcript is expressed constituitively (39). Similar findings have been reported for the nodABCD genes of R. trifolii (44), R. leguminosarum (45), and B. japonicum (Banfalvi and Stacey, unpublished results). The nodABC inducer found in plant exudates has been identified as a member of the flavonoid family of plant secondary metabolites (46-48). Each plant produces a family of flavonones, one of which acts as the primary inducer of the nodABC genes of the compatible rhizobia. For example, the inducer for the nodABC genes of R. meliloti has been identified as luteolin (5,7,3',4' tetrahydroxy flavonone) (46); 2,4-dihydroxy flavonone is the primary inducer of the genes in R. trifolii (47). There is now some evidence to suggest that the specificity for different flavonones is determined by the nodD gene (48). That is, the transcription of nodABC is induced to a differing degree by a spectrum of flavonones if different nodD genes are present (48). This fact may explain why some rhizobia possess more than one copy of nodD. For example, some strains of R. meliloti possess at least two apparently functional nodD genes (49). In contrast, R. leguminosarum appears to possess only one nodD gene (50). Recent work by Horvath et al. (51) shows that chimeric nodD genes constructed by recombinant DNA methods from the nodD genes of R. meliloti and R. sp. MPIK3030 interact with different inducers depending on which Rhizobium donated the C-terminal portion of the protein. This is the strongest evidence to support the notion that the nodD gene product directly interacts with the inducing chemical.

The DNA sequence of the promoter regions of the nod genes has now been determined in many cases (31-38). These data allow the comparison of sequence and the presumptive identification of conserved controlling regions. Rostas et al. (52) have recently performed these comparisons between the promoter regions of the various nod gene transcripts of R. meliloti. The promoter sequences all have a 47 bp sequence 200-250 bp upstream of the translational start. Deletion of this region abolished nodABC function (52). Within this 47 bp region, there is a highly conserved 25 bp sequence which has been termed the Nod Box. A Nod Box sequence has been found associated with most of the nod genes sequenced from both Rhizobium and Bradyrhizobium. This sequence has been postulated to be essential for nodD-dependent gene induction (52).

Synthesis of the nod gene inducers by plants takes place via the general phenyl propanoid metabolic pathway (Figure 2). This pathway appears to be intimately involved in plant-microbial interactions. For example, in addition to producing the inducer of nod genes in rhizobia, acetosyringone, produced by a branch metabolic pathway leading to lignin biosynthesis, has been shown to be an inducer of the vir genes in Agrobacterium tumefaciens (53, 54). The expression of vir genes is essential for the establishment of crown gall disease by A. tumefaciens. The phenyl propanoid pathway also leads to the synthesis of flavonoid derived phytoalexins in many plants (reviewed in 55). Phytoalexins are

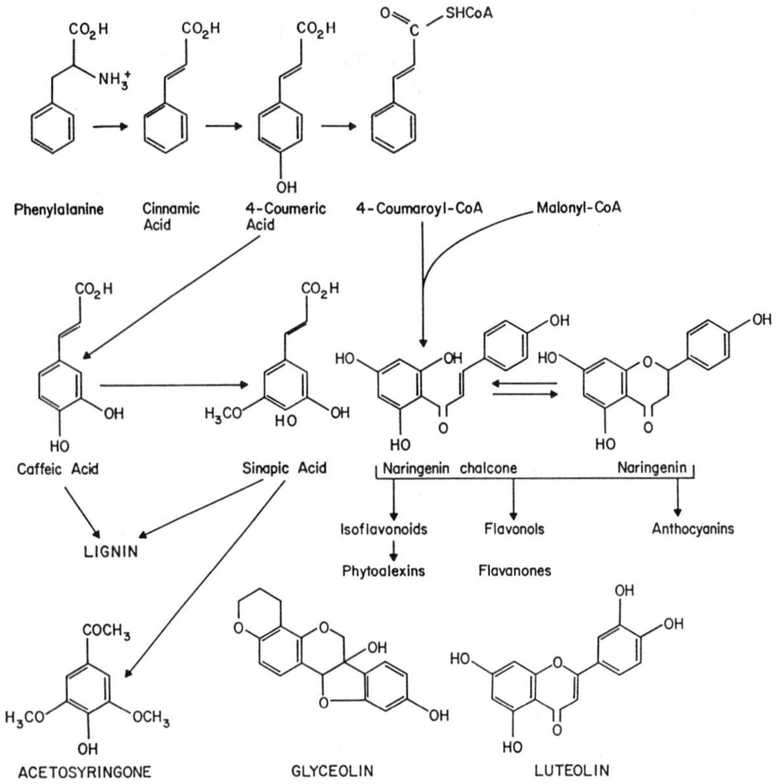

Figure 2. General phenyl propanoid pathway for secondary metabolism in plants. The pathway has been abbreviated and those products involved in plant-microbial interactions highlighted (see text).

antimicrobial chemicals produced by plants in response to pathogen attack (reviewed in 56). Phytoalexins are thought to be important in the plant defense response to pathogen invasion.

Genetics of Host Specificity. R. meliloti, R. trifolii, R. leguminosarum. In addition to the common nodulation genes discussed above, it is apparent that other genes for host specificity and additional functions are required for nodulation. Transfer of the cloned nodABCDIJ genes and surrounding DNA to Agrobacterium and Sym plasmid-cured rhizobia often results in nodule morphogenesis, whereas transfer of the nodABCDIJ genes alone does not (for example, 57, 58). These results indicate that the other genes necessary for nodule formation are closely linked to the common nodulation genes. Further genetic analysis located the genes necessary for host specificity (hsn) near the common nod genes in R. meliloti, R. trifolii, and R. leguminosarum. In R. meliloti, four genes have been identified as being involved in host specificity, nodE, F, G, H (59). Mutations in nodE, F, and G result in a phenotype showing a delay in nodule formation (59). Mutations in nodH result in a Nod$^-$ phenotype. These results would apparently indicate that only nodH is absolutely essential for nodulation. The functions of the nodEFGH gene products are unknown. However, in R. meliloti, this cluster of genes imposes host specificity on the root hair curling function encoded by the common nodulation genes (60). In the absence of nodEFGH functions, R. meliloti can induce root hair curling on non-host plants; in the presence of nodEFGH function, only alfalfa root hairs are curled by R. meliloti (60). Thus, the common nod genes and the host specificity genes interact. This interaction appears to be post-transcriptional in that the nodEFGH genes do not appear to influence the transcription of nodABC (61).

Genes that are likely analogous to nodEFGH have been reported in R. leguminosarum (31, 62) and R. trifolii (63, 64). Presently, there is a need to clarify the relationships between nod genes isolated from different species. The host specificity genes of R. leguminosarum and R. trifolii are immediately adjacent to the hac genes; unlike R. meliloti where the hac and hsn genes are separated by 7-8 kb (Figure 3). The regulation of the host specificity genes appears similar to that of the nodABC transcript. That is, transcription is nodD-dependent and inducible by plant-produced flavonones (44). At least ten genes are involved in the nodulation of alfalfa, peas, and clover by their respective symbionts (ie. nodA, B, C, D, E, F, G, H, I, and J or equivalents). These genes or perhaps a subset are all that are necessary to allow nodule formation when transferred into A. tumefaciens or Sym-plasmid-cured strains of rhizobia (57, 58, 65, 66). Considering the complexity of the nodulation process it is surprising that so few genes are essential. The picture is changing, however, to indicate that several other genes may affect nodule formation.

Rhizobia Possessing Broad Host Range. Some Rhizobium species and several Bradyrhizobium species have the capability to nodulate a wide array of plant hosts. For example, Rhizobium sp. strain MPIK3030 (a derivative of strain NGR234) can induce nodules on Vigna unguiculata, Macroptilium atropurpureum, Psophocarpus

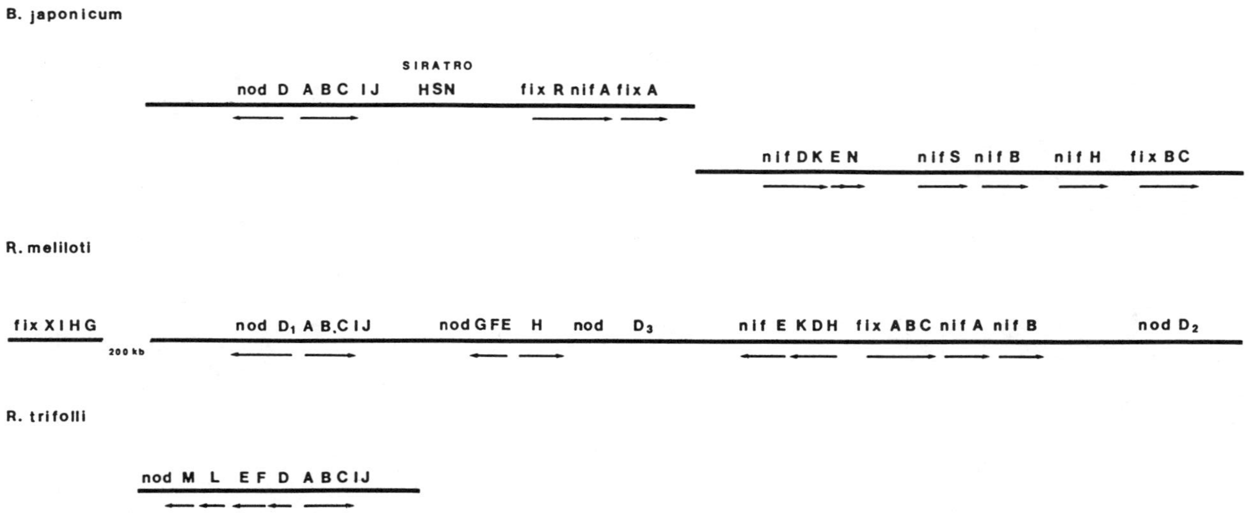

B. japonicum

SIRATRO

nod D A B C I J HSN fix R nif A fix A

nif D K E N nif S nif B nif H fix B C

R. meliloti

fix X I H G 200 kb nod D_1 A B C I J nod G F E H nod D_3 nif E K D H fix A B C nif A nif B nod D_2

R. trifolii

nod M L E F D A B C I J

Figure 3. Genetic maps of the *nod*, *nif*, and *fix* genes of *Bradyrhizobium japonicum* and *Rhizobium meliloti* and the *nod* genes of *Rhizobium trifolii*. The linkage between the *nod* and *nif KDH* genes of *B. japonicum* has not been established.

tetragonolobus, Glycine max, Medicago sativa, and a few others (67). Recently, a gene locus from R. sp. MPIK3030 (68) and R. sp. NGR234 (69) was cloned and shown to encode specificity for nodulation of siratro (M. atropurpureum). This locus is one of three host specificity regions identified by Broughton et al. (70) from strain MPIK3030. Each of the three loci identified appear to determine specificity for a different set of hosts. That is, this strain contains multiple host specificity loci which allow nodulation of a broad range of hosts.

A similar situation likely exists in Bradyrhizobium species. In general, Bradyrhizobium species have a broader host range than Rhizobium species. Nieuwkoop et al. (37) recently found homology in B. japonicum to the siratro host specificity genes isolated from R. sp. MPIK3030. Mutations in this region of the B. japonicum genome prevent nodulation of siratro but did not affect nodulation of soybean. Therefore, it is likely that B. japonicum possesses at least two distinct host specific loci for soybean and siratro, respectively.

Examples of Other Genes that Affect Nodulation. A considerable amount of information now points to the importance of the rhizobial cell surface and cell surface polysaccharides to successful nodulation (reviewed in 6, 56). This literature is too voluminous to review here. A number of recent mutants, however, have been isolated and partially characterized that appear to be affected in their cell surface. For example, Leigh et al. (71) and Finan et al. (72) have characterized mutants of R. meliloti that do not stain with the fluorescent dye calcofluor. These mutants have a phenotype in which the plant produces a nodular outgrowth of the root but the rhizobia are unable to induce infection thread formation. Bacteria are only found in intercellular spaces in the outermost layers of the root. These mutants have been classified as Fix^- but are obviously affected in early nodulation functions. This is an example where the distinction between fix and nod genes may not be clear. The defect in these mutants is apparently due to a mutation on the second (non-Sym) plasmid in R. meliloti (29, 30).

Mutants of an apparently similar phenotype to those of R. meliloti mentioned above have been found independently by Dylan et al. (73) and in R. phaseoli by Vander Bosch et al. (74). The mutants of R. meliloti isolated by Dylan et al. (73) were obtained by mutating a DNA region that showed homology to the chvB gene region of Agrobacterium tumefaciens. In A. tumefaciens, mutations in the chromosomal chvB locus result in an avirulent phenotype (75, 76). A. tumefaciens mutants defective in chvB function are defective in attachment to the plant surface and cannot synthesize cyclic B-2 glucan (77). Mutants of R. meliloti in the chvB locus appear to have a similar phenotype (73). Cyclic B-2 glucans have only been reported to be synthesized by members of the Rhizobiaceae (eg. Agrobacterium and Rhizobium). A report by Abe et al. (78) implicates cyclic B-2 glucan in infection thread formation. Treatment of clover roots with B-2 glucan resulted in an increase in the number of infection threads formed by the R. trifolii inoculant.

Genetics of Nitrogen Fixation. There are two general classes of genes (nif, fix) in Rhizobium and Bradyrhizobium which affect nitrogen fixation. Operationally, any mutant that forms a nodule but cannot fix nitrogen is termed Fix^- and the affected gene termed a fix gene. Subsequent analysis of this gene may show its analogy to a gene previously identified in K. pneumonia (nifQBALFMVSUXNEYKDHJ, reviewed in 79-81) and the gene is then signified by its nif designation. The reason for this distinction between nif and fix should be clear from the previously discussed exopolysaccharide mutants. These Fix^- mutants form a nodule but are obviously not mutated in genes specific for nitrogenase construction.

Nif Genes in Klebsiella Pneumoniae. The nifKD and nifH genes identified in K. pneumoniae are the structural genes for component I and II, respectively, of the nitrogenase enzyme (reviewed in 79-81). This enzyme contains an iron-molybdenum cofactor (FeMoCo) which is essential for enzymatic activity (82, 83). Several genes (nifQBNEV) are involved in the synthesis of FeMoCo (79-81). The nifM and nifSU gene products are necessary for component II and component I synthesis, respectively (80-81). The mode of action of these genes is unknown. The nifJ gene encodes a pyruvate:flavodoxin oxidoreductase which transfers electrons from pyruvate to the nifF gene product, flavodoxin (84, 85). The complete electron transfer chain from pyruvate to component I of nitrogenase has been reconstituted in vitro using purified components (85). A function has not been ascribed to nifX or Y.

The nifA gene product acts as a positive activator of transcription of all of the other nif genes except itself and nifL (86, 87). The nifL gene is transcribed with nifA and acts as a repressor of nif gene expression in the presence of O_2 (88). In addition to the nifA and L regulatory genes, regulation of the nif operon in K. pneumoniae requires two additional unlinked genes. Under conditions of NH_4^+ starvation, the ntrC and ntrA gene products are necessary to activate nifAL transcription (89-92).

The Nif and Fix Genes of Rhizobia. The nitrogenase genes (nif) of Rhizobium were first identified by DNA-DNA hybridization to the previously cloned genes from Klebsiella pneumoniae (e.g., 93, 94). Similar to the nod genes, the nif genes of Rhizobium species are found on plasmids; whereas in Bradyrhizobium species they are assumed to be chromosomally encoded (reviewed in 3, 95).

In Rhizobium species, the nifKDH genes are often, but not always, found to be transcribed as one unit (95). In some strains of R. phaseoli and R. fredii multiple copies of nifH have been found (96-98). In Bradyrhizobium japonicum, the nifKD and nifH genes are made on two distinct mRNAs (99). This also appears to be the case for B. parasponia (100) and a cowpea-nodulating Bradyrhizobium species (101). The separation of the nifDK and nifH genes may be a distinction between Bradyrhizobium and Rhizobium species.

The regulation of the nif genes in Rhizobium and Bradyrhizobium appears to be similar to that in Klebsiella pneumoniae. A nifA-like regulatory gene homologous to that in Klebsiella pneumoniae has been identified in R. meliloti (102-104), R. leguminosarum (105), B. japonicum (106), and R. trifolii (107). Sundaresan et al. (108)

showed that activation of the *R. meliloti nifH* promoter was dependent on *nifA*. A similar situation exists for the *nif* and *fix* genes identified in *B. japonicum* (106, 109). In contrast to *K. pneumoniae*, a functional *ntrC* gene does not appear to be essential for symbiotic nitrogen fixation in *R. meliloti* in that *ntrC*$^-$ strains remain Fix$^+$ (110).

In addition to the *nifKDH* and *nifA* genes, analogous genes to *K. pneumoniae nifB*, *S*, *E*, and *N* have been identified in *Rhizobium* and *Bradyrhizobium* species (111-117). In all cases, these genes have been identified by DNA sequence comparison. In addition to these genes of known or assumed function, a number of *fix* genes have been identified (i.e., *fixABCGHIX*, 110-117). However, not all of the *fix* gene regions identified by mutagenesis have been assigned letter designations (example 104, 118, 119). The function of most of the identified *fix* genes is unknown. Recent evidence, however, suggests that the *fixABC* genes may encode functions necessary for nitrogen fixation in all aerobic nitrogen-fixing bacteria (120). Hennecke and coworkers (120) have found homology to the *fixABC* genes of *B. japonicum* in DNA from *Azotobacter* and other aerobic diazotrophs. A possible function for the *fixABC* gene products is the synthesis of proteins necessary for coupling electron transport to nitrogenase.

Acknowledgments

The work of the author's laboratory reviewed here was supported in part by the U.S. Department of Agriculture grant 86-CRCR-1-2120 and by Public Health Service grant 1-R01 GM 33494-01A1 from the National Institutes of Health.

1. Roberts, G. P.; Brill, W. J. *Annu. Rev. Microbiol*. 1981, *35*, 207.
2. Jordan, D. C. *Int. J. Syst. Bacteriol*. 1982, *32*, 136.
3. Prakash, P. K.; Atherly, A. G. *Int. Rev. Cytol*. 1985, *104*, 1.
4. Allen, O. N.; Allen, E. K. *The Leguminosae*; The University of Wisconsin Press: Madison WI, 1981.
5. Trinick, M. J. *Nature* 1973, *244*, 459.
6. Hodgson, A. L. M.; Stacey, G. *CRC Crit. Rev. Biotechnol*. 1986, *4*, 1.
7. Long, S. R. *In Plant-Microbe Interactions*; Nester, E.; Kosuge, T., eds; MacMillan Press: New York, 1984; Vol. 1, p. 265.
8. Long, S. R.; Buikema, W.; Ausubel, F. M. *Nature* 1982, *298*, 485.
9. Evans, H. J.; Bottomley, P. J.; Newton, W. E. *Nitrogen Fixation Research Progress*; martinus Nijhoff Publ.: Dordrecht, The Netherlands, 1985.
10. Marvel, D. J.; Kuldau, G.; Hirsch, A.; Richards, E.; Torrey, J. G.; Ausubel, F. M. *Proc. Natl. Acad. Sci*. 1985, *82*, 5861.
11. Noti, J. D.; Dudas, B.; Szalay, A. A. *Proc. Natl. Acad. Sci*. 1985, *82*, 7379.
12. Russell, P.; Schell, M. G.; Nelson, K. K.; Halverson, L. J.; Sirotkin, K. M.; Stacey, G. *J. Bacteriol*. 1985, *164*, 1307.
13. Scott, D. B.; Chua, K.-Y.; Jarvis, B. D. W.; Pankhurst, C. E. *Mol. Gen. Genet*. 1985, *201*, 43.

14. Brewin, N. J.; Beringer, J. E.: Buchanan-Wollaston, A. V.; Johnston, A. W. B.; Hirsch, P. R. J. Gen. Microbiol. 1980, 116, 261.
15. Buchanan-Wollaston, A. J.; Beringer, J.; Brewin, N. J.; Hirsch, P. R.; Johnston, A. W. B. Mol. Gen. Genet. 1980, 178, 195.
16. DeJong, T. M.; Brewin, N. J.; Phillips, D. A. J. Gen. Microbiol. 1981, 124, 1.
17. Hirsch, P. R.; Van Montagu, M.; Johnston, A. W. B.; Brewin, N. J.; Schell, J. J. Gen. Microbiol. 1980, 120, 403.
18. Hooykas, P. J. J.; Snijdewit, F. G. M.; Schilperoot, B. A. Plasmid 1982, 8, 73.
19. Johnston, A. W. B.; Beringer, J.E .; Beynon, J. L.; Brewin, N.; Buchanan-Wollaston, A. V.; Hirsch, P. R. In Plasmids: Medical--Environmental, and Commercial Importance; Timmis, K. N.; Puhler, A., eds.; Elsevier: N. Holland, 1979.
20. Knol, A. J. M.; Hontelez, J. G. J.; Vanden Bos, R. C.; Van Kamnen, A. Nucl. Acids Res. 1980, 8, 4337.
21. ________________. J. Gen. Microbiol. 1982, 128, 1839.
22. Tichy, H. V., Lotz, W. FEMS Microbiol. Lett. 1981, 10, 203.
23. Banfalvi, Z.; Randhawa, G. S.; Kondorosi, E.; Kiss, A.; Kondorosi, A. Mol. Gen. Genet. 1983, 189, 129.
24. Buikema, W. J.; Long, S. R.; Brown, S. E.; Vanden Bos, R. C.; Earl, C.; Ausubel, F. M. J. Mol. Appl. Genet, 1983, 2, 249.
25. Burkhart, B.; Burkhart, H. J. J. Mol. Biol. 1984, 175, 213.
26. Djordjevic, M. A.; Zukos, W.; Shine, J.; Rolfe, B. G. J. Bacteriol. 1983, 156, 1035.
27. Kondorosi, A.; Kondorosi, E.; Pankhurst, C. E.; Broughton, W. J.; Banfalvi, Z. Mol. Gen. Genet. 1983, 188, 433.
28. Banfalvi, Z.; Kondorosi, E.; Kondorsi, A. Plasmid 1985, 13, 129.
29. Finan, T. M.; DeVos, G. F.; Signer, E. R. In Nitrogen Fixation Research Progress; Evans, H. J.; Bottomley, P. J.; Newton, W. E., eds.; Martinus Nijhoff Publ.: Dordrecht, The Netherlands, 1985; p. 135
30. Hynes, M. F.; Simon, R.; Muller, P.; Niehaus, K.; Labes, M.; Puhler, A. Mol. Gen. Genet. 1986, 202, 356.
31. Hombrecher, G.; Gotz, R.; Dibb, N. J.; Downie, A J.; Johnston, A. W. B.; Brewin, N. J. Mol. Gen. Genet. 1984, 194, 293.
32. Jacobs, T. W.; Egelhoff, T. T.; Long, S. R. J. Bacteriol. 1985, 162, 469.
33. Long, S. R.; Egelhoff, T. T.; Fischer, R. F.; Jacobs, T. W.; Mulligan, J. T. In Nitrogen Fixation Research Progress; Evans, H. J.; Bottomley, P. J.; Newton, W. E., eds.; Martinus Nijhoff Publ.: Dordrecht, The Netherlands, 1985; p. 87.
34. Rossen, L; Johnston, A. W. B.; Downie, J. A. Nucl. Acids Res. 1984, 12, 9497.
35. Torok, I.; Kondorosi, E.; Stepkowski, E.; Posfai, J.; Kondorosi, A. Nucl. Acids Res. 1984, 12, 9509.
36. Scott, K. F. Nucl. Acids Res. 1986, 14, 2905.
37. Nieuwkoop, A. J.; Banfalvi, Z.; Deshmane, N.; Gerhold, D.; Schell, M. G.; Sirotkin, K. M.; Stacey, G. J. Bacteriol. 1987, 169, 2631.
38. Egelhoff, T. T.; Long, S. R. J. Bacteriol. 1985, 164, 591.

39. Mulligan, J. T.; Long, S. R. Proc. Natl. Acad. Sci. 1985, 82, 6609.
40. Schmidt, J.; John, M.; Kondorosi, E.; Kondorosi, A.; Wieneke, U.; Schroder, G.; Schroder, J.; Schell, J. EMBO J. 1984, 3, 1705.
41. Schmidt, J.; John, M.; Wieneke, V.; Knussman, H.-D.; Schell, J. Proc. Natl. Acad. Sci. 1986, 83, 9581.
42. Higgins, C. F.; Hiles, I. D.; Salmond, G. P. C.; Gill, D. R.; Downie, J. A.; Evans, I. J.; Holland, I. B.; Gray, L.; Buckel, S. D.: Bell, A. W.; Hermondson, M. A. Nature 1986, 323, 448.
43. Van Brussell, A. A. N.; Zaut, S. A. J.; Canter-Cromers, H. C. J.; Wijffelman, C. A.; Pees, E.; Tak, T.; Lugtenberg, B. J. J. J. Bacteriol. 1986, 165, 517.
44. Innes, R. W.; Kuempel, P. L.; Plazinski, J.; Canter-Cremers, H.; Rolfe, B. G.; Djordejevic, M. A. Mol. Gen. Genet. 1985, 201, 426.
45. Rossen, L.; Shearman, C. A.; Johnston, A. W. B.; Downie, J. A. EMBO J. 1985, 4, 3369.
46. Peters, N. K.; Frost, J. W.; Long, S. R. Science 1986, 233, 977.
47. Redmond, J. WJ.; Bathey, M.; Djordjevic, M. AM.; Innes, R. W.; Kuempel, P. L.; Rolfe, B. G. Nature 1986, 323, 632.
48. Wijffelman, C. A.; Zaat, B. A. J.; Van Brussel, A. A.; Spaink, H. P.; Okker, R. J. H.; Pees, E.; de Maagel, B. A.; Lugtenberg, B. J. J. In Recognition in Microbe-Plant Symbiotic and Pathogenic Systems; Lugtenberg, B. J. J., ed.; Springer-Verlag: The Netherlands, 1986; p. 123.
49. Gottfert, M.; Horvath, B.; Kondorosi, E.; Rodriguez-Quinones, F.; Kondorosi, A. J. Mol. Biol. 1986, 191, 411.
50. Downie, J. A.; Knight, C. D.; Johnston, A. W. B.; Rossen, L. Mol. Gen. Genet. 1985, 198, 255.
51. Horvath, B.; Bachem, C. W. B.; Schell, J.; Kondorosi, A. EMBO J. 1987, 6, ______.
52. Rostas, K.; Kondorosi, E.; Horvath, B.; Simoncsits, A.; Kondorosi, A. Proc. Natl. Acad. Sci. 1985, 83, 1757.
53. Stachel, S. E.; Nester, E. W.; Zambayski, P. C. Proc. Natl. Acad. Sci. 1986, 83, 379.
54. Stachel, S. E.; Mossens, E.; Van Montagu, M.; Zambryski, P. C. Nature 1986, 318, 624.
55. Smith, D. A.; Banks, S. W. Phytochem. 1986, 25, 979.
56. Halverson, L. J.; Stacey, G. Microbiol. Rev. 1986, 50, 193.
57. Hirsch, A. M.; Drake, D.; Jacobs, T. W.; Long, S. R. J. Bacteriol. 1985, 161, 223.
58. Hirsch, A. M.; Wilson, K. J.; Jones, J. D. G.; Bang, M.; Walker, V. V.; Ausubel, F. M. J. Bacteriol. 1984, 158, 1133.
59. Kondorosi, E.; Banfalvi, Z.; Kondorosi, A. Mol. Gen. Genet. 1984, 193, 445.
60. Truchet, G.; Debello, F.; Vasse, J.; Terzaghi, B.; Garnerone, A.-M.; Rosenberg, C.; Batut, J.; Marllet, F.; Demarie, J. J. Bacteriol. 1985, 164, 1200.
61. Innes, R.; Djordejevic, M.; Rolfe, B.; Derarie, J.; Kuempel, P. L. In Molecular Genetics of Plant-Microbe Interactions; Verma, D. P. S.; Brisson, N., eds; Martinus Nijhoff Publ.: Dordrecht, The Netherlands, 1987; p. 229.

62. Downie, J. A.; Hombrecher, G.; Ma, Q.-S.; Knight, C. D.; Wells, B.; Johnston, A. W. B. Mol. Gen. Genet. 1983, 190, 359.
63. Djordejevic, M. A.; Schofield, P. R.; Rolfe, B. G. Mol. Gen. Genet. 1985, 200, 463.
64. Schofield, P. R.; Ridge, R. W.; Rolfe, B. G.; Shine, J.; Watson, J. M. Plant Mol. Biol. 1984, 3, 3.
65. Truchet, G.; Rosenburg, C.; Vasser, J.; Juillet, J.-S.; Canut, S.; Denairie, J. J. Bacteriol. 1984, 145, 1063.
66. Wong, C. H.; Pankhurst, C. E.; Kondorosi, A.; Broughton, W. J. J. Cell Biol. 1983, 97, 787.
67. Trinick, M. J. J. Appl. Bacteriol. 1980, 49, 39.
68. Bachem, C. W. B.; Banfalvi, Z.; Kondorosi, E.; Schell, J.; Kondorosi, A. Mol. Gen. Genet. 1986, 203, 43.
69. Bassam, B. J.; Rolfe, B. G.; Djordjevic, M. A. Mol. Gen. Genet. 1986, 203, 49.
70. Broughton, W. J.; Wong, C. H.; Lewin, A.; Samrey, U.; Myint, H.; Meyer, Z. A., H.; Dowling, D. N.; Simon, R. J. Cell Biol. 1986, 102, 1173.
71. Leigh, J. A.; Signer, E. R.; Walker, G. C. Proc. Natl. Acad. Sci. 1985, 82, 6231.
72. Finan, T. M.; Hirsch, A. M.; Leigh, J. A.; Johansen, E.; Kuldau, G. A.; Deegan, S.; Walker, G. C.; Signer, E. R. Cell 1985, 40, 869.
73. Dylan, T.; Ditta, G.; Kashyap, L.; Douglas, C.; Yanofsky, M.; Nester, E.; Helinski, D. In Nitrogen Fixation Research Progress; Evans, H.; Bottomley, P. J.; Newton, W. E., eds.; Martinus Nijhoff Publ.: Dordrecht, The Netherlands, 1985; p. 134.
74. Vandenbosch, K. A.; Noel, K. D.; Kaneko, Y.; Newcomb, E. H. J. Bacteriol. 1985, 162, 950.
75. Douglas, C. J.; Staneloni, R. J.; Rubin, R. A.; Nester, E. W. J. Bacteriol. 1985, 161, 850.
76. Douglas, C. J.; Helperin, W.; Nester, E. W. J. Bacteriol. 1982, 152, 1265.
77. Puvanesarajah, V.; Schell, F. M.; Stacey, G.; Douglas, C. J.; Nester, E. W. J. Bacteriol. 1985, 164, 102.
78. Abe, M.; Anumara, A.; Higashi, S. Plant Soil 1982, 64, 315.
79. Dixon, R. A.; Buck, M.; Drummond, M.; Hawkes, T.; Khan, H.; MacFarlane, S.; Merrick, M.; Postgate, J. R. Plant Soil 1986, 90, 225.
80. Orme -Johnson, W. H. Ann. Rev. Biophys. Chem. 1985, 14, 419.
82. Shah, V. K.; Brill, W. J. Proc. Natl. Acad. Sci. 1977, 74, 3249.
83. Shan, V. K.; Chisnell, J. R.; Brill, W. J. Biochem. Biophys. Res. Commun. 1978, 81, 232.
84. Nieva-Gomez, D.; Roberts, G. P.; Klevickis, S.; Brill, W. J. Proc. Natl. Acad. Sci. 1980, 77, 2555.
85. Shah, V. K.; Stacey, G; Brill, W. J. J. Biol. Chem. 1983, 258, 12064.
86. Buchanan-Wollaston, V.; Cannon, M. C.; Beynon, J. L.; Cannon, F. C. Nature 1981, 294, 776.
87. Roberts, G. P.; Brill, W. J. J. Bacteriol. 1980, 144, 210.
88. Merrick, M., Hill, S.; Hennecke, H.; Hahn, M.; Dixon, R.; Kennedy, C. Mol. Gen. Genet. 1982, 185, 75.

89. Alvarez-Morales, A.; Dixon, R.; Merrick, M. EMBO J. 1984, 3, 501.
90. Drummond, M.; Clements, J.; Merrick, M.; Dixon, R. Nature 1983, 301, 302.
91. Merrick, M. EMBO J. 1983, 2, 39.
92. Ow, D. W.; Ausubel, F. M. Nature 1983, 30, 307.
93. Hennecke, H. Nature 1981, 291, 354.
94. Ruvkin, G. B.; Ausubel, F. M. Proc. Natl. Acad. Sci. 1980, 77, 191.
95. Ausubel, F. M. In Microbial Development; Losick, R.; Shapiro, L., eds.; Cold Spring Harbor, New York, 1984; p. 275
96. Prakash, R. K.; Atherly, A. G. J. Bacteriol. 1984, 160, 785.
97. Quinto, C.; de la Vega, H.; Flores, M.; Leemans, J.; Cevalos, M. A.; Pardo, M. A.; Ma, R. A.; Girand, D. L.; Calva, E.; Palacios, R. Proc. Natl. Acad. Sci. 1985, 82, 1170.
98. Quinto, C.; de la Vega, H.; Flores, M.; Fernandez, L.; Ballate, T.; Soberon, G.; Palacios, R. Nature 1982, 299, 724.
99. Kaluza, K.; Fuhrman, M.; Hahn, M.; Regensburger, B.; Hennecke, H. J. Bacteriol. 1983, 155, 915.
100. Scott, K. F.; Rolfe, B. G.; Shine, J. DNA 1983, 2, 141.r
101. Yun, A. C.; Szalay, A. A. Proc. Natl. Acad. Sci. 1984, 81, 7358.
102. Szeto, W. W.; Zimmerman, J. L.; Sunderasan, V.; Ausubel, F. M. Cell 1984, 36, 1035.
103. Weber, G.; Reilander, H.; Puhler, A. EMBO J. 1985, 4, 2751.
104. Zimmerman, J. L.; Sezto, W. W.; Ausubel, F. M. J. Bacteriol. 1983, 156, 1025.
105. Downie, J. A.; Ma, Q.-S.; Knight, C. D.; Hombrecher, G.; Johnston, A. W. B. EMBO J. 1983, 2, 947.
106. Fischer, H.-M.; Alvarez-Morales, A.; Hennecke, H. EMBO J. 1986, 5, 1165.
107. Scott, D. B.; Court, C. B.; Ronson, C. W.; Scott, K. F.; Watson, J. M.; Schofield, P. R.; Shine, J. Arch. Microbiol. 1984, 151, 157.
108. Sundaresan, V.; Jones, J. D.; Ow, D. W.; Ausubel, F. M. Nature 1983, 301, 728.
109. Alvarez-Morales, A.; Hennecke, H. Mol. Gen. Genet. 1985, 199, 306.
110. Ausubel, F. M.; Buikema, W. J.; Earl, C. D.; Klingensmith, J. A.; Nixon, B.; Szeto, W. W. In Nitrogen Fixation Research Progress; Evans, H. J.; Bottomley, P. J.; Newton, W. E., eds.; Martinus Nijhoff Publ.: Dordrecht, The Netherlands, 1985; p. 165.
111. Hennecke, H.; Fischer, H.-M.; Ebeling, S.; Guber, M.; Thony, B.; Gottfert, M.; Lamb, J.; Hahn, M.; Ramseier, T.; Regensberger, B.; Alvarez-Morales, A.; Studer, D. In Molecular Genetics of Plant-Microbe Interactions; Verma, D. P. S.; Brisson, N., eds.; Martinus Nijhoff Publ.: Dordrecht, The Netherlands, 1987; p. 191.
112. Fuhrman, M.; Fischer, H.-M.; Hennecke, H. Mol. Gen. Genet. 1985, 199, 315.
113. Hennecke, H.; Alvarez-Morales, A.; Betancourt-Alvarez, M.; Ebeling, S.; Filger, M.; Fischer, H.-M.; Gubler, M.; Hahn, M.; Kaluza, K.; Lamb, J.

W.; Meyer, L.; Regensburger, B.; Studer, D.; Weber, J. In Nitrogen Fixation Research Progress; Evans, H. J.; Bottomley, P. J.; Newton, W. E.; eds.; Martinus Nijhoff Publ.: Dordrecht, The Netherlands, 1985; p. 157.
114. Norel, F.; Desnoues, N.; Emerich, C. Mol. Gen. Genet. 1985, 199, 352.
115. Noti, J. D.; Yun, A. C.; Folkerts, O.; Szalay, A. A. In Molecular Genetics of Plant Microbe Interactions; Verma, D. P. S.; Brisson, N., eds.; Martinus Nijhoff Publ.: Dordrecht, The Netherlands, 1987; p. 202.
116. Rossen, L.; Ma, Q.-S.; Mudd, E. A.; Johnston, A. W. B.; Downie, J. A. Nucl. Acids Res. 1984, 12, 7123.
117. Weber, G.; Aquillar, O. M.; Gronemeier, B.; Reilander, H.; Puhler, A. In Advances in Molecular Genetics of the Bacteria-Plant Interaction; Szalay, A. A.; Legocki, R. P., eds.; Cornell Univ. Publ.: Ithaca, New York, 1985; p. 13.
118. Rostas, K.; Sista, P. R.; Stanley, J.; Verma, D. P. S. Mol. Gen. Genet. 1984, 197, 230.
119. So, J.-S.; Hodgson, A. L. M.; Haugland, R.; Leavitt, M.; Banfalvi, Z.; Nieuwkoop, A. J.; Stacey, G. Mol. Gen. Genet. 1987, 207, 15.
120. Gubler, M.; Hennecke, H. FEBS Lett. 1986, 200, 186.

RECEIVED July 19, 1987

Chapter 23

Genetic Engineering of Nuclear-Encoded Components of the Photosynthetic Apparatus in *Arabidopsis*

Michael P. Timko [1,2], Lydia Herdies [3], Eleonor de Almeida [3], Anthony R. Cashmore [2,4], Jan Leemans [3], and Enno Krebbers [3]

[1]Department of Biology, University of Virginia, Charlottesville, VA 22903
[2]Laboratory of Cell Biology, Rockefeller University, New York, NY 10021
[3]Plant Molecular Biology Laboratory, Plant Genetic Systems, NV, Ghent, Belgium

Light regulates the coordinate transcription of nuclear genes involved in the development of the photosynthetic apparatus. We have been investigating the mechanism involved in the tissue-specific and photoregulated expression of genes encoding the small subunit (SSU) polypeptides of ribulose-1, 5-bisphosphate carboxylase in the crucifer Arabidopsis thalliana. We demonstrate that the SSU polypeptides are encoded by a small family within the nuclear genome of this plant and describe the structure and organization of these genes. The results of initial experiments using the promoter of one of these genes (rbcS 1A) to express chimaeric genes in transgenic plants are presented. These studies provide the basis for future study the nature and location of nucleotide sequences required for properly regulated expression and demonstrate a system for the expression of altered genetic information in plant cells.

Over the past decade considerable advancement has been made in our understanding of the molecular and genetic basis for the control of plant cellular growth and differentiation. A major contributory factor to this increased knowledge has been the development of recombinant DNA tools for the identification, isolation and manipulation of plant genes. These techniques coupled with the refinement of procedures for the introduction of foreign, or altered, genetic information into plant genomes and the ability to regenerate intact transgenic plants from single cells has permitted the in depth analysis of many important physiological processes. Among the areas of research which have received particular attention are examinations of the control

[4]Current address: Plant Science Institute, Department of Biology, University of Pennsylvania, Philadelphia, PA 19104

0097-6156/88/0362-0279$06.00/0

mechanisms active in the regulation of gene expression involved in the development and function of the photosynthetic apparatus. The focus of these studies has been the development of experimental systems for identifying the biochemical and molecular genetic factors which define phenotypically assayable photosynthetic parameters. Their goal is the later manipulation of such factors for the improvement of the photosynthetic apparatus at a genetic level. The potential benefits to man from the successful manipulation of plant genomes are both varied and numerous.

Photoregulation of plant gene expression

The development and sustained growth of most higher plants is dependent on the quantity and quality of light. Light serves as both an activator and modulator of plant morphogenic responses (1-2). The basis for such responses has been correlated to selective alterations in gene expression, as well as effects on cellular activities at the translational and post-translational level. One of the various photomorphogenic responses, the light-dependent transformation of a proplastid-etioplast into a mature photosynthetically competent chloroplast has intrigued researchers. The dramatic changes observed during plastid photomorphogenesis are controlled by at least three distinct photoreceptors: protochlorophyll(ide), phytochrome, and a poorly characterized blue-light receptor (2-4). By what mechanism the activities of these various photoreceptors are coordinated, and to what degree each contribute to plastid photomorphogenesis is unknown. These interrelationships are further complicated by the fact that in addition to multiple photoreceptors, plastid photomorphogenesis is dependent upon the coordinate expression of genes located in both the plastid and nuclear genomes (5).

Ribulose-1,5-bisphosphate carboxylase (RuBPCase) is the bifunctional enzyme which catalyzes the initial carbon dioxide fixation step in the Calvin cycle and participates as an oxygenase in photorespiration. It is a multimeric protein consisting of eight small subunit (SSU) and eight large subunit (LSU) polypeptides of 14 and 55 kD, respectively, and it is located exclusively in the chloroplast stroma. Among the major polypeptide products of light-induced nuclear gene expression are the SSU polypeptides. These polypeptides are formed on cytoplasmic polysomes as higher molecular mass precursors containing amino-terminal extensions termed transit peptides (6-8). The precursors are then imported into the chloroplast by an energy-dependent mechanism (9) subsequent to which the transit peptide is proteolytically removed (8). Following processing, SSU polypeptides are assembled into the RuBPCase holoenzyme with LSU polypeptides which are encoded in the chloroplast genome and are synthesized on chloroplast ribosomes (5).

A number of independent laboratories have demonstrated that the SSU polypeptides are encoded by multigene families within the nuclear genome in both dicot and monocot species (10-18, 21). These gene families (designated rbcS) vary to a considerable degree in size (2 to 12 members) and organization among the various plant species. In some plant species, the linkage and chromosomal location of the various members of this gene family have now been determined (16, 19-20). Comparisons of the nucleotide sequences and predicted amino acid sequences of the various rbcS gene family members have revealed substantial intraspecific conservation at the nucleotide and amino acid sequence level within the coding portion of these genes, as well as interspecific homologies at the protein sequence level (14, 16, 21).

Transcription of rbcS gene family members is regulated by light and in a cell-specific manner. The rbcS genes are expressed in photosynthetic leaf tissue, stem and floral parts, but not in roots or other storage tissues (22). Little or no translatable mRNAs encoding these polypeptides are detected in dark-grown, or etiolated plant tissues. Exposure to light, however results in a rapid accumulation of mRNAs encoding these polypeptides (see ref. 2). Photoregulted rbcS gene expression is mediated by phytochrome (23). In addition, the involvement of unidentified blue-light receptors active in the regulation of rbcS gene expression at some developmental stages and in certain cell types has been reported (27-28). Transcriptional run-off experiments using isolated nuclei indicate that light-regulation of rbcS gene expression occurs at the level of transcription (24-25).

The transcriptional control of rcS gene expression was examined and the nucleotide sequences required for photoregulated and cell-specific expression were identified within the promoters of these genes. These studies used gene transfer systems based upon the T-DNA of *Agrobacterium tumefaciens* (21, 36-38, 42). Furthermore, in the case of the pea rbcS genes, this expression involves enhancer-like elements residing various distances upstream of the TATA-box (40-41).

We have been investigating the mechanism and factors necessary for the photoregulation and cell-specific expression of rbcS genes in the crucifer *Arabidopsis thaliana*. We report preliminary evidence that in this crucifer, the SSU polypeptides are encoded by a small family within the nuclear genome. We describe the structure and organization of these genes. Finally, we discuss the results of initial studies in which the promoter of one of these genes was used in the construction and expression of chimaeric genes in plant cells. These experiments are directed toward establishing a system to investigate the sequences necessary for regulated rbcS gene expression, and to establish a method for the regulated expression of altered, or substitute gene products in plant cells.

Development of Arabidopsis as a system for studying rbcS gene expression

Current trends in plant molecular biology have led to the necessity of developing systems which allow the rapid identification, isolation and analysis of genetic traits. The crucifer *Arabidopsis* has recently gained increasing favor among plant biologists as a tool for the study of gene structure and regulation (see ref. 29). Of the many advantages *Arabidopsis* provides to the study of gene regulation, its small genome size relative to other plant species, its relatively low content of repetitive DNA, and the availability of well defined linkage groups assigned to each of its five chromosomes are most notable. Its ease of culture in the laboratory and short generation time permit the rapid analysis of mutants. Furthemore, this crucifer lends itself to Agrobacterium-mediated cell transformation (30-31), therefore allowing the study of introduced foreign genetic traits.

In order to characterize the rbcS gene family in Arabidopsis, a genomic DNA library prepared from A. thaliana var. Columbia consisting of restriction endonuclease EcoRI partial digestion products cloned in the lambda phage Sep 6 (obtained from E. Meyerowitz, CalTech) was screened by plaque hybridization (32) using a ^{32}P-labeled PstI insert of pSS15, a cDNA encoding most of the pea preSSU polypeptide (33). From our initial screening of approximately four genomic equivalents (34), seven independent recombinant phages were isolated and plaque purified. DNA was prepared from each recombinant phage and restriction endonuclease cleavage maps were generated. Each phage was found to contain approximately 10 - 11 Kb of cloned genomic DNA, and cloned DNA inserts in each independent phage were nearly identical. Southern blot hybridization studies of the phage DNA revealed the presence of two EcoRl restriction fragments of approximately 2.0 kb which contained sequences hybridizing specifically to radioactively-labeled pea rbcS cDNA sequences. The two hybridizing EcoRl restriction fragments were purified from a single phage (designated λ ats 2) and ligated into pUC 9 linearized by EcoRl, yielding plasmids pATS-3 and pATS-22, respectively. Subsequent restriction endonuclease cleavage and Southern blot hybridization analysis demonstrated that the 5'-most portion of the gene was encoded in sequences contained in pATS-3 and the 3' portions of the gene were encoded in pATS-22. This gene, designated 1A, was structurally characterized, its nucleotide sequence determined, and subsequently used in further studies described below.

The complete nucleotide sequence of the rbcS 1A gene and its immediate 5'- and 3'- flanking regions and the predicted amino acid sequence of the encoded polypeptide are shown in Figure 1. The rbcS 1A gene contains an open reading frame beginning with methionine that encodes a polypeptide corresponding to 180 amino acids in length; the first 55 amino acids comprise the transit peptide portion of the preSSU protein. The Arabidopsis rbcS 1A gene reading frame is interupted apparently by two intervening sequences of 106 bp (Intron I) and 136 bp (Intron II). The location of these introns have been defined by comparison to nucleotide and predicted amino acid sequences of other known rbcS genes (10-18) and by comparing intron boundry sequences with consensus splice site sequences observed in other plant genes (26). The positioning of the two introns within the Arabidopsis 1A gene is similar to that observed in rbcS genes of pea (11), soybean (15), and tomato (16), but differs from the rbcS genes reported in Nicotiana tabacum (17) and N. plumbaginifolia (21) which contain a third intron, and wheat (14) which lack the second intron. The first exon of rbcS 1A encodes the entire transit

```
ACCAGGCAAGTAAAATGAGCAAGCACCACTCCACCATCACACAATTTCACTCATAGATAACGATAAGATTCATGGAATTA

TCTTCCACGTGGCATTATTCCAGCGGTTCAAGCCGATAAGGGTCTCAACACCTCTCCTTAGGCCTTTGTGGCCGTTACCA

AGTAAAATTAACCTCACACATATCCACACTCAAAATCCAACGGTGTAGATCCTAGTCCACTTGAATCTCATGTATCCTAG

ACCCTCCGATCACTCCAAAGCTTGTTCTCATTGTTGTTATCATTATATATAGATGACCAAAGCACTAGACCAAACCTCAG

                          MET ALA SER SER MET LEU SER SER ALA THR MET VAL ALA
TCACACAAAGAGTAAAGAAGAACA  ATG GCT TCC TCT ATG CTC TCT TCC GCT ACT ATG GTT GCC

SER PRO ALA GLN ALA THR MET VAL ALA PRO PHE ASN GLY LEU LYS SER SER ALA ALA PHE
TCT CCG GCT CAG GCC ACT ATG GTC GCT CCT TTC AAC GGA CTT AAG TCC TCC GCT GCC TTC

PRO ALA THR ARG LYS ALA ASN ASN ASP ILE THR SER ILE THR SER ASN GLY GLY ARG VAL
CCA GCC ACC CGC AAG GCT AAC AAC GAC ATT ACT TCC ATC ACA AGC AAC GGC GGA AGA GTT

ASN CYS MET GLN
AAC TGC ATG CAG GTCATTTATATTTCTTCTTTCACTTTTTTATTATTCCATATGATTTTTTTCGGTTCTTTCTTC

                                        VAL TRP PRO PRO ILE GLY LYS LYS LYS
GAATCTACATAAACTAATATCATTGGAAAAATCGAAAAAATAG GTG TGG CCT CCG ATT GGA AAG AAG AAG

PHE GLU THR LEU SER TYR LEU PRO ASP LEU THR ASP THR GLU LEU ALA LYS GLU VAL ASP
TTT GAG ACT CTC TCT TAC CTT CCT GAC CTT ACC GAT TCC GAA TTG GCT AAG GAA GTT GAC

TYR LEU ILE ARG ASN LYS TRP ILE PRO CYS VAL GLU PHE GLU LEU GLU
TAC CTT ATC CGC AAC AAG TGG ATT CCT TGT GTG GAA TTC GAG TTG GAG GTAATTAAACAAAAT

TTAAACATCTATATAAACTAGCTAGATCTTAGGAAAATTTGGTTTAATATATTAGGATCTTGATTTATATAAACATGTT

                                          HIS GLY PHE VAL TYR ARG GLU HIS GLY
CAAAATGTTATCTGAGTGGTTTGTAACATGTGGTTTGTATAG CAC GGA TTT GTG TAC CGT GAG CAC GGT

ASN SER PRO GLY TYR TYR ASP GLY ARG TYR TRP THR MET TRP LYS LEU PRO LEU PHE GLY
AAC TCA CCC GGA TAC TAT GAT GGA CGG TAC TGG ACA ATG TGG AAG CCT CCC TTG TTC GGT

CYS THR ASP SER ALA GLN VAL LEU LYS GLU VAL GLU GLU CYS LYS LYS GLU TYR PRO ASN
TGC ACC GAC TCC GCT CAA GTG TTG AAG GAA GTG GAA GAG TGC AAG AAG GAG TAC CCC AAT

ALA PHE ILE ARG ILE ILE GLY PHE ASP ASN THR ARG GLN VAL GLN CYS ILE SER PHE VAL
GCC TTC ATT AGG ATC ATC GGA TTC GAC AAC ACC CGT CAA GTC CAG TGC ATC AGT TTC GTT

ALA TYR LYS PRO PRO SER PHE THR GLY
GCC TAC AAG CCA CCA AGC TTC ACC GGT TAATTTCCCTTTGCTTTTGTGTAAACCTCAAAACTTTATCCCC

CATCTTTGATTTTATCCCTTGTTTTTCTGCTTTTTTCTTCTTTCTTGGGTTTTAATTTCCGGAGTTAACGTTTGTTTTC

CGGTTTGCGAGACATATTCTATCGGATTCTCAACTGTCTGATGAAATAAATATGTAATGTTCTATAAGTCTTTCGATTT

GATATGCATATCAACAAAAAGAAAATAGGACAATGCGGCTACAAATATGAAATTTACAAGTTTAAGAACCATGAGTCGC
```

Figure 1. Nucleotide sequence and predicted amino acid sequence of the Arabidopsis rbcS 1A gene.

peptide and the first two amino acids of the mature protein. The two introns of the *Arabidopsis* rbcS 1A gene are unrelated in nucleotide sequence to each other and show no homology with the introns of other rbcS genes from other plant species.

Since there is considerable information available on the structure of SSU polypeptides from other organisms, it was of interest to compare the amino acid sequence of the rbcS 1A polypeptide with those of other plant species (Table I). In each of the pair-wise comparisons listed, the predicted amino acid sequence for the transit peptide and mature portion of the protein are compared separately. In cases where more than one amino acid sequence was available for a given species and used in the comparisons, the upper and lower homology levels are presented. The predicted amino acid sequence of the mature polypeptide encoded by rbcS 1A gene exhibits a moderate degree (approximately 70% or greater) of homology to mature SSU polypeptides from most other dicot species. As might be expected, considerably less homology is observed with monocot species. Within short domains of the mature protein, conservation in interspecific comparisons of 95% or higher are found. The relevance of such conservation remains to be elucidated. It has been suggested previously that such conservation may reflect a possible role of these regions in the interaction of SSU and LSU in the RuBPCase holoenzyme (14). Comparisons of transit peptides reveals considerably less conservation of amino acid sequence among species than observed within the mature protein. Transit peptides from all species show a strict conservation at the processing site of the precursor and an overall conservation in charge distribution and organization within the transit peptide (51).

Table I. Comparison of amino acid sequence differences between the *Arabidopsis* SSU polypeptide encoded by rbcS 1A and those of selected plant species

	% Divergence (transit peptide/mature protein)
Pisum sativum (43)	33.3/ 26.8
Nicotiana tabacum (17)	30.9/ 27.6
Nicotiana plumbaginifolia (22)	35.0/ 28.4
Lycopersicon esculentum (16)	31.5/ 29.3-30.8
Glycine max (20)	26.1/ 19.5-21.1
Lemna gibba (13)	39.6/ 27.2
Triticum aestivum (14)	47.2/ 35.2
Silene pratensis (53)	32.1/ 32.2

Sixty-four nucleotides upstream of the initiating methionine codon is the sequence CATTATATATAG. This sequence is highly conserved among all rbcS genes thus far examined (see ref. 36), and includes the 'TATA box' thought to specify the start site of transcription. Based upon the positioning of this regulatory element and by S1 nuclease mapping experiments we have defined an untranslated leader sequence of 23 nucleotides for the transcript derived from this gene. In addition to this well defined regulatory element, within the 5'-flanking portion of the rbcS 1A gene are several short, highly conserved elements previously noted within the upstream regions of rbcS genes of various plant species (12, 20-21, 41, 43). Since these genetic elements lie in regions demonstrated to be involved in conferring proper cell-specific and photoregulated expression in pea (40-41, 44), the presence of similar conserved sequence elements residing in the immediate 5'-flanking sequences of the Arabidopsis rbcS 1A gene and other rbcS genes suggests that they have functional significance. Further detailed studies such as those recently initiated and described below should be useful in determining their importance.

Previous studies of the complexity of rbcS gene families in other organisms have shown the SSU polypeptide to be encoded by small gene families ranging in size from 2-12 (10-18, 21). Based upon a comparison of the complexity of certain gene families in Arabidopsis (34-35) with their counterparts in other plant species it has been suggested that the small size of the Arabidopsis genome may be in part due to a relative reduction in the size of multigene families. It was therefore of considerable interest to us, to determine the number of genes encoding the SSU polypeptide in this organism. Southern hybridization analysis of total leaf genomic DNA was used to estimate the number of rbcS genes present within the Arabidopsis genome (Figure 2). Total DNA prepared from leaf tissues of light-grown plants was digested to completion with various restriction endonucleases, the cleaved DNA fragments were separated by electrophoresis through 1% agarose and blotted to nitrocellulose. Restriction fragments from either the 5' -most coding portions or 3' -most coding region of the rbcS 1A gene, radioactively-labeled to a high specific activity with ^{32}-P, were hybridized to DNA bound on the filters. Based upon the number and size of the individual hyridizing fragments observed in our experiments, we interpreted the results of these hybridization studies as indicating that the rbcS gene family in Arabidopsis consists of approximately four members.

Since the initial screening of the phage lambda genomic library resulted in the isolation of only one of these genes, a second genomic library consisting of restriction endonuclease Sau3A partial digestion products cloned in the lambda phage Charon 35 (kindly provided by Dirk Inze, Rijksuniversiteit Gent) was screened by plaque hybridization using the rbcS gene 1A coding sequences as hybridization probe. From this second series of screenings, we have successfully isolated phage containing the remaining three members of the Arabidopsis rbcS gene family, as well as reisolated phage containing the rbcs 1A gene. All three of the remaining rbcS genes were isolated on a contiguous piece of cloned genomic DNA, approximately 17 kb in length (Figure 2). The three additional rbcS genes contained on this fragment have been designated 1B, 2B and 3B. The intergenic regions are approximately 2.4 kb and 2.7 kb, respectively and all three genes are encoded in the same orientation. The three rbcS genes in this cluster are separated by at least 10 kb from the 1A gene; and at

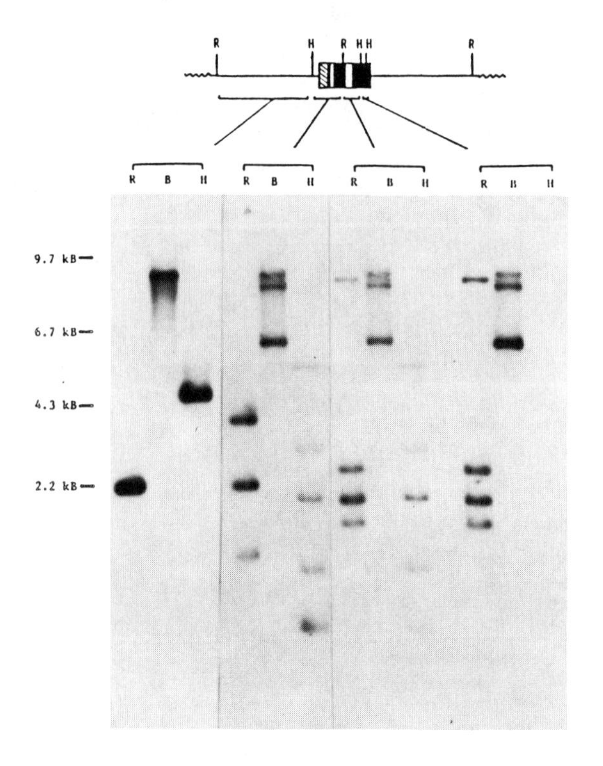

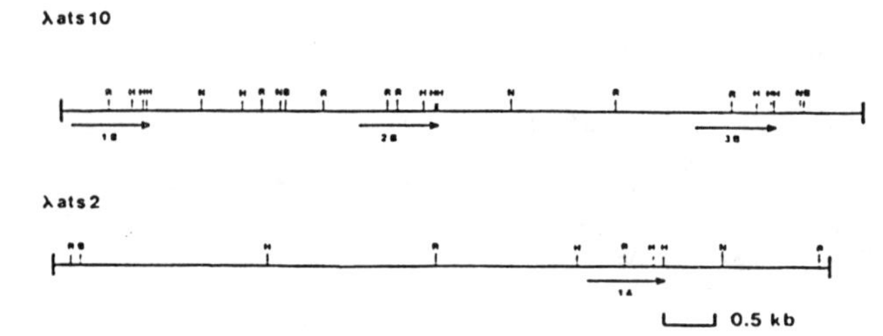

Figure 2. The organization and complexity of the rbcS gene family in Arabidopsis. Panel A shows an autoradiograph of Southern blot hybridization analysis of leaf total genomic DNA probed with radioactively-labeled restriction endonuclease fragments from 5' - and 3' - regions of the rbcS 1A gene. Bound Arabidopsis DNA was digested with the restriction endonucleases indicated (R - EcoRI; N - NDEI; H - HindIII; B - BamHI). The lower panel shows the partial restriction endonuclease cleavage maps of phage containing the four rbcS genes of Arabidopsis.

present we do not have any information supporting their linkage in the Arabidopsis genome. The hybridizing restriction fragments within the recombinant phage now isolated account for all of the hyridization bands observed in our genomic Southern hybridization analysis (see Figure 2).

Individual fragments of phage DNA were subcloned into pUC vectors and the nucleotide sequence of relevant portion of the genes were determined. Analysis of the nucleotide sequences of the rbcS 1B, 2B and 3B genes revealed that the three genes are distinct in nucleotide sequence from each other, as well as from the rbcS 1A gene (E. Krebbers, J.L. and M.P.T., manuscript in preparation). Pair-wise comparisons of the nucleotide sequences within the coding portions of rbcS 1B, 2B and 3B revealed that these genes are more similar to each other, than any individual gene from this cluster is to rbcS 1A. A notable difference among the two classes of genes (rbcS 1B - 3B versus 1A) is the length of the encoded protein. Genes rbcS 1B, 2B and 3B each encode a mature SSU polypeptide which is one amino acid residue longer at the carboxy-terminus than the polypeptide encoded by 1A. The amino acid sequence of the transit peptide encoded by the four Arabidopsis rbcS genes differs at one or more residues depending on which two sequences are compared.

Whether all four rbcS genes present within the Arabidopsis genome are expressed and to what extent each member contributes to the overall levels of rbcS mRNA levels present in various photosynthetic tissues is not known at this time. The four Arabidopsis rbcS genes show regions of considerable homology within the immediate 5' and 3' untranslated flanking portions of the individual genes; however, the presence of small insertions and/or deletions and duplications within these portions of the four genes will make it possible to distinguish the individual transcripts if present. We are currently investigating this point. It is clear, however, from our transcript mapping studies discussed briefly above, that gene 1A is expressed in light grown leaf tissues.

The use of the Arabidopsis rbcS 1A gene promoter for the expression of chimaeric genes

The future successful manipulation of plant gene expression requires both detailed information on the nature and location of the nucleotide sequences required for proper transcriptional regulation and the ability to use such information to express altered, or foreign genetic information in plant cells. We have examined the potential of the rbcS 1A gene promoter as model for the study of transcriptional regulation and for the properly regulated, high level expression of novel genetic information in transgenic plants.

In order to investigate these questions, we adopted an approach previously successfully employed both in our, as well as other laboratories (21, 36-38, 42). Chimaeric genes were constructed in which sequences from the 5' -flanking region of the Arabidopsis rbcS 1A gene were fused to the coding region of the bacterial neomycin phosphotransferase II (npt II) gene from the transposon Tn5. In our initial studies, two chimaeric gene fusions were prepared. In the first construction, designated pGS1400, an approximately 1550 bp fragment including 5' -flanking sequences, the untranslated leader sequence and the initiating methionine codon of the preSSU encoded by the rbcS gene 1A are fused in frame to the coding region of the npt II polypeptide. In the second construction, designated pGS1401, the same _Arabidopsis_ sequences as in pGS1400 plus sequences encoding the transit peptide of the preSSU are fused in frame to the npt II coding region. The fusion protein encoded by this chimaeric gene contains a regenerated CYS-MET processing site at the fusion point between the transit peptide and the npt II coding region (E. de Almeida, et al., manuscript in preparation).

These chimaeric genes were inserted into Ti-plasmid based plant transformation vectors (45) and _Agrobacteria_ containing these chimaeric genes were used to transform tobacco cells (_N. tabacum_ var. SR1) by the leaf-disc method (39). Plantlets arising from calli selected on the basis of kanamycin resistance were excised, rooted, and the expression characteristics examined. In a parallel series of experiments, chimaeric genes in which the npt II coding region was fused to the promoter sequences from the nopaline synthase gene of the _Agrobacterium_ Ti plasmid, the 35S promoter from cauliflower mosaic virus, or the TR 1' promoter from the TR-DNA of an octopine type Ti plasmid (46) were introduced into plant cells. Transcription directed from all three promoters is constitutive in plant cells, and the CaMV 35S promoter has been previously demonstrated to confer high level expression in leaf tissue (41). The absolute level of expression of the various chimaeric gene constructs in several independently-derived transgenic tobacco plants was examined (Figure 3). Total RNA was prepared from leaves of light-grown plants, and equivalent amounts of RNA from the individual transformants was fractionated on agarose gels, blotted to nitrocellulose, and then probed with radioactively-labeled sequences complementary to npt II RNA. It can be seen that the levels of RNA are significantly higher in all of the plants expressing npt II under the control of the _Arabidopsis_ rbcS 1A promoter than in plants expressing these same sequences under the control of any of the other three promoters.

The high levels of RNA encoding the npt II protein observed in the leaf tissues of plants expressing the chimaeric genes encoded on pGS1400 and pGS1401 result in similarly high levels of both npt II protein and enzymatic activity. This point was demonstrated through Western blot analysis as shown in Figure 3. Equivalent amounts of leaf protein extract from individual transformants expressing pGS1400 and pGS1401 were fractionated by SDS-polyacrylamide gel electrophoresis and the fractionated protein was blotted to nylon filters. Filters were incubated with antibody raised against the npt II protein and immunoreactive protein bands visualized using the procedures of Towbin, et. al. (48). Plants expressing pGS1401 (npt II fused to the rbcS 1A promoter and transit peptide) were observed to contain significantly higher levels of npt II transcripts (data not shown) and total npt II protein than plants expressing pGS1400 (fusions minus the transit peptide coding sequences) (Figure 3). The basis

NPT II expression under the control of different promoters in transgenic tobacco plants

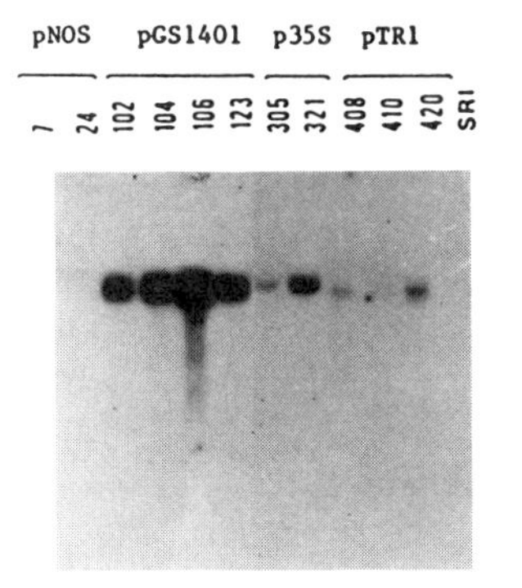

Western blot analysis of NPT II protein levels in leaf tissue of transformed tobacco plants

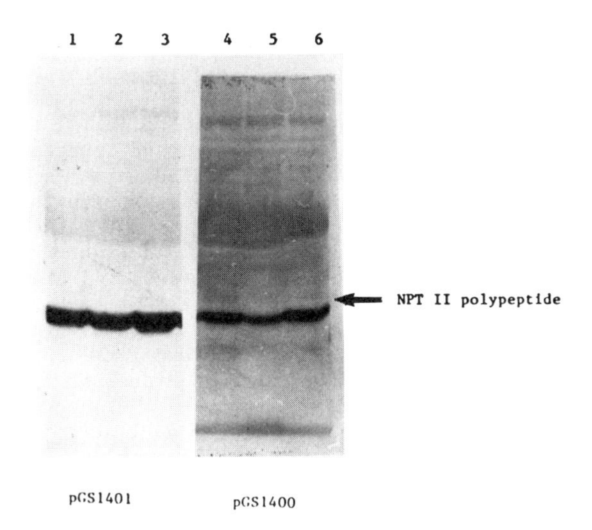

Figure 3. Comparison of absolute expression levels of npt II genes under the control of various promoters in transgenic tobacco plants. The upper panel shows the results of Northern blot analysis in which equivalent amounts of total RNA isolated from light-grown leafs of transgenic plants containing the constructs noted in the Figure were probed with radioactively-labeled sequences complementary to npt II RNA. See text for description of chimaeric genes. In the lower panel, Western blots of protein extracts from leaves of three independently derived transgenic plants expressing chimaeric genes encoded on plasmids pGS1400 and pGS1401 are shown.

for the increased levels of both total RNA encoding npt II and npt II protein in plants expressing pGS1401 versus pGS1400 is not known. It is possible that nucleotide sequences within the coding portion of the transit peptide of preSSU are involved in mRNA stability and/or translational efficiency. What role if any the nature of the encoded protein (i.e., npt II) has in this effect is also unknown. Further studies of the role of these sequences are clearly necessary.

It was also of interest to us to determine whether the two chimaeric gene constructs (pGS1400 and pGS1401) were being properly expressed in a tissue-specific and photoregulated fashion in the heterologous tobacco genomic background. Tissue was excised from leaves, stems, roots, and floral parts (sepals, petals, anthers and stigmas) of transgenic plants expressing pGS1400 and pGS1401 and npt II activity was measured in crude protein extracts prepared from these tissues (Figure 4). Highest levels of npt II expression are observed in leaves and sepals; lower levels of activity are observed in stems and other foral parts. Npt II activity was absent, or detectable at reduced levels in comparison to photosynthetic tissues in the roots of some plants. Qualitatively similar patterns of npt II expression are observed for plants transformed with pGS1400 (data not shown).

To determine whether the nucleotide sequences contained within the 1550 bp 5' -flanking portion of the *Arabidopsis* rbcS 1A gene are capable of directing high levels of photoregulated expression, plants transformed with either pGS1400 (data not shown) or pGS1401 were placed in darkness for 48 hrs. Following this dark treatment, young leaves were excised and crude protein extracts prepared. The plants were then placed in white light and protein extracts prepared from leaves at various time points for periods extending up to 48 hrs. The amount of npt II activity present in extracts prepared from dark-adapted and light-grown leaf tissues was then measured. These studies (Figure 4) show that following extended dark periods, there is little or no detectable npt II activity in leaves. Upon exposure to light, there is a rapid induction and expression of npt II under the control of the rbcS 1A promoter as evidenced by the appearance of npt II enzymatic activity.

These data demonstrate that the nucleotide sequences necessary for the proper tissue-specific and photoregulated expression of the *Arabidopsis* rbcS 1A gene are located within the approximately 1550 bp 5' to the coding portion of the gene. Our present findings are consistent with previous observations (21, 36-38, 42) demonstrating that tissue-specific and photoregulated rbcS gene expression is mediated by sequences located within the 5'-flanking region of the gene. These studies provide the basis for a more rigorous examination of the transcriptional regulatory mechanism(s) involved in rbcS gene expression in *Arabidopsis*.

Discussion

We have demonstrated that in the crucifer *Arabidopsis thaliana*, the SSU polypeptides of RuBPCase are encoded in the nuclear genome by a multigene family consisting of four members, three of which

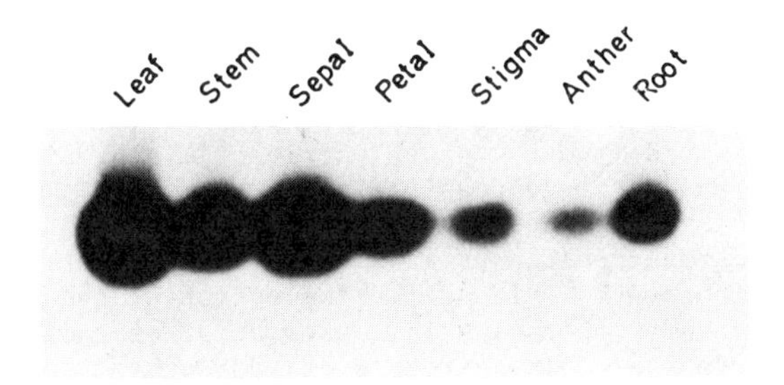

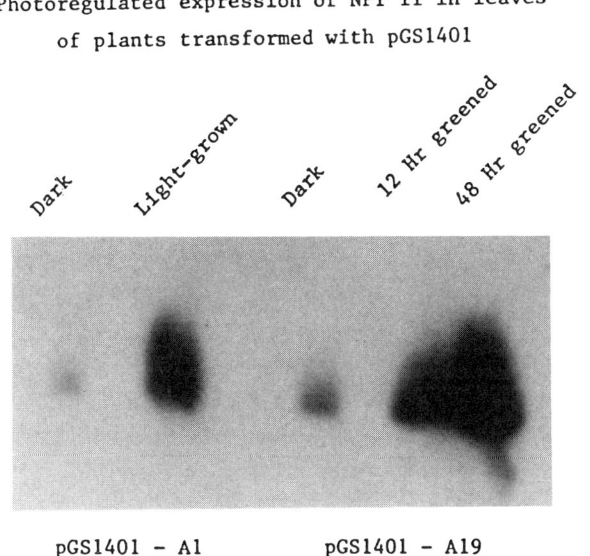

Figure 4. The tissue-specific and photoregulated expression characteristics of the chimaeric gene contained on plasmid pGS1401 in transgenic tobacco. Crude protein extracts were prepared from the various tissues indicated (upper panel) or from the leaves of light-grown, or dark-adapted plants (lower panel) and Npt II activity was measured (52).

are tightly clustered along the chromosome. The level of complexity and organization of the *Arabidopsis* rbcS gene family is not dissimilar to that observed for this gene family in other plant species (11-20).

Previous comparisons of gene families in *Arabidopsis* with their counterparts in other plant species (47-49) led to the suggestion that the reduced size of the *Arabidopsis* genome may in part reflect a reduction in the complexity of multigene families within this organism. This appears not to be the case for the rbcS family.

Tightly linked and highly conserved rbcS gene family members have been described in other plant species (11, 12, 16, 20). In most cases it has not been possible to determine whether the observed high levels of intraspecific homology among rbcS gene family members reflects recent gene duplication events or gene correction. In tomato, evidence for selection at the protein level as a mechanism operating in conservation of nucleotide sequences within the coding portions of rbcS genes has been presented (16). The high degree of conservation among the three clustered members of the *Arabidopsis* rbcS gene family at the nucleotide sequence level and in comparisons of the amino acid sequence of the encoded polypeptide suggests that these genes may be the products of fairly recent gene duplication events. Our observations with the rbcS gene family of Arabidopsis are similar in many respects to those made with a second unrelated gene family in this organism, that encoding the constituent polypeptides of the light-harvesting chlorophyll a/b- protein complex (47). The cab gene family consists of three tightly linked genes encoding identical mature polypeptides approximately 96% homologous within their coding regions, and possibly a fourth less related gene.

The results of experiments directed toward establishing *Arabidopsis* as a system for the study of cell-specific and photoregulated gene expression involved in the development of the photosynthetic apparatus have also been presented. We demonstrated that nucleotide sequences within the 5' -flanking region of the rbcS 1A gene are capable of directing proper cell-specific and photoregulated expression in transgenic tobacco plants. These observations are consistent with previous examinations of the nature and location of critical *cis*-regulatory elements in rbcS gene promoters from other plant species (21, 36-38, 42). They further support the suggestion that the mechanism for cell-specificity and photoregulation of rbcS gene expression is conserved among higher plant species, thus allowing the regulatory sequences and factors involved in these processes to function interspecifically. Previous studies of transgenic expression of rbcS genes and chimaeric genes containing the promoters of rbcS genes have noted effects on the level and cell-specificity of expression, likely resulting from position effects as a consequence of sites of integration along the chromosome. The magnitude of variation in the specificity and level of expression among independently derived transformants is currently being analyzed. We would also like to know whether the same nucleotide sequences necessary for cell-specificity and photoregulation in transgenic tobacco are required in the homologous genetic background. We expect that cell-specific and photoregulated rbcS gene expression in Arabidopsis is mediated by discrete enhancer-like elements similar to those recently described in the flanking regions of rbcS genes from pea (40-41). Experiments addressing this question are currently in progress.

For future biotechnological applications, a reliable system for the properly regulated, high level expression of novel genetic information in plant cells is necessary. The demonstration that the rbcS 1A promoter is capable of directing significantly high levels (in comparison to other test promoters) of apparently correctly regulated expression of at least one genetically engineered gene in transgenic plants is clearly one of the more potentially important findings of this present study. Work is presently underway to determine the general usefulness of Arabidopsis rbcS 1A promoter-based vectors for the high level expression of foreign genetic information in plant cells. In conclusion, we have demonstrated that it is possible to successfully manipulate the expression of genes involved in the development and functioning of the photosynthetic apparatus of higher plants. These studies open the door to a more detailed dissection and manipulation of the photosynthetic machinery. Such studies are a step toward the production of useful products for medicine, pharmacy, industrial and agricultural commodities. Whether this potential will be realized in the develompment of new and improved commercial varieties which will provide more economical and less labor-intensive yields for the farmer remains to be seen.

Acknowledgments

This work was supported by funds from the Pratt Foundation awarded to MPT. MPT is indebted to ARC for his support during the initial stages of these studies. E. de A. was supported by a grant from the International Institute for Education. We thank Dirk Inze for providing us with a genomic library and for helpful discussions, and Tamela Davis for preparing the manuscript.

Literature Cited

1. Mohr, H. In Encyclopedia of Plant Physiology; Shropshire, W.; Mohr, H., Eds.; Springer-VerLag: Berlin. 1983; Vol. 16, New Series, p. 336.
2. Tobin, E.M.; Silverthorne, J. 1985. Ann. Rev. Plant Physiol. 1985, 36, 569-593.
3. Anderson, J.M. Ann. Rev. Plant Physiol. 1986, 37, 93-136.
4. Harpster, M.; Apel, K. Physiol. Plant. 1985, 64, 147-152.
5. Ellis, R.J. Ann. Rev. Plant Physiol. 1981, 32, 111-137.
6. Dobberstein, B., Blobel, G.; Chua, N.H. Proc. Natl. Acad. Sci. USA 1977, 74, 1082-1085.
7. Cashmore, A.R.; Broadhurst, M.K.; Gray, R.E. Proc. Natl. Acad. Sci. USA 1978, 75, 655-659.
8. Highfield, P.E.; Ellis, R.J. Nature 1978, 271, 420-424.
9. Grossman, A.R.; Bartlett, S.G.; Chua, N.H. Nature 1980, 285, 625-628.

10. Coruzzi, G.; Broglie, R.; Cashmore, A.R.; Chua, N.H. J. Biol. Chem. 1983, 258, 1399-1402.
11. Cashmore, A.R. In Genetic Engineering of Plants - An Agricultural Perspective; Kosuge, T.; Meredith, C.P.; Hollaender, A., Eds.; Plenum Press: New York. 1983, p. 29-38.
12. Dean, C.; van den Elzen, P.; Tamaki, S.; Dunsmuir, P.; Bedbrook, J. Proc. Natl. Acad. Sci. USA 1985, 82, 4964-4968.
13. Stiekema, W.S.; Wimpee, C.F.; Tobin, E.M. Nucl. Acids Res. 1983, 11, 8051-8061.
14. Broglie, R.; Coruzzi, G.; Lamppa, G.; Keith, B.; Chua. N.H. Biotechnology 1983, 1, 55-61.
15. Berry-Lowe, S.L.; McKnight, T.D.; Shah, D.M.; Meagher, R.B. J. Molec. Appl. Genet. 1982, 1, 483-498.
16. Pichersky, E.; Bernatzsky, R.; Tanksley, S.D.; Cashmore, A.R. Gene 1986, 40, 247-258.
17. Mazur, B.J.; Chui, C.F. Nucl. Acids Res. 1985, 13, 2373.
18. Pinck, L.; Fleck, J.; Pinck, M.; Hadidance, R.; Hirth, L. FEBS Lett. 1984, 145, 145-148.
19. Polans, N.O.; Weedens, N.F.; Thompson, W.F. Proc. Natl. Acad. Sci. USA 1985, 82, 5083-5087.
20. Grandbastien, M.A.; Berry-Lowe, S.; Shirley, B.W.; Meagher, R.B. Plant Molec. Biol. 1986, 7, 451-465.
21. Coruzzi, G.; Broglie, R.; Edwards, C.; Chua, N.H. EMBO J. 1984, 3, 1671-1679.
22. Poulsen, C.; Fluhr, R.; Kauffman, J.M.; Butry, M.; Chua, N.H. Mol. Gen. Genet. 1986, 205, 193-200.
23. Tobin, E.M. Plant Mol. Biol. 1981, 1, 34-51.
24. Gallagher, T.F.; Ellis, R.J. EMBO J. 1982, 1, 1493-1498.
25. Silverthorne, J.; Tobin, E.M. Proc. Natl. Acad. Sci. USA. 1984, 81, 1112-1116.
26. Brown, J.W.S. Nucl. Acids Res. 1986, 14, 9549-9559.
27. Richter, G. Plant Mol. Biol. 1984, 3, 271-276.
28. Fluhr, R.; Chua, N.H. Proc. Natl. Acad. Sci. USA 1986, 83, 2358-2362.
29. Meyerowitz, E.M.; Pruitt, R.M. Science 1985, 229, 1214-1218.
30. Horsch, R.G.; Fry, J.E.; Hoffmann, N.L.; Eichholtz, D.; Rogers, S.G.; Fraley, R.T. Science 1985, 227, 1229-1231.
31. An, G.; Watson, B.D.; Chiang, C.C. Plant Physiol. 1986, 81, 301-305.
32. Benton, W.D.; Davis, R.W. Science 1977, 196, 180-182.
33. Broglie, R.; Bellemare, G.; Bartlett, S.G.; Chua, N.H.; Cashmore, A.R., Proc. Natl. Acad. Sci. USA 1981, 78, 7304-7308.
34. Leutwiler, L.S.; Hough-Evans, B.; Meyerowitz, E.M. Mol. Gen. Genet. 1984, 194, 15-23.
35. Chang, C.; Meyerowitz, E.M. Proc. Natl. Acad. Sci. USA 1986, 83, 1408-1412.
36. Broglie, R.; et. al Science 1984, 224, 838-843.
37. Herrera-Estrella, L.; Van den Broeck, G.; Maenhaut, R.; Van Montagu, M.; Schell, J.; Timko, M.; Cashmore, A.R. Nature 1984, 310, 115-120.
38. Morelli, G.; Nagy, F.; Fraley, R.T.; Rogers, S.G.; Chua, N.H. Nature 1985, 315, 200-204.

39. Horsch, R.B.; Fraley, R.T.; Rogers, S.G.; Sanders, P.R.; Lloyd, A.; Hoffmann, N. Science 1984, 223, 496-498.
40. Timko, M.P.; et. al. Nature 1985, 318, 579-582.
41. Fluhr, R.; Kuhlemeier, C.; Nagy, F.; Chua, N.H. Science 1986, 232, 1106-1111.
42. Faciotti, D.; O'Neill, J.K.; Lee, S.; Shewmaker, C.K. Biotechnology 1985, 3, 241-246.
43. Timko, M.P.; Kausch, A.P.; Hand, J.M.; Herrera-Estrella, L.; Van den Broeck, G.; Van Montagu, M.; Cashmore, A.R. In Molecular Biology of the Photosynthetic Apparatus; Arntzen, C.; Bogorad, L.; Bonitz, S.; Steinback, K. Eds. Cold Spring Harbor Laboratory: Col Spring Harbor; 1985, p. 381-396.
44. Nagy, F.; Morelli, G.; Fraley, R.T.; Rogers, S.G., Chua, N.H. EMBO J. 1985, 4, 3063-3068.
45. De Blaere, R.; Bytebier, B.; De Greve, H.; Deboeck, F.; Schell, J.; Van Montagu, M.; Leemans, J. Nucl. Acids Res. 1985, 13, 4777-4788.
46. Velten, J.; Velten, L.; Hain, R.; Schell, J. EMBO J. 1984, 3, 2723-2730.
47. Leutwiler, L.S.; Meyerowitz, E.M.; Tobin, E.M., Nucl. Acids Res. 1986, 14, 4051-4064.
48. Towbin, et al. Proc. Natl. Acad. Sci. USA 198, 76, 4350-4354.
49. Chaboute, M.E.; Chaubet, N.; Phillipps, G.; Ehling, M.; Gigot, C. Plant Molec. Biol. 1987, 8, 179-191.
50. Pruitt, R.E.; Meyerowitz, E.M. J. Mol. Biol. 1986, 187, 169-183.
51. Karlin-Neumann, G.A.; Tobin, E.M. EMBO J. 1986, 5, 9-13.
52. Van den Broeck, G.; Timko, M.P.; Kausch, A.P.; Cashmore, A.R.; Van Montagu, M.; Herrera-Estrella, L. Nature 1985, 313, 358-363.
53. Smeekens, S.; van Oosten, J.; de Groot, M.; Weisbeek, P. Plant Molec. Biol. 1986, 7, 433-440.

RECEIVED July 8, 1987

Chapter 24

Application of Biotechnology to Improvement of Plant Food Properties

Role of Competing Strategies and Impediments to Progress

T. J. Orton and A. A. Reilley

Western R&D Station, DNA Plant Technology Corporation, 182 Lewis Road, Watsonville, CA 95076

Biotechnological techniques offer several strategies for food quality improvement. The procedures of most promise are genetic transformation and somaclonal variation. Once technical and legal constraints are overcome, plant transformation will dramatically impact food properties by altering raw material inputs. Techniques such as somaclonal variation are already having an impact on the commercial implementation of biotechnology in the food industry.

One of the most significant steps in the modern evolution of <u>Homo sapiens</u> was the transformation from a hunting/gathering to a sedentary agricultural society. This development enabled man to control the organisms upon which his subsistence depended rather than to remain totally subject to the vagaries of his environment. During succeeding millenia, reaping and sowing spawned the art of plant breeding, the science of genetics, and now the powerful technologies of molecular and cellular biologies. This short contribution assesses the current impact of these so-called new biotechnologies on plant food properties, and surveys future prospects.

<u>Commercial Context</u>

As technologies have been successfully developed and commercial potential recognized, efforts have been made to bring them to bear. Like any tool or method, biotechnology can only work within a set of narrow windows, or viable product concepts. A product is only viable when a number of criteria are adequately satisfied, e.g.

0097-6156/88/0362-0296$06.00/0

1. Product has measurable demand.
2. Profit potential of the product creates adequate incentive to components of supply system (e.g. seed companies, farmers, crude processors).
3. Barriers exist to the entry of competitors.

Because biotechnology is expensive, and usually long-term, weighing these and other criteria becomes extremely critical in deciding among product alternatives. Furthermore, biotechnology is dependent upon genetics and plant breeding as a vehicle to reach finished products (see Figure 1). Because the totality of organisms is the result of complex interaction of thousands of genes and fluctuating environments, it will be some time before molecular biology will be capable of providing this information based on arrays of nucleotide sequences. Plant breeding uses a combination of science and intuitive skills to visualize the essence of these sequences based on overall phenotype. With this knowledge the breeder can mold the genetic structure of the finished product which will hopefully bear the specific gene or genes from biotechnological inputs.

The reader is directed to several earlier reviews for a description of the various biotechnological inputs listed in Figure 1: Sharp et al. 1984, Evans and Sharp 1986, Moshy 1986, Kachetourians and McCurdy 1987, Roels and Thijssen 1987.

There are over 8,000 different food products available to the consumer at the average North American supermarket. Each product has a unique blend of raw materials, formulations, process engineering, and packaging (Moshy 1986). Up to now, R & D activities in the food industry have been focused entirely on the latter three areas. Biotechnology will make it possible to add functional attributes to raw materials, thus allowing product claims not previously possible (e.g. 'natural'), improved cost-effectiveness, and ultimately, entirely new product concepts. Examples of raw material attributes which affect or alter food product quality are listed in Table I.

Positive or negative food properties can envisioned for increases or decreases of metabolites, including:

1. Amino acids and proteins
2. Carbohydrates
3. Fatty acids
4. Vitamins and structural precursors
5. Deletion of toxic compounds (by either blocking synthesis or introduction specific catabolic enzymes)
6. Compounds which promote or retard spoilage

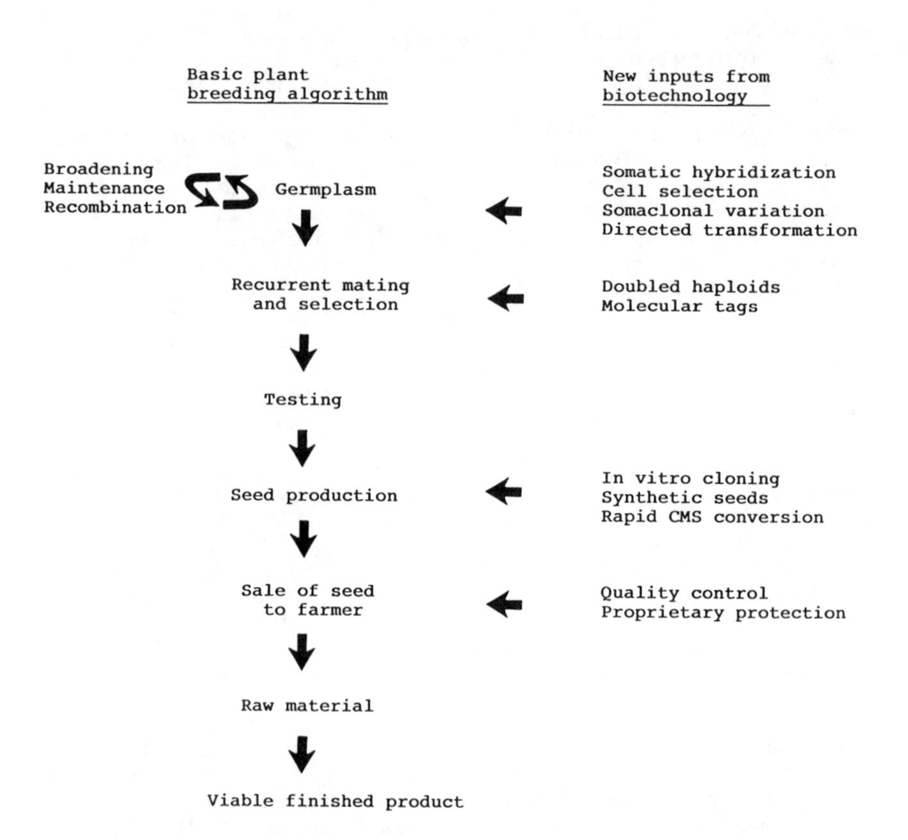

Figure 1. The basic plant breeding process and associated biotechnological inputs.

TABLE I. - Crop plant properties which impact on food products

Property	Probable mode of inheritance
Processing quality:	
Soluble solids	moderately complex
Shelf stability	simple to complex
Uniformity (size, constitution)	complex
Nutrition:	
Proteins and amino acids	simple to moderately complex
Carbohydrates	
nutritive	simple to moderately complex
dietary fiber	simple to complex
Fatty acids	simple to moderately complex
Vitamins	simple
Aesthetic:	
Flavor/aroma	moderately complex
Texture	moderately complex
Color	simple to moderately complex
General:	
Yield	complex
Stress tolerance	complex
Pest or disease resistance	simple

The presence or absence of metabolic intermediates, their over-or underproduction, and thier conversion to new intermediates or end-products not normally present are the largest collective targets for biotechnologies. This is true because many of these technologies work best at the level of one or few genes. Matabolic pathways are mediated by the presence and activity of catabolic and anabolic enzymes, each of which is encoded by one or few genes. By deleting, adding, or modulating these genes, the corresponding molecular make-up of plant tissues can be altered. Food properties constitute an important example where in relatively simple changes in the plant genome can result in superior consumer products.

Useful Technologies

A compilation of demonstrated or conceptualized food properties made possible by biotechnology is provided in Table II. As discussed in the previous section, nearly all of the food properties under consideration are under very simple genetic control. With respect to the new technological inputs depicted in Figure 1, the two approaches which stand out are 1) broadening the base of germplasm through directed transformation with cloned genes or modulators or existing genes and 2) somaclonal variation which is defined as variability derived in conjunction with a cell or tissue culture process.

Plant cell culture techniques are essential to both genetic transformation and somaclonal variation (Figure 2). These procedures are relatively straightforward and have been developed for a wide range of species. Explants are placed onto defined nutrient media under controlled environmental conditions. Plant growth regulators are added to the media to mediate formation of multiple buds, callus, or single cells. These cultures may be used for transformation experiments. Whole plants are regenerated from cell cultures by further manipulation of plant growth regulators.

It has been known for some time that mutations occur spontaneously in culture, but the actual mechanisms by which they arise are not understood. Patterns of somaclonal mutations among plants regenerated from cell and tissue cultures among diverse species share certain similarities. An extraordinarily high proportion of regenerated plants contain mutations in a small subset of genes, resulting in recessive mutations and reductions in fertility, although there are numerous exceptions to this pattern. Manifestations of somaclonal variation can be altered by genotype, culture medium, age of culture, and method of culture. The approach is, therefore, unlike mutation breeding, wherein random mutations are induced by exposure to chemical mutagens or ionizing radiation resulting in larger numbers of different mutations in relatively small proportions.

TABLE II - Demonstrated or conceptualized changes in food properties through biotechnology

Property	Food Value	Biotechnology	Reference (s)
Biological decaffeination	Natural or more cost-effective alternative to decaffeination of coffee, teas, etc.	Directed transformation with cloned microbial caffeinase gene	Weetall et al. 1983
Elevated pools of essential amino acids	Balanced basic meal for animal or human diet (corn, wheat, rice, soybean, etc.)	Cell selection for feed-back insensitivity of allosteric biosynthetic enzymes, e.g. aspartyl kinase (threonine, methionine)	Green 1982
Elevated levels of proteins containing high proportions of essential amino acids	Balanced basic meal for animal or human diet (corn, wheat, potatoes, legumes, etc.)	a) Directed transformation of storage protein gene and/or its regulatory sequences b) Recovery from pool of somaclonal variants exhibiting changes in peptide makeup	Murai et al. 1983 Larkin 1987 Larkin et al. 1983
Modulation of lip-oxygenase activity (and other catalytic enzymes)	a) Down-modulation to increase stability in vegetable oils b) Up-modulation to alter flavor characteristics in fruit tissues (e.g. tomatoes, stone fruits)	a) Recovery of somaclonal somaclonal variant b) Site-directed muta-genesis for down-modulation or trans-formation with active cloned gene for up-modulation	Sha et al. 1969 Chase 1979 Neidleman 1986
Increased soluble solids	Energy savings in leading to lower costs for processing, e.g. tomatoes	Recovery of somaclonal variants	Evans et al. 1984

Continued on next page.

TABLE II. (Continued)

Property	Food Value	Biotechnology	Reference(s)
Altered fatty acid composition, e.g. increased or decreased saturation, new specialty fatty acids	Health benefits from decrease in serum cholesterol, among others	a) Transformation with genes or existing biosynthetic genes with more active promotors b) Site-directed mutagenesis to down modulate undesirable fatty acid biosynthetic genes c) Recovery of somaclonal variants with desired mutations	Neidleman 1986
Enhanced sweetness	a) Natural or more cost-effective alternative to refined sugar, e.g. high fructose corn syrup b) Enhanced flavor and organoleptic quality in fruits and vegetables	a) Recovery of somaclonal variants with mutations affecting carbohydrate metabolism b) Transformation with genes encoding carbohydrate metabolic enzymes	Moshy 1986
Reduced or enhanced aliinase activity in onions	Reduce bitterness or increase characteristic odor	a) Recovery of somaclonal variant or site-directed gene inactivation for down-modulation b) Transformation with active gene for up-modulation	Neidleman 1986
Resistance to Lepidopterous larval pests	Reduce raw material costs and eliminate pesticide residues	Transformation with toxin gene from *Bacillus thuringiensis*	Adang et al. 1986

Donor Plant

Explants are taken from the donor plant, surface sterilized, and placed onto nutrient media.

Media contains growth regulators which induce multiple bud, callus, or shoot formation.

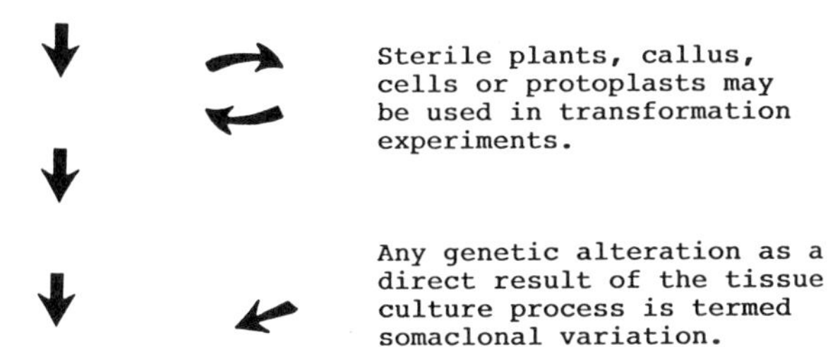

Whole plants are regenerated from tissue culture by further manipulation of growth regulators.

Regenerated plants are screened for genetic improvement.

Improved Plant

Figure 2. General procedures of plant tissue culture.

The attractiveness of somaclonal variation as a genetic technique lies in its extreme simplicity, involving only the extablishment of a culture, the regeneration of plants from it, and then sorting through the population of regenerate for the desired trait or traits. Thus, it is relatively cheap, quick, and requires only minimal technical capabilities. There are no known barriers to traits which can or cannot be targeted. Moreover, mutations arising from somaclonal variation have been regarded as an acceleration of natural evolution and do not fall under recombinant DNA safety guidelines. The main drawback of somaclonal variation lies in our present lack of predictive capabilities. Since a multitude of variables are seemingly out of control the *a priori* probability of achieving a given result becomes incalculable.

The second major arena for complementing traditional plant breeding involves the development of transgenic plant via microbial vectors. Directed transformation in plants is made possible by the capabilities to 1) cut and splice DNA, and 2) mobilize DNA sequences from the test tube into the functioning genetic apparatus of whole plants (Goodman et al. 1987). This provides the theoretical possibility to transform any plant with any existing or conceivable gene. The obvious scientific and commercial potential of this has stimulated tremendous activity in this area world wide. Necessary steps include cloning DNA sequences in *E. coli*, identifying clones which carry a gene of interest, modifying the gene of interest, modifying the gene in a desired fashion, affixing appropriate regulatory sequences, integrating and sequence into a vector such as the Ti plasmid of *Agrobacterium tumifaciens* (Schell and Van Montagu, 1983) which can deliver it into the genome of the target plant cell, and finally regenerating whole plants from the transformed cell. Although remarkable progress has been made, each of these steps is time-consuming, expensive, and still frought with technical hurdles. Practical limitations render it difficult to work with any traits which are not dominant and simple in inheritance. And then after successful introduction of a gene into a food plant the F.D.A., U.S.D.A., E.P.A., or other Federal agency must scrutinize it's impact on health and public welfare. All these considerations make it impossible to predict with any degree of assurance whether or not a particular product of biotechnology will ever come to market.

Summary

New technologies continue to impact the evolution of food product development. Directed transformation will make a huge impact on agriculture, but mitigating factors will protract the time frame. In the mean time, more immediately available strategies, such as somaclonal variation, will pave the way for commercialization of biotechnology in the food industry.

Literature Cited

Adang, M.J., Firoozabady, E., Deboer, D.L., Klein, J., Merlo, D.J., Murray, E., Rocheliau, T., Rashka, K., Staffield, G., and Stock, C. 1986. Expression of a Bacillus thuringiensis crystal protein in Nicotiana tabaccum. In: VI International Congress of Plan Tissue and Cell Culture, Univ. of Minnesota Press, Minneapolis, p 404.

Chase, T. 1974. Flavor enzymes. Adv. Chem. Ser. 136:241-266.

Evans, D.A., Sharp, W.R., and Medina-Filho, H.P. 1984 Somaclonal and gametoclonal variation. Amer, J. Bot. 71:759-774.

Evans, D.A., and Sharp, W.R., 1986. Potential applications of plant cell culture. In: Biotechnology in Food Processing. ed. S.K. Harlander and T.P. Labuza Noyes Publ., Park Ridge, NJ, pp 133-143.

Goodman, R.M., Hauptli, H., Crossway, A., and Knauf, V.C. 1987. Gene transfer in crop improvement. Science 236:48-54.

Green, C.E. 1982. Inheritance and expression of lysine and threonine resistance selected in maize tissue culture. In: Variability in Plants Regenerated from Tissue Culture. ed. E.D. Earle and Y. Demarly. Praeger, New York, pp 188-201.

Khachatourians, G.C. and McCurdy, A.R. 1987. Biotechnology: applications of genetics to food production. In: Food Biotechnology, ed. D. Knorr, Marcel Dekker, Inc. New York, pp 3-19.

Larkin, P.J., Ryan, S.A., Brettel, R.I.S., and Scowcroft, W.R. 1983. Heritable somaclonal variation in wheat. Theoret. Appl. Genet. 67:443-455.

Larkins, B.A. 1987. Modification of proteins encoded by seed storage genes. In: Tailoring Genes for Crop Improvement. ed. by G. Bruening, J. Harada, T. Kosuge, and A. Hollaender, Plenum, New York, pp 163-176.

Moshy, R. 1986. Biotechnology: its potential impact on traditional food processing, In: Biotechnology in Food Processing. ed. S.K. Harlander and T.P. Labuza, Noyes Publ., Park Ridge, NJ., pp 1-14.

Murai, N., Sutton, D.W., Murray, M.G., Slightom, J.L., Merlo, D.J. Reichert, N.A. Sengupta-Gopalan, C., Stock. C.A., Baker, R.F., Kemp, J., and Hall, T.C. 1983.

Phaseolin gene from bean is expressed after transfer to sunflower via tumor-inducing plasmid vectors. Science 222:476-481.

Neidleman, S. 1986. Enzymology and food processing. In: Biotechnology in Food Processing. ed. S.K. Harlander and T.P. Labuza, Noyes Publ., Park Ridge, NJ. pp 37-56.

Roels, J.A. and Thijssen, H.A.C. 1987. Impact of biotechnology on the food industries. In: Food Biotechnology, ed. by D. Knorr, Marcel Dekker, New York, pp 21-35.

Schell, J. and Van Montagu, M. 1983. The Ti plasmids as natural and as practical gene vectors for plants. Bio/Tech 1:175-180.

Sha, F.M., Salunkhe, D.K. and Olsen, L.E. 1969 effects of ripening processes on chemistry to tomato volatiles. J. Amer. Soc. Hort. Sci. 94:171-176.

Sharp. W.R., Evans, D.A., and Ammirato, P.V. 1984. Plant genetic engineering: designing crops to meet food industry specifications. Food Technol. 38:112-119.

Weetall, H.H. and Zelko, J.T. 1983. Applications of microbiol enzymes for production of food-related products. Dev. Ind. Microbiol. 24:71-77.

RECEIVED September 21, 1987

Chapter 25

Biotechnology in Livestock Production

David E. Reed

Molecular Genetics, Inc., 10320 Bren Road East, Minnetonka, MN 55391

The application of biotechnological advances to food animal production has been far more difficult and costly than was expected. This paper concerns technological approaches to improving veterinary vaccines and preventatives. In particular, some of the successes and failures will be discussed based upon considerations of technology, cost, and intended product use. Examples of biotechnologically derived products currently produced, are a monoclonal antibody product for prevention of calf diarrhea, a genetically modified porcine herpesvirus vaccine, and a subunit vaccine for prevention of calf diarrhea. Products not yet available but in final testing include bovine somatotropin, a subunit vaccine for prevention of pseudorabies in swine, and diagnostic tests which can distinguish infected animals from animals which have been vaccinated.

For the purposes of this paper, the term "biotechnology" is defined here as recombinant DNA (rDNA) and hybridoma or monoclonal antibody (MAB) technologies. Translation of biotechnology into products has been far more difficult than most of us imagined. A comprehensive list of the biotechnologically derived products which are being developed for veterinary use has been published (1). This publication states that an impressive number of veterinary biotechnology products including dozens of vaccines and numerous preventatives, growth promotants, and diagnostics will be available between 1988 and 1996. In most cases, however, the problems in translating the technology into products have been both underestimated and understated.

The purpose of this paper is to give a general overview of current progress in applying biotechnology to improve vaccines, growth promotants, and therapeutics used in the production of livestock. This discussion is not meant to be an inclusive list of

0097–6156/88/0362–0307$06.00/0

the products now in development but will focus on the concepts and, using examples, will site approaches which have failed as well as those which have succeeded.

Early Success of rDNA Approach:

One of the early successes of rDNA technology in livestock production was the cloning of bovine somatotropin (BST). Expression systems which used the bacterium *E. coli* allowed production of nearly limitless quantities of BST as insoluble aggregates of protein in the bacterium. After initial problems were solved by the use of reducing agents and strong chaotropic agents to solubilize the aggregated protein, the *E. coli* produced BST was found to be equal in potency to naturally derived BST (2). It appears that the remaining research problems with BST involve development of novel delivery systems which will obviate the need for daily injections.

Extending the rDNA Approach to Vaccine Development

Subunit vaccines. Researchers in several commercial laboratories, buoyed with the rDNA successes of BST and other small (≈20,000 Dalton MW) peptides, predicted similar success in producing viral subunit proteins in limitless quantities. Success in cloning and expressing a number of viral surface proteins has been achieved. In most cases, the immunologically important surface polypeptides were identified and the genes specifying those peptides were isolated, sequenced, and placed into *E. coli* expression systems. Unfortunately, in almost all the cases, the immunizing potential of *E. coli* produced polypeptides was significantly less than that of the native viral polypeptide and vaccines formulated from these polypeptides generally were not efficacious. The exact reasons for these failures were difficult to determine. It is known that denaturation of polypeptides by solubilization processes (e.g. 10 *M* urea and 0.1 *M* dithiothreitol treatment) can destroy confirmation-dependent antigenic sites. In addition, unlike mammalian cell systems in which mammalian viruses are produced, certain post-transcriptional or post-translational processing events such as cleavage or glycosylation do not occur or occur differently in *E. coli* expression systems. In our laboratory, we were successful in expressing in *E. coli* canine and porcine parvovirus structural proteins. These proteins, when used to immunize animals, would produce antibody which was recognized immunologically by authentic virus protein from infected mammalian cells but, unlike authentic virus protein, failed to produce protective or neutralizing antibody (3). In a porcine herpesvirus system, our laboratory compared the efficacy of vaccines prepared from *E. coli* produced viral protein with that of vaccines made from authentic viral protein (L'Italien, Zamb, Robbins, Marshall, unpublished data). Compared to authentic viral protein, approximately 1000 times more *E. coli* produced viral protein than authentic viral protein was required to immunize a pig.

The killed *E. coli* - BPV recombinant bacterin is more successful example of rDNA technology. This product has been tested in extensive immunogenicity trials in cattle and found to be highly efficacious in preventing warts (4).

If rDNA approaches to subunit vaccines are to be successful, there are some possibilities to increase the efficacy. A straightforward approach is to introduce the genes for the immunogenic proteins into mammalian cell expression systems. Unfortunately, this has not been easy and products produced by this method have not yet been commercialized.

We can conclude that it will be, technically feasible to produce viral subunit vaccines using rDNA technology. However, the cost of the research necessary to complete development of many of the veterinary vaccines will make it economically not feasible. The development of the first rDNA vaccine for for use in animals, foot & mouth disease (FMD) of cattle, required that the immunizing subunit be defined, and the gene cloned and expressed. This took a number of years and the investments of several research groups. In order to make the rDNA vaccine for FMD a practical vaccine, the research, development, and clinical testing must be repeated for each type of FMD virus and possibly for each subtype because the final FMD vaccine product needs to be multivalent. This will present a rather fundamental dilemma when the cost of the research begins to exceed the potential revenues for the product.

The subunit vaccines seem to be highly dependent upon adjuvants in order to be efficacious. Unfortunately, the adoption of new adjuvants has been slow. One important reason for this involves the high cost of taking a product through the Food Safety Inspection Service approval process. Because the mechanisms are so varied by which adjuvants potentiate immunologic response to an injected antigen, adjuvants still are picked by largely empirical means. Additionally, the species-specificity of many adjuvants makes it difficult to extrapolate results obtained in laboratory animals to results expected in food animals.

Live Virus Vaccines. There are two areas in live virus vaccine development where the rDNA research has been successful. That is in production of live virus vaccines in which a virulence gene has been deleted and production of live virus vaccines which are genetic recombinants between vaccinia virus and the immunizing subunit(s) genes of another virus.

The traditional empirical methods of attenuation (mutagenesis, multiple cell culture passage, temperature selection, and passage in non-host animals), eventually should be replaced by the more exacting rDNA methods whereby a virulence gene is identified and deleted. An example of this is the commercially available porcine herpesvirus vaccine in which the viral thymidine kinase (TK) gene has been deleted (5). The TK^- vaccine is safer than the parent TK^+ vaccine by being less neurovirulent. Live virus vaccines, however, are less safe than the non-living vaccines because the efficacy of the live virus vaccine requires that the vaccine virus replicate in the host animal. In addition, the live virus vaccines, in comparison to the killed vaccines, are much more likely to be contaminated with passenger viruses or mycoplasmas.

Research success with live virus vaccines which are recombinants with vaccinia has been reported commonly. For example, a recombinant vaccinia virus carrying the Rift Valley fever virus (RVFV) G1 and G2 glycoprotein genes conferred 90 - 100% protection of mice challenged with virulent virus (6). In contrast, a RVFV subunit vaccine produced by rDNA methods in E. coli conferred only 56 - 70% protection.

Unfortunately, because of the vaccinia pathogenicity for humans, the recombinant vaccinia approach to veterinary vaccine development is burdened with safety risks beyond that of conventional live virus vaccines. It is unclear whether or not U.S. Department of Agriculture regulatory clearance will be forthcoming for vaccinia recombinants.

Monoclonal Antibody (MAB) Technology

MABs for Treatment or Prevention of Disease. One MAB product currently being sold for the prevention of animal disease is a MAB against the K99 pilus of E. coli (7). Newborn calves infected with the $K99^+$ E. coli often die of diarrhea. The product is a single oral dose of MAB given to calves within 12 hours of birth. The mechanism of action is not precisely determined but presumably the MAB blocks the attachment of piliated $K99^+$ E. coli to the intestinal wall. The efficacy of this product in field use appears to be very high. An additional MAB product has been developed for conferring passive immunity to young pigs in the face of an outbreak of porcine herpesvirus. This product is intended to reduce the death losses from pseudorabies.

MABs for Producing Subunit Vaccines. Monoclonal affinity chromatography technology has provided the tools to make vaccines of unprecedented purity. Our company is developing a vaccine for porcine herpesvirus using affinity chromatography technology. The product is prepared by extracting a single surface glycoprotein from infected cells. Beyond the obvious safety benefits of a highly pure product, the product has an additional advantage which is compatibility with a serologic test for pseudorabies. Because pseudorabies is a controlled disease, any animal which is serologically positive is subject to restrictions on sale or shipment. A diagnostic test which detects serologic response to a surface glycoprotein not included in the vaccine will detect infected pigs but will not detect pigs vaccinated with the affinity purified vaccine.

Summary

In hindsight it appears that scientists and administrators in both the public and private sectors have expected too much and too soon from rDNA technology. It is time to reassess both the technology and the needs. The research goals must be linked to products to improve livestock performance and not linked to a particular technology. For example, we must first ask what is needed in a foot & mouth disease (FMD) vaccine before we decide that rDNA technology can improve the product. One of the great promises of the rDNA technology is reduced cost of vaccines. With FMD, cost of the vaccine has not been a major problem.

The most pressing problems of FMD vaccines have been in safety (allergic reactions and incomplete inactivation of the virus). Other technologies besides rDNA can be used to solve safety problems.

Biotechnology has just begun to impact food animal production. However, because the costs of biotechnology are great and the markets in veterinary biologicals (vaccines, diagnostics, and antibody products) are small relative to the human market, adoption of the rDNA and MAB technology for the improvement of food animal products is likely to be slow.

Literature Cited

1. Emerging Developments in Veterinary Biotechnology. PB86-222379. U.S. Food and Drug Administration, Rockville, MD. U.S. Department of Commerce, National Technical Information Service: Springfield, VA. July, 1986.
2. George, H. J.; L'Italien, J. J.; Pilacinski, W. P. DNA 1985, 4, 273-281.
3. Halling, S.M. & Smith, S. Gene 1984, 29, 263-269.
4. DeLorbe, W.; Pilacinski, W. P.; Lum, M. A.; et al. in Vaccines 87 Modern Approaches to New Vaccines: Prevention of Aids and other Viral, Bacterial, and Parasitic Diseases; Chanock, R. M.; Lerner, R. A.; Brown, F.; Ginsberg, H. (Eds.) Cold Springs Harbor Laboratory Publications: Cold Springs Harbor, NY, 1987, pp 431-434.
5. Kit, S.; Kit, M.; Pirtle, E. C. Am. J. Vet. Res. 1985, 46, 1359-1367.
6. Collett, MS.; Keegan, K.; Hu, S.-L.; et al. in The Biology of Negative Strand Viruses; Mahy, B.; Kolakofsky, D. (Eds.) Elsevier: New York, 1987, pp 321-329.
7. Sherman, D. M.; Acres, S. D.; Sadowski, P. L.; et al. Infection & Immunity 1983, 42, 653-658.

RECEIVED August 17, 1987

Chapter 26

Biotechnology for Agriculture and Food in the Future

Ralph W. F. Hardy

Boyce Thompson Institute for Plant Research at Cornell, Tower Road, Ithaca, NY 14853 and BioTechnica International, Inc., 85 Bolton Street, Cambridge, MA 02140

Biotechnology products and processes are expected to provide necessary inputs for agriculture and food including increased productivity, higher value-in-use food products and new products for non-food markets. Several examples of recent major science, technical, and policy developments are highlighted: N_2 fixation, microbial pesticides, plant cell culture regeneration and somoclonal variation, genetic engineering of plants, monoclonal antibodies and DNA probes, microbial production of animal products, genetic engineering of animals, reproduction technology, genetically-engineered yeast, designed genes, regulation of gene expression, chemical agriregulators, proprietariness, and regulation and field testing. Types of products and processes for each of crops, animals, and foods are identified with indicated sequence of commercialization. World sales of $5 to $10 billion are projected by 2000.

Biological science and technology for agriculture and food has been in accelerated advance in recent years. From these biological advances are expected significant products and processes to meet major needs of agriculture and food in developed and developing countries. The needs are increased production efficiency, increased value-in-use of animal and plant food products, and new agricultural products for major non-food markets.

These innovations are expected to come from the new biotechnology. Biotechnology has been used in agriculture for centuries. Traditional biotechnology used genetic engineering at the organismal level with selection and breeding as the key techniques. The new biotechnology enables genetic engineering at the cellular and molecular level. This new biotechnology will be

0097-6156/88/0362-0312$06.00/0

the focus of my comments. These new biotechnology products and processes are expected to have significant impact on agriculture in the 1990's and beyond. Surveys of informed public- and private-sector researchers and farmers suggest that the new biotechnology is expected to be the major source of innovation by the beginning of the twenty-first century.

Biotechnology products and processes will include chemicals, diagnostics, microbes, seeds, and animals. Examples of major anticipated products are self nitrogen-fertilizing crops that would eliminate the need for the $20 billion annual worldwide expenditure for fertilizer nitrogen and genetically-engineered meat-producing animals with major improvements in efficiency of production and in dietary quality of the meat.

These advances in science and technology are being driven by the traditional agricultural research laboratories but to a much greater extent by a few private universities, research institutes, and a major investment by established and development-stage companies focused on generating the science and technology and converting it to useful products and processes.

Bioscience, Biotechnology and Related Advances for Agriculture and Food

Several key advances will be highlighted below. Collectively, they document the strength of the growing base of relevant knowledge and related policy for biotechnology products and processes. The knowledge base in agricultural science trails, to a major extent, that in human health science and explains why the initial impact of the new biotechnology on agriculture will trail that in human health care.

Nitrogen Fixation. The best-understood gene system of agricultural significance is *nif*, the nitrogen-fixation complex. These seventeen genes have been characterized, and their regulation is understood so as to enable their restructuring for increased activity. Genes involved in the symbiotic relationship between legume plants and rhizobial bacteria are being identified. A recent advance is the recognition that a plant flavone, luteolin, signals the expression of rhizobial genes for nodulation. Although the N_2-fixation system is complex, several of the technical hurdles have been crossed on the way to create nitrogen-fixing higher plants. The route is defined for achieving this major technical advance. Laboratory N_2-fixing plants are expected now by the early 1990's, a much earlier date than projected a few years ago.

Microbial Pesticides. A variety of formulations of *Bacillus thuringiensis* have been used for several years as pesticides. Molecular biotechnology is enabling modification of *Bacillus thuringiensis* for improved efficacy. The toxin gene has been incorporated into *Pseudomonas*, a rhizosphere inhabitant, to enable protection against soil insects. In addition, there are a large number of other known viral and fungal pathogens of insects, diseases, and weeds. Molecular and cellular biotechnology is expected to generate useful products based on these other pathogens. One of the largest collections of fungal pathogens for insects--about 750 accessions--is located at the Boyce Thompson Institute.

Cell Culture Regeneration and Somoclonal Variation. The number of different plants that can be regenerated from cell cultures is increasing. Beyond the traditional tobacco, carrot, and other early examples, it is now possible to regenerate soybean and most recently rice as the first example of a major cereal crop. This latter success was facilitated by Rockefeller-Foundation support focussed exclusively on rice. It is expected within the next few years that plant regeneration from cell cultures will become standard practice for most crop plants.

Plant cells in culture undergo a process called somoclonal variation which leads to greater expressed genetic diversity in regenerated plants than in the parental plant. Somoclonal variation is proving useful in crops such as sugar cane and maybe wheat and potato.

Genetic Engineering of Plants. Prior to 1983 it was impossible to genetically engineer plants at the molecular level in contrast to microbial and animal capabilities. Since that time various techniques have been used to engineer successfully a wide variety of plants. These techniques include the use of *Agrobacterium tumefaciens* and its modified T_i plasmid in dicotyledonous plants and electroporation and micro-injection of naked DNA in various plants. Successful examples of plant genetic engineering include incorporation of single foreign genes to detoxify or resist selected antibiotics and herbicides. Herbicides to which plants have been genetically engineered with resistance include the old herbicide atrazine, the middle-aged herbicide glyphosate, and the new high potency herbicides, sulfonylureas and imidazolinones. In most cases a single-base change in the DNA converts an enzyme from susceptible to resistant.

The majority of genetically engineered plants to date are single gene additions to provide resistance to toxic chemicals. More recently, disease and pest resistance are being molecularly engineered into plants. Earlier this year the gene for a coat protein of tobacco mosaic virus was incorporated into tomato and other plants with expression of protection against TMV by the engineered plant. In another case, a gene for the *Bacillus thuringiensis* toxin was incorporated into a tobacco plant enabling this plant to resist certain insect pests.

A recent exciting example is the incorporation of the genes for bacterial and firefly luciferase into higher plants. For bacterial luciferase, this is the first example of the incorporation into a plant of two genes whose coordinate expression is required for the formation of a functional enzyme, luciferase in this case. These plants with the luciferase gene emit light. This two-gene example suggests that more complex genes will be engineered genetically into plants in the near future.

Monoclonal Antibodies and DNA Probes. DNA probes and monoclonal antibodies are proving useful for diagnosis of disease in plants and animals. DNA probes are also useful for latent viral disease and are being used commercially within the last year for identification of certain human diseases, food contaminants, and infectious bacteria associated with periodontal disease. In addition, one of the early products in animal biotechnology was a monoclonal product named Genecol 99 for reduction of scours in new-born calves.

Microbial Production of Animal Products. Advances in biotechnology for human health have provided the basis for microbial products for the animal industry. These products include vaccines, lymphokines, and hormones. Lymphokines such as interferon have been produced in microbes and may be useful in animal health. Bovine growth hormone produced by microbes significantly increases milk production per animal and production efficiency; porcine growth hormone increases the rate of gain by swine, decreases the amount of feed needed for a pound of gain, and increases substantially the leanness of pork. Microbial vaccines have been produced for foot and mouth disease as well as for pseudo rabies and other viruses. A synthetic subunit vaccine for foot and mouth disease has been constructed using two separate but small parts of the virus gene, an example of how biotechnology knowledge will direct chemical synthesis.

Genetic Engineering of Animals. In 1981 functional foreign genes were first incorporated into animal embryos. The most dramatic example of transgenic engineering in animals was the 1983 introduction of multiple copies of growth hormone into rodent embryos. The resultant rodents grew up to double the size of their litter mates who did not have the additional growth hormone genes. Similar covers in the early 1980's of Nature and Science, showing transgenic super rodents, are the most visual representations of the power of biotechnology. Much experimental work has sought to extend these rodent experiments to domestic animals with no successful report to date. However, one expects molecular genetic engineering of domestic animals to succeed and thereby improve health, efficiency of production, and value-in-use of animals.

Reproduction Technology. Significant advances have occurred over the last decade in reproductive technology. Embryo transfer has become commercial enabling major increases in progeny of genetically superior females through the use of superovulation and transfer of fertilized embryos to surrogate mothers. *In vitro* fertilization is another useful technique. Advances have also occurred in sperm sexing to enable production of the economically-desired animal sex.

Genetically-Engineered Yeast. Industrial yeasts are involved in the production of many beverages, foods, and some industrial products. These products include cheese, bread, beer, wine, spirits, and industrial alcohol. Molecular genetic engineering techniques have been developed in recent years to enable the engineering of industrial yeasts. One example is the engineering of these multi-ploidy yeasts for the production of light beer--beer with reduced calories by reduction in residual starch. Normal yeasts are unable to completely convert starch to alcohol because of their inability to degrade the starch beyond its branch points. A debranching-enzyme gene has been engineered into the industrial yeast so that it can completely convert starch to alcohol, resulting in the production in a single natural step of light beer. In another case, an industrial yeast has been genetically engineered with the incorporation of a gene to enable lactose utilization, a by-product of cheese manufacture, in the production of high concentrations of ethanol, eliminating a potential pollutant and producing a useful source of energy. Other genes could be incorporated into industrial yeasts to improve the efficiencies of manufacture of cheese, bread, wine, and spirits and to make other useful products.

Designed Genes. Prior to the early 1980's mutagenesis of genes was a random, unpredictable procedure using chemical mutagens or radiation. In the early 1980's genes were first mutagenized in a directed manner by a process called site-specific mutagenesis (SSM), an application of chemistry to biotechnology. One of the earliest demonstrations of SSM was the change of a gene for the penicillin-inactivating enzyme, β-lactamase--a serine at the active site was replaced by a cysteine. The product of the designed gene was a thio-β-lactamase, a novel enzyme with activity altered from the natural β-lactamase. Since that time, site-specific mutagenesis has become a major chemical-biotechnology activity. This SSM technique is expected to design genes in a directed way so that their products are more useful for production of crops, animals, foods, or other areas. One opportunity for major impact on agriculture from site-specific mutagenesis is the redesign of the carbon dioxide-fixing enzyme, the most abundant enzyme in nature, to eliminate its wasteful oxygenase activity and increase its useful carboxylase activity. The outcome could be a doubling in the yield of most crop plants with minimal additional input.

Regulation of Gene Expression. Introduced genes must be expressed at the appropriate time to be effective. Understanding of the regulation of gene expression is critical. Recent work has identified genetic elements involved in light regulation of gene expression. In addition, studies of so-called signal sequences associated with chromosomal genes whose products are transported either outside the cell or to organelles within the cell have advanced.

Chemical Agriregulators. Genes are expressed at defined times during the growth and development of an organism. As the knowledge of the regulation of gene expression advances, it is expected that synthetic chemical molecules will be designed for regulation of gene expression. These agriregulators will cause expression of plant protectants at the desired time, control of growth and development, and alteration of the composition of the harvested product as a few examples. Herein lies a major opportunity for future agrichemical products.

Proprietariness. In 1980 a Supreme Court action extended the patent act to include a novel, living organism--a microbe that digested oil. Life forms were no longer denied patent protection. In 1986 a novel seed containing an elevated tryptophane level was awarded a patent. This extension of patent proprietariness to a seed will encourage commercial investment in the development of improved plants. Similar extensions may be anticipated for engineered domesticated animals.

Regulation and Field Testing. In the 1970's there was a major concern that recombinant DNA, the basic process for molecular genetic engineering, might lead in the laboratory to unsafe products. This fear has not been substantiated, and in fact, the risk to the health of humans has been negligible to zero from the products of recombinant DNA. In the 1980's a major concern has been expressed about deliberate release or introduction of molecularly genetically engineered organisms into the environment.

In agriculture, field research is a necessary and normal extension of laboratory research. On June 26, 1986 the Federal Register announced the first regulatory steps for agricultural products. Emphasis was placed on product risk/benefit rather than on process representing significant progress over an earlier proposal.

An excessively cautious approach and an unwillingness to recognize the centuries of successful and safe agricultural experience with traditional biotechnology has produced major delays in field research on organisms and processes produced by molecular genetic engineering. To date a dead genetically-engineered *Bacillus thuringiensis*, three genetically-engineered tobaccos, and some animal vaccines have been field evaluated. All agree that product not process should be assessed for risk. Yet in the current U.S. regulatory environment, a product made by the process of recombinant DNA is subjected to much evaluation prior to the initial field test while an equivalent product made by traditional or cellular genetic engineering requires no such evaluation. Such illogical thinking is constraining and, to some extent, constipating field research with biotechnology products and processes produced by molecular genetic engineering techniques. Already at least one test of an agricultural product has been done with approval outside the country (in New Zealand) possibly because of the unrealistic U.S. regulatory policy.

Bioproducts and Bioprocesses for Food and Agriculture

It is very early in the development of products and processes from the new biotechnology for commercial use in agriculture and food. World sales of such products are probably not more than $100 million in 1986. Conservative projections suggest that they might be $5 to $10 billion by the turn of the century with considerable growth thereafter. These products and processes are expected to significantly lower unit costs of production of commodity crops and animals. In addition, they are expected to provide significant additional value-in-use to existing commodity agricultural products. An early example was the conversion by Canadian scientists of rapeseed, an industrial oil seed crop, to canola, an edible oil seed crop. This biotechnology conversion has made canola the major source of edible oil in Canada. Canola has just been recognized as GRAS, generally recognized as safe, in the U.S.

Cellular and molecular biotechnology should provide many additional changes that will increase the value-in-use of crops and animals. In addition, the need for processing of animal and plant food may, in part, be replaced by genetic changes in the animal or crop to eliminate the need for some of these expensive chemical and engineering processes. Some examples of suggested products and processes will be listed below.

Products and Processes for Crop Production. Diagnostics will enable the early detection of disease pests and resistant pests so as to provide more effective crop protection. Diagnostics may also provide tools to identify crops with improved value-in-use. In addition, diagnostics may be used for facile identification of proprietary genetic material. Restriction maps of genetic material are being evaluated for fingerprint proprietary bacterial and crop material.

Microbial products will become increasingly important in insect and disease protection. Viruses, bacteria, and fungi are already being used in various parts of the world for insect control. Genetically-engineered rhizobia with improved nitrogen fixation and yield capabilities in certain legumes are expected to be in field trial in 1987. Such improved rhizobia are anticipated to increase nutrient input and yield of crops such as alfalfa and soybeans. Genetically-engineered microbes may also be useful in post-harvest processing of crops such as silage making for improved nutritional value of the silage. Undoubtedly, microbes will become more important in crop production, decreasing the need for chemical pesticide and fertilizer use.

Genetically-engineered seeds will be the major impact of the new biotechnology on crop production. In the near term, these seeds will enable plants to resist or detoxify herbicides, thereby increasing the breadth of crops on which effective and/or low-cost herbicides can be used as well as enabling the farmer to rotate crops. The residues from previous herbicide treatments have constrained many such rotations in recent times. Such herbicide-safe crops probably will be the first generation of genetically engineered plants. Shortly thereafter, plants with disease resistance, especially for viruses and for insects, should become available. In addition to biotic stresses, plants possessing resistance to abiotic stresses are expected to be developed. One of the most costly inputs for crop production is fertilizer nitrogen. In the early 1990's plants genetically engineered to fix their own nitrogen are expected to be developed in the laboratory. Possibly before the turn of the century, self nitrogen-fertilizing crops would become initially available for commercial use.

Beyond these cost reductions, genetically-engineered plants with capabilities beyond those of present plants are expected. Such plants will have their harvested parts modified in quantity and/or quality so as to more closely match consumer needs. Herein lies one of the major opportunities of the new biotechnology for crop agriculture and food. In addition, existing or new crops will be produced by biotechnology to provide materials useful for the non-food market. These may include industrial chemicals, specialty chemicals, polymers, and other materials of use to society. A renewable competitive energy source may be the largest impact of biotechnology on agriculture enabling the use of our excess crop production capacity beyond that needed for food to be used to meet some of our non-food needs.

Chemicals will be designed based on knowledge generated by the new biotechnology. These biodesigned chemicals will prove useful as agriregulators for the crop production process and may represent the future role of the agrichemical industry.

Animal Biotechnology Products and Processes. Diagnostics and vaccines for improved animal health care are already being developed and to a limited extent utilized. Therapeutics based on natural biological molecules, such as the lymphokine interferon, will improve animal health. Diseases are estimated to reduce U.S. livestock and poultry productivity by at least 20% annually, an estimated economic loss of $14 billion per year. Biotechnology products should greatly decrease these significant losses due to disease.

Food additives will improve the efficiency of animal production. Molecules based on natural biological products will decrease the need for synthetic chemical feed additives and the attendant consumer concern with such products. The potential for growth hormones in milk and pork production are the earliest such examples. It is interesting to note that supplemental ovine growth hormone is not expected to be of major significance in poultry production. The intensive poultry-breeding programs in recent decades utilizing traditional or organismal genetic engineering may have achieved the same result over a long period of time, as now possible with molecular genetic engineering over a short period of time. Some of these bioregulator molecules will improve the quality of the food product such as the indicated reduced fat content of pork produced by porcine growth hormone. In the longer term animals modified by genetic engineering will have a complement of genes to minimize disease loss and maximize efficiency of production and quality of the product.

Food Biotechnology Products and Processes. Research and development work on food products and processes is less advanced than that in the plant and animal area. For the most part, research to decrease process costs is just beginning. Undoubtedly, biotechnology will enable improvement in important consumer and health associated aspects of food. These may include longer shelf life, improved appearance, improved flavor, and increased perceived healthfulness of the food among others. The light-beer example noted earlier is one of the few completed products or processes in the food area.

Summary

The progress made in biological science and technology during the last five years has been impressive, providing a strong base for new biotechnology products and processes for agriculture and food. This science and technology base is much more advanced than most recognize. The next major impact of biotechnology, after that initially already being felt in human health diagnostics and therapeutics, will be in the agricultural and food area.

RECEIVED September 3, 1987

CHEMICAL MARKETING

Chapter 27

Commercial Biotechnology: An Overview

Peter Hall

SRI International, 333 Ravenswood Avenue, Menlo Park, CA 94025

During the past ten years, biotechnology has offered a unifying concept for viewing the commercial benefits to be derived from life processes. As the term is now used, biotechnology refers to the application of basic scientific disciplines involving the life sciences, chemistry and engineering to develop a range of powerful tools with the ability to provide commercial products and processes encompassing a range of industries. As time has progressed, the term biotechnology has become so broad as to include any life science associated activity with either real or remote commercial potential. While the media and much of the scientific community continue to use biotechnology (or generic engineering) to refer to state-of-the-art manipulations of genes in prokaryotes and eukaryotes (including plant cells) the definition has become much broader. The term often refers to both established and state-of-the-art endeavors as diverse as aquaculture, production of industrial enzymes, development of microbial agents for cleaning up oil spills and protecting crops, microbial fermentation to produce specialty chemicals and even the brewing of beer.

The scope and interactions of the various components of biotechnology are summarized in Figure 1.

The commercialization of biotechnology continues to be driven by:

- o the desire of entrepreneurial scientists to realize a financial return for their laboratory research developments;
- o the infusion of investment capital from private and corporate investors;
- o the desire of established, market-led pharmaceutical, chemical, agribusiness, food product and

0097-6156/88/0362-0322$06.00/0

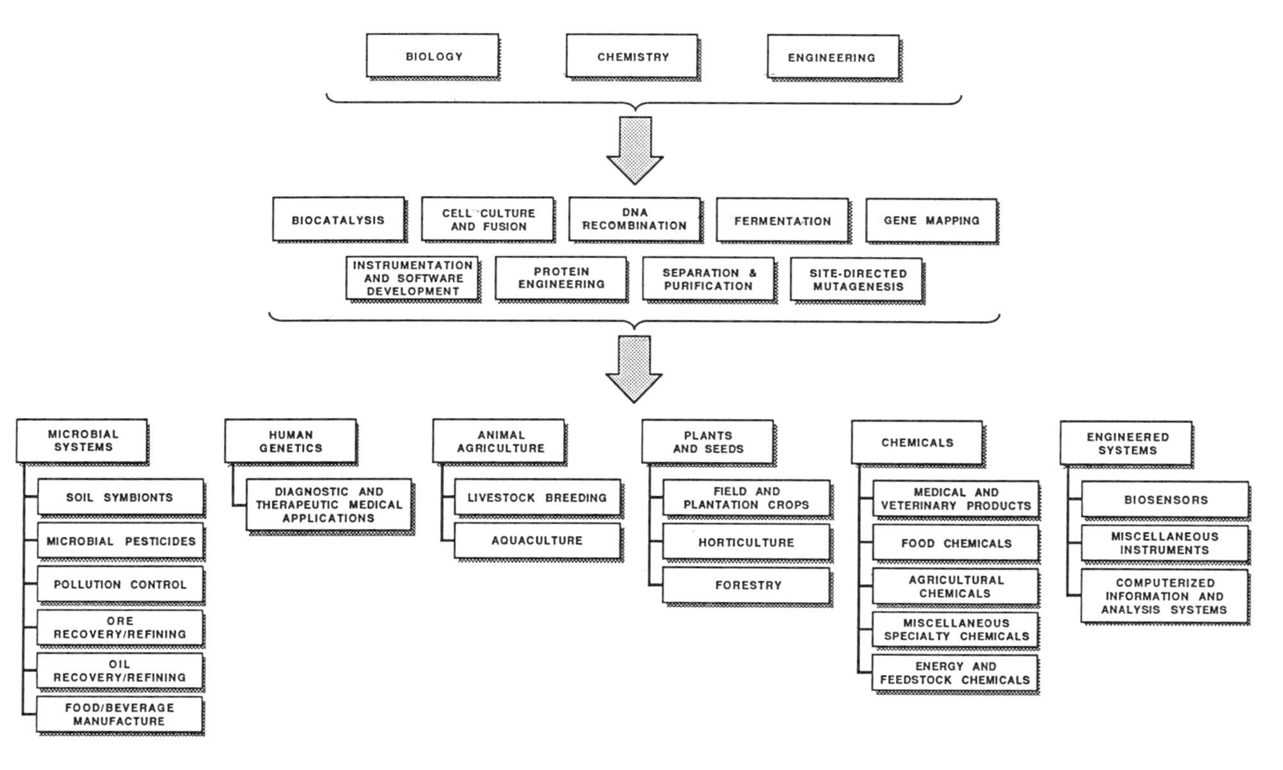

FIGURE 1 SCOPE AND INTERACTIONS OF BIOTECHNOLOGY

other diversified companies to protect their existing business and exploit new, technology-driven business opportunities;

- and the desire of regional, national and supranational governmental bodies to stimulate industrial development.

Worldwide commitment to biotechnology is illustrated by the high levels of government and corporate R&D funding involved--currently in the neighborhood of $4.5 billion. This is shown in Table 1. Like the pharmaceutical industry, with which it intersects, the biotechnology industry is increasingly seen as a valuable national asset by countries around the world. National and regional governments and development agencies are assuming a greater role in fostering the transfer of technology from academia to industry. Industry/university collaborations have become increasingly common and in most cases protect the interests of faculty members (e.g., freedom to publish) while providing industry with ample opportunities for establishing patent protection in the areas they are funding.

The basic underpinning technologies in most cases seem to present no significant barriers to biotechnology commercialization. For instance, the major therapeutic human proteins that have been targeted by industry have been cloned, with expression limits in many cases approaching theoretical limits.

The present barriers to the commercial development of biotechnology are threefold: (1) Selection of viable product targets--with all its successes, biotechnology is still often "technology looking for markets". (2) Limited market potential--creation of new markets or significant expansion of existing markets will not occur until a larger number of unique and proven compounds have been developed; for biotechnology, it will be difficult to create new markets. The greatest new market opportunities may be tissue plasminogen activator (TPA), human and animal growth hormones, a successful lymphokine treatment for acquired immune deficiency syndrome (AIDS) and other immunodeficiency-related disorders, diagnostics for screening for AIDS and cancer, and environmental monitoring, as well as new instruments and software packages. In the long term, new genetically engineered varieties of higher plants and microbial plant symbionts will provide unique new market opportunities. (3) Processing costs--production, separation, and purification costs are in danger of making biotechnologies non-competitive with existing processes and products. For most commodity and even specialty chemical products, yield improvements using

Table 1
CURRENT ANNUAL EXPENDITURES ON BIOTECHNOLOGY RESEARCH AND DEVELOPMENT
(Millions of U.S. Dollars)

COUNTRY/REGION	SOURCE		TOTAL
	Government	Industry (est.)	
United States	525	1,500-2,000*	2,000-2,500
Japan	55	1,000	<1,100
Europe			
Belgium	166		
Fed. Rep. of Germany	76		
France	186		
Italy	45		
Netherlands	14		
Sweden	5		
United Kingdom	12		
Other	56		
Subtotal	560	>500	<1,100
TOTAL	1,140	3,000-3,500	4,200-4,700

*Of this total, about $500 million is accounted for by dedicated (start-up) biotechnology companies.

Source: SRI International

biotechnology will be difficult to achieve and costly in relation to returns.

Product Development Status

Table 2 summarizes the current and projected markets for products arising from biotechnology. The market projections and other estimates included here are intended only as rough estimates. Note that for many of these market estimates, SRI has taken a more conservative, and we hope more realistic, position than is evident in many industry projections. A complicating factor in comparing the many market estimates and projections that have been generated is that they differ in the definitions of biotechnology used and the products and categories included.

Figure 2 summarizes the commercialization timetable for selected biotechnology product categories.

Biotechnology will have its great impact on markets for medical products and specialty chemicals after 1995, when second- and third-generation products and the products of protein engineering begin to make major inroads. For therapeutic compounds, practical, cost-effective new drug delivery methods will be imperative.

Cost will continue to be a major barrier to the implementation of biotechnology and the development of new products. Therapeutic products will encounter the same cost-containment pressures that are now increasingly being felt by the rest of the pharmaceutical industry. Early indications are that health care providers may already have some resistance toward some of these products, especially those for which alternative, lower cost therapies are available from traditional sources.

For most specialty chemicals and especially for commodity chemicals and energy sources, process economics will usually favor traditional fermentation or chemical synthesis routes.

Even with the numerous barriers to the commercial development of biotechnology products, a growing number of medical diagnostic and therapeutic products are gaining market approval around the world. Since 1983, more than 150 monoclonal antibody-based immunodiagnostic products have been approved in the United States alone. During 1985-1986, a number of important biotechnology health care products gained regulatory approval:

Novel Diagnostic Products

o In May 1985, Abbott received approval for its "Corzyme" enzyme immunoassay (EIA) and "Corab"

Table 2
ESTIMATED MARKETS FOR BIOTECHNOLOGY PRODUCTS
(Millions of Constant Dollars)

MARKETS	1985	1990	1995	Average Annual Growth (%)** 1985-1995
Health Care Products				
Therapeutics	100	1,400	4,250	39
Diagnostics	250	1,500	4,500	37
Vaccines	- -	100	1,000	(59)
Subtotal	350	3,000	9,750	39
Chemicals				
Industrial enzymes	- -	25	50	(15)
Water-soluble polymers	- -	25	100	(32)
Subtotal	- -	50	150	(25)
Feed and Food				
Additives and ingredients*	1,500	2,000	4,000	10
Flavors and fragrances	- -	100	200	(15)
Animal feed additives (incl. growth hormones)	- -	50	500	(59)
Subtotal	1,500	2,150	4,700	12
Waste Treatment and Decontamination	- -	500	1,000	(15)

MARKETS	1985	1990	1995	Average Annual Growth (%)** 1990-1995
Agriculture				
Seeds/Plants (by recombinant methods)	- -	100	1,000	(59)
Microbial soil inoculants	- -	50	500	(59)
Microbial pesticides	- -	50	150	(25)
Subtotal	- -	200	1,650	(53)
Instrumentation				
Laboratory bioreactors	100	180	320	12
Biosensors	- -	25	100	(32)
Computer systems	50	100	200	15
Laboratory instrumentation	275	675	1,700	20
Downstream processing equipment	130	200	270	8
Production plants	60	100	130	8
Subtotal	615	1,280	2,720	16

	1985	1990	1995	
TOTAL - All Markets	2,465	7,180	19,970	23

* High fructose corn syrup (HFCS) produced by means of immobilized enzymes.
** Growth rates in parentheses are from 1990-1995.
Source: SRI International

	MAJOR IMMEDIATE HURDLE			
	Market Acceptance	Process Economics	Technical Breakthrough	Initial Market Entry
Pharmaceuticals				
Hormones	■			1983
Interferons and Lymphokines	■			1984
Enzymes	■			1987
Vaccines	■			1986
Antibiotics		■		1990
Diagnostics	■			1983
Agriculture (Animal)				
Vaccines		■		1983
Growth Promotants		■		1988
Gene Transfer			■	1995
Agriculture (Plant)				
Improvement of Desired Characteristics			■	1984[1]
Plant Products		■		1983
Nitrogen Fixation			■	2000
Microbial Insecticides		■		1987
Specialty Chemicals				
Amino Acids		■		1983
Industrial Enzymes	■			1975
Vitamins		■		1990
Single Cell Protein		■		1984[2]
Biopolymers[3]		■		1990
Steroids[3]		■		1990
Environment				
Pollution Control		■		1988
Microbial Mining			■	1990
Microbial Enhanced Oil Recovery		■		1990
Energy				
Chemicals from Biomass		■		1990

[1] The first commercialized varieties of agronomic importance have been produced through plant cell culture.
[2] ICI's Preteen operation based in Billingham, England doesn't employ a recombinant organism.
[3] Some products currently on the market employ microbial biotransformation.
Source: SRI International

Figure 2 COMMERCIAL APPLICATION OF NEW BIOTECHNOLOGIES

radio immunoassay (RIA) assays, the first DNA probe products to be approved and marketed.

- In August 1985, Integrated Genetics received approval for its "Gene-Trak" DNA probe salmonella test for food.
- In September 1985, Enzo Biochem received approval for DNA probe tests for herpes. They are marketed by Ortho (Johnson & Johnson).
- In July 1986, Integrated Genetics received approval for its "Gene-Trak" DNA probe Campylobacter test.

Therapeutic Products

- On October 11, 1985, human growth hormone (Protropin) from Genentech was approved for the treatment of hypopituitary dwarfism. Protropin is Genentech's first product. Orphan drug status has been granted in the United States for growth hormone deficiency and Turner's syndrome. Other indications are under consideration. Approval is expected soon throughout Western Europe.
- On June 4, 1986, alpha-interferon from Hoffmann-La Roche Inc./Genentech and Schering-Plough/Biogen was approved for the treatment of hairy cell leukemia. Schering's Intron-A is being marketing in a number of countries. It has been registered in 11 countries altogether, with approval expected soon throughout Western Europe, as well as in Japan. The U.S. approval is the first for Roche's Roferon-A.
- On June 19, 1986, Ortho Pharmaceutical Corp. received approval for its Orthoclone OKT 3 monoclonal antibody-based product for therapeutic use. This product is used to treat acute allograft rejection in renal transplant patients.

Vaccines

- In June 1986, hepatitis B vaccine from recombinant yeast developed by SmithKline was first introduced in Singapore and Malasia as Engerix-B. This was the first market introduction of a recombinant human vaccine.
- On July 23, 1986, recombinant hepatitis B vaccine from Chiron Corp. was first approved for marketing in the United States. It will be marketed by Merck & Co. as Recombivax HB.

Miscellaneous

- In 1986, the U.S. Food and Drug Administration (FDA) granted orphan drug status to Wellcome's monoclonal antibody-based FAb digitalis antidote product, Digibind. Digibind is available on a limited basis in the U.K.

Perhaps the single most important medical product now under development using biotechnology is TPA. This product has a ready market and unless it runs into unexpected regulatory difficulties, could have worldwide sales of $1,000 million by 1995.

A number of promising product developments outside the health care field were also announced:

- In January 1986, the U.S. Department of Agriculture (USDA) approved pseudorabies vaccine for use on swine from TechAmerican Group Inc. This was the first approved genetically engineered live virus vaccine.
- An instrument for automated sequencing of DNA is now being marketed by Applied Biosystems Inc.

Diagnostics is the medical industry segment that was the first to develop, has grown more rapidly than any other, and will be the most radically transformed by biotechnology. Since its inception in 1983, the monoclonal antibody-based immunodiagnostic assay business has grown rapidly, and in 1985 sales in this segment totaled about $250 million worldwide. This market will now be joined by assays based on DNA probes, primarily for use in screening for genetic disorders. By 1990, biosensors will begin to be introduced for use in medical diagnostic applications. The search for effective AIDS diagnostics has provided a tremendous impetus to these areas over the past 2 years.

It is highly unlikely that biotechnology will have a significant impact on the commodity chemicals business by the year 2000; however, biotechnology will start to play a role in the development of industrial and research enzymes and water-soluble polymers by 1990. In addition, biotechnology will play a role in the development of some flavors and fragrances and of animal feed additives. It is unlikely that biotechnology will significantly replace standard fermentation processes, except in those cases in which an immobilized enzyme system is already in place. There is also little chance that vitamin production by means of a recombinant organism will be commercialized before 1995.

Biotechnology will have the following impacts on agriculture through the development of improved plant varieties between now and 1995:

- Development of recombinant microbial plant symbionts will enhance the uptake of nitrogen, phosphates, and other nutrients. Eventually, plants will be engineered with the ability to fix their own nitrogen.

- Development of microbial pesticides, ice-minus bacteria, and other microbial products will enhance plant growth.
- New crop varieties with improved processing characteristics and consumer appeal will be developed.
- Herbicide-resistant plants will be developed, and perhaps plants with resistance to various pests will be developed if the trait is readily identified and involves a single gene.

Agricultural biotechnology developments at the cellular level are already resulting in commercialized products. Gene transfer will begin to result in commercialized products after 1990, but it is doubtful that these products will begin to have a major impact until 1995. The greatest single barrier to the commercialization of new crop varieties based on plant biotechnology is the lack of basic understanding of cytogenetics.

The controversy surrounding environmental release could become a barrier to the development of new microbial systems, but it is likely to be a long-term problem. So far, the field testing of new genetically engineered plant varieties seems to have generated little concern from consumer and enviromental advocates.

Any significant development involving biotechnology in agriculture for the developed world is likely to raise questions regarding the impact of this technology on farm economics and the diminishing role of the small family farm.

The following milestones involving the commercialization of plant biotechnology for agriculture occurred in 1985/1986:

- In February 1985, Calgene announced the successful expression of glyphosate-tolerant genes in regenerated tobacco plants.
- In November 1985, Applied Genetics Sciences (AGS) received federal approval to conduct field tests of recombinant ice-minus *Pseudomonas flourescens* and *P. syringae* (Frostban) on strawberry plants in California.
- In February 1986, the U.S. Environmental Protection Agency (EPA) suspended AGS's experimental use permit to conduct field tests of Frostban and fined the company for violation of FIFRA regulations.
- On May 13, 1986, researchers at the University of California at Berkeley received Federal approval to conduct field tests of recombinant ice-minus *P. syringae* on potato plants in Northern California. Field testing commenced on August 6 following delays caused by local opposition.

- o On May 30, 1986, Agracetus began the first outdoor field trials involving genetically engineered plants (tobacco modified to resist crown gall disease).
- o On June 30, 1986, Ciba-Geigy received regulatory approval to begin the first field trials involving genetically engineered herbicide resistance in plants (atrazine resistance in tobacco).

Regulatory and Public Policy Issues

By far, the most important regulatory/public policy issue confronting the biotechnology industry during 1985 and 1986 has been environmental release. The controversy gradually intensified pitting such of the scientific community against activists such as Jeremy Rifkin. Since September 1983, Rifkin has brought eight biotechnology-related lawsuits against governmental agencies and individuals. He has so far won three and lost one; four are still pending. The real issues revolve around risk and how it should be managed in the face of uncertainty. Rifkin and his allies have successfully used public fears (some would say they have actually created these fears) regarding the potential adverse effects of a genetically engineered organism on the environment to force the U.S. regulatory and scientific communities to become more thorough in evaluating the potential risks. Recent industrial accidents involving new technology (e.g., Chernobyl, Three Mile Island, and even Bhopal) have crystallized the public's distrust of regulatory agencies and their ability to adequately protect the public's interest, and have increased concern about allowing technology to be used before all the risks involved have been thoroughly assessed. The recent findings that the biotechnology company, AGS violated EPA regulations by conducting unauthorized tests outside the greenhouse and that the USDA approved an animal vaccine without first consulting recombinant DNA experts within the agency and in other government agencies have understandably added to the public's concern and skepticism.

On June 18, 1986, President Reagan signed a U.S. federal regulatory guidelines package compiled by the Office of Science and Technology Policy (OSTP) in conjunction with federal regulatory agencies. This package was the culmination of 2 years of work by OSTP on developing an overall framework for U.S. federal regulatory responses to the many facets of biotechnology. The underlying principles followed in developing these guidelines have been that biotechnology should and can be regulated under existing regulatory statutes and provision and that products developed using biotechnology should be evaluated on the same basis as those produced by any other production method. Although the Administration would like to use the same mechanism for regulating biotechnology that is used for other products, it has allowed for the creation of numerous scientific advisory panels to assure that each application is adequately evaluated.

Industry seems to be largely in favor of these guidelines, hoping they will forestall legislation now pending in the House and Senate to deal with regulatory issues. Critics of the guidelines are primarily concerned about their overall flexibility and failure to define certain critical terms such as environmental release. Rifkin's Foundation on Economic Trends has filed suit against the government, claiming that the new guidelines have no scientific basis and fail to adhere adequately to already established procedures regarding the filing of environmental impact statements.

U.S. regulatory policy has been and continues to be the primary model for biotechnology regulatory policy outside the United States. Thus recent developments here will have far-reaching effects in other countries.

Biotechnology Industry Developments

In many ways, 1985 and 1986 have heralded a complete turnaround in the fledgling biotechnology industry, at least based on the reaction of Wall Street. Substantial gains in biotechnology company stocks have been linked with a number of significant "headline-making" events, including:

- The reduction or elimination of financial losses among biotechnology start-up companies that begin in 1985.
- Intensification of acquisition and joint-venture activities--primarily the announcement by Lilly in late September 1985 of an agreement-in-principle to acquire Hybritech and the announcement in October 1985 of the acquisition of Genetic Systems by Bristol-Myers.
- Regulatory approval by the FDA in October 1985 of Genentech's recombinant human growth hormone product, Protropin--the first recombinant therapeutic agent to be developed, manufactured, and marketed by a biotechnology start-up firm. This approval was followed in 1986 by the approval of a number of other products developed by start-up companies.

- Publication of a National Cancer Institute (NCI) study of interleukin-2 in a variety of cancers--giving a tremendous boost to Cetus, Immunex, Collaborative Research, and Biogen.
- Early reports on the efficacy of recombinant TPA as a thrombolytic agent.
- Promising results in the diagnosis and therapy of AIDS.
- The success of the monoclonal antibody-based immunodiagnostic market in the United Stated, currently about $150 million.

Biotechnology start-up companies are gaining more business acumen as demonstrated by the hiring of top executives and other employees from established, market-led companies; the start-ups are no longer relying solely on the leadership of scientific researchers turned entrepreneurs.

They are also learning the necessity of selecting unique product targets for which they have a proprietary position. Achieving this goal is particularly difficult in biotechnology, a field in which there are more companies than product targets and in which patent protection is often questionable. Companies are focusing on narrower product areas in which they have known expertise. The minor shakeout involving biotechnology companies producing L-phenylalanine using improved fermentation and biocatalytic methods in 1985 illustrates the difficulties inherent in focusing on a product target simply because the product has a large and growing market and proprietary technology can be developed to reduce production costs.

Biotechnology start-up companies are becoming more integrated, encompassing manufacturing, marketing, and distribution capabilities in addition to technology capabilities alone. Biotechnology companies are increasingly finding it necessary to form alliances with established, market-led companies in order to gain new capabilities. The established companies are using collaborations with start-up companies as part of their own biotechnology business strategies.

An increasing number of start-up companies are making the transition from being "technology boutiques" to being fully integrated product companies. None of these companies will forsake their strong technology underpinnings because the biotechnology industry will always remain technology-driven to a large extent. However, the developers of monoclonal antibody-based immunodiagnostics have demonstrated that biotechnology can address real market needs and that biotechnology companies can successfully sell products. Genentech is currently the only start-up company based on the development of recombinant DNA products that has made the transition, but others such as Cetus and Amgen are sure to follow. Some companies, such as Celltech and Damon Biotech, are realizing their goal of becoming primarily contract manufacturers, using their own proprietary technology to produce bulk products for other companies. These companies are successful and are building large-scale production facilities.

Although biotechnology start-up companies will continue to play a leading role in the development and commercialization of biotechnology, the established, market-led companies are ahead in many ways and are gaining momentum daily. The established companies have the financial stability and market presence needed and are also beginning to approach, and in some cases to overtake, the start-up companies in terms of technology. Established companies have been able to gradually buy the technology by funding R&D contracts with biotechnology start-up companies, obtaining an equity position in those companies, forming joint ventures to pursue product development goals, and outright acquisition. The desirability of biotechnology start-up companies as acquisition candidates has become one measure of their success and future potential.

In addition, some established, market-lead companies are successfully building their own internal biotechnology R&D capabilities. Outstanding examples of companies using that approach include DuPont, Eastman Kodak, and Monsanto. Japanese companies from the pharmaceutical, chemical, food and beverage, and heavy manufacturing industries now commonly include biotechnology as a component of their long-term business development strategies.

RECEIVED August 19, 1987

Chapter 28

Selected Applications of Bioprocesses for Chemicals

Acrylamide, Vitamin C, and Phenylalanine

Jerry L. Jones[1], W. S. Fong[1], P. Hall[2], and S. Cometta[1]

[1]SRI International, 333 Ravenswood Avenue, Menlo Park, CA 94025

Although biotechnology offers real commercial opportunities for improving the production of commodity, specialty, and fine chemicals and for producing new products, the near-term timing and extent of these opportunities are difficult to predict. This paper summarizes the techniques and process economics of three new bioprocessing routes for chemicals: acrylamide production by enzymatic transformation of acrylonitrile to replace an existing chemical catalysis process, ascorbic acid production using a recombinant microorganism to replace several steps in the current process, and phenylalanine production by fermentation or by two alternative enzymatic conversion routes. These processes illustrate the complexity, time, and cost of evaluating new applications for biotechnology in chemicals production. Finding these applications will require multidisciplinary teams, advanced facilities, and considerable prior knowledge about the bioconversion steps to be used.

New markets resulting from biotechnology have been forecast to represent tens of billions of dollars in annual sales by the year 2000. Most people expect the major markets for biotechnology in the next five to ten years to be for producing new diagnostic and therapeutic products for human health care. These opinions are supported by the publically reported investments to date in biotechnology. The Policy Research Corporation International (PRCI) recently published estimates of total investments in biotechnology between 1978 and 1985 in the United States (1). Of the total of ~ $4 billion dollars,

---

[2]Current address: Resources Development Foundation, Mount View, CA 94042

0097-6156/88/0362-0336$06.00/0

- Almost 75% has been devoted to health care.
- About 60% has been allocated to work on therapeutic products with 70% of that for anticancer products.
- 18% has been applied toward products for agriculture with only 4% of the total for agricultural chemicals.
- 4% was targeted for specialty chemicals.

Major market opportunities exist in crop agriculture, including new and improved microbial pesticides, microbial soil inocula for nitrogen fixation and increased release of phosphorus fertilizer, and improved plants in terms of pest and disease resistance or pesticide resistance, their impacts are likely to be mid- to long-term (10 to 25 years). Therapeutic products for animal agriculture also represent major market opportunities and are more likely to be near- to mid-term.

The near-term outlook is certainly not as clear or bright for the use of biotechnology to improve the production of commodity, specialty, and fine chemicals (nonprotein chemicals) or to produce new products. Microbial fermentations and enzymatic transformations are already used in the production of a range of products, including

- Commodity Products: ethanol, monosodium glutamate, high fructose corn sweetener, lysine, and recently acrylamide.
- Specialty and Fine Chemicals: industrial enzymes, xanthan gums, citric acid, vitamin C, vitamin B_{12}, miscellaneous low-volume amino acids, and recently the amino acid phenylalanine and sweetener aspartame.
- Whole Cell Products: microbial pesticides and recently recombinant organisms for various agricultural applications.

From numerous conversations with the staff of chemical companies worldwide over the last several years, we conclude that few individuals, including ourselves, have been able to identify precisely where to focus bioprocessing R&D work for chemicals. In fact, numerous companies are struggling to find enough good commercial targets to justify further growth of biotechnology programs directed specifically toward chemicals production.

In this paper we examine three examples of new bioprocessing routes (at various stages of development or use) for chemicals. In selecting the examples, we limited our scope of biotechnology applications to recombinant DNA use in microbial fermentation processes or for microbial production of enzymes for biochemical transformations.

In looking for market opportunities one can consider various areas:

- Existing chemical products and markets
  - Those produced entirely (or with only slight chemical modification) by extraction from natural plant sources (e.g., guar gum, alginates).
  - Those produced entirely by chemical synthesis from oil, gas, or coal-derived feedstocks.
  - Those produced directly by fermentation (aerobic or anaerobic) of sugars.
  - Those produced by a combination of fermentation and chemical synthesis steps.
  - Those produced by a combination of chemical synthesis and enzymatic transformation steps.

- Modifications of existing products for existing markets
- New products for existing markets
- New products for new markets.

Although all areas offer business opportunities, to be specific and somewhat quantitative, we selected three examples from the existing chemical products and markets category:

- Acrylamide production by enzymatic transformation of acrylonitrile to replace an existing chemical catalysis process.
- Ascorbic acid production using a recombinant microorganism to replace several steps in the current process.
- Phenylalanine production by fermentation or by two alternative enzymatic conversion routes.

In our discussion of the last example, we compare three bioprocess routes to a single product to illustrate the complexity of selecting a new bioprocessing route.

Acrylamide

The first example chosen is for a process with an unusual origin. Microbial processes are used at chemical production facilities worldwide to treat chemical waste constituents. Research on biotransformation of waste streams in Japan led to a new enzymatic process for one of the smaller tonnage commodity products--acrylamide from acrylonitrile.

Acrylamide's two reactive centers (vinyl and amide group) can be reacted sequentially to form a variety of homopolymers and copolymers for use in applications such as water treatment, pulp and paper production, and mineral processing. Major producers include American Cyanamid, Dow Chemical, Nalco Chemical, Mitsubishi Chemical, and Mitsui Toatsu. Worldwide production is about 100,000 metric tons (mt).

Acrylamide is produced by the hydration of acrylonitrile. Until the 1970s, the route generally followed consisted of reacting acrylonitrile with 85% sulfuric acid at about 90°C to form acrylamide sulfate, followed by neutralization with ammonia at about 43°C and separation of product by crystallization. More recently, acrylamide has been produced by a catalytic process in which acrylonitrile in solution is passed over a catalyst such as Raney copper at approximately 100°C, followed by purification and concentration to a 50% solution. Although solid acrylamide can be obtained from the catalytic process, the market is generally supplied by the acrylamide solution, which is a lower cost product. Acrylamide in solution has a value of roughly $2/kg, and worldwide production represents a value of about $200 million.

Nitto Chemical, a Japanese producer of acrylonitrile, has been using microorganisms to eliminate the residual acrylonitrile in the waste streams from its production operation. As an extension of this technology, Nitto developed an enzymatic process for producing acrylomide and built a small plant with a capacity of 4,000 mt/yr. The hydration is catalyzed by nitrile hydralase, an enzyme produced by microorganisms such as *Bacillus*, *Bacteridium*, *Corynebacterium*, *Micrococcus*, *Nocardia*, and *Pseudomonas*:

$$CH_2{=}CHCN + H_2O \xrightarrow{\text{Nitrile Hydralase}} CH_2{=}CHCONH_2$$

The microorganisms for the hydration of acrylonitrile are typically cultivated aerobically. After harvesting, they are immobilized in acrylamide gels for use in an immobilized enzyme reactor. The process was recently evaluated at SRI International by Fong (2), and selected excerpts from that study are presented below.

The reaction must be conducted between the freezing temperature of the mixture and about 59°F (15°C) at a pH of 7-9. The use of this pH range minimizes the formation of by-products (such as acrylic acid) from further hydrolysis of the acrylamide formed by the hydration. The concentration of acrylonitrile in the reaction medium must also be kept low (e.g., 2 to 5 wt%) because acrylonitrile is toxic to the microorganisms. Under proper conditions, the hydration of acrylonitrile is essentially complete, yielding acrylamide at close to 100% of theory and producing only very small amounts of by-products such as acrylic acid.

From the SRI process evaluation, based primarily on Nitto patents, it was concluded that the enzymatic route has the potential to be economically viable in plants with capacities approaching 10,000 mt/yr or greater. The process economics could be improved with longer useful life of immobilized cells (1000-hour base case assumption by SRI) and/or higher enzyme activity per cell. The latter improvement can potentially be achieved by rDNA techniques.

Vitamin C - Ascorbic Acid

Building on the extensive previous work in applied microbiology, biochemistry and enzymology related to existing commercial fermentation products also offer potential for new processes. For example, various bacteria have been identified with particularly attractive traits in terms of specific chemical transformations (e.g., substrate specificity, conversion efficiency). Until rDNA technology became available, the only way to use the traits from several species for a given synthesis was in a mixed culture process (several species of organisms present) or in a tandem fermentation process using two pure culture steps in series. One good example of a potential application for combining specific biotransformation capabilities from different existing microbes into one organism is a new process proposed by Genentech for producing vitamin C.

Worldwide, the consumption of vitamins during the early 1980s exceeded 70,000 mt and represented sales of well over \$1 billion. Vitamin C was the dominant product and represented over 50% of the total production with a sales value of $\geqslant$ \$10/kg and a value worldwide of about \$400 million. Major producers include Hoffman-LaRoche, Merck, and Takeda.

Currently, only two vitamins, B_{12} and B_1, are produced entirely by fermentation. Vitamin C production includes, as one initial step, the conversion of D-sorbitol to L-sorbitol by *Acetobacter suboxydans*. The remainder of the process entails all chemical synthesis steps.

The production economics of the current process for producing vitamin C at the 4500 mt/yr scale are roughly as follows (3)(as a percent of plant gate revenue requirement, excluding G&G, sales, and research charges):

Materials purchases	37%
Labor	6
Utilities	4
Fixed costs	7
Capital-related charges	46
	100%

Somewhat less than 10% of the plant revenue can come from sale of by-products. Because of the multistep nature of the current complex process, plant investment costs as well as operating costs are high. Total investment costs for a 4500 mt/yr plant would be about $100 million today. A simplified process with fewer steps could offer significant potential savings. A direct single-step fermentation would be ideal. However, biosynthetic pathways for vitamins are complex and each pathway uses many enzymes. As a result, improvements in direct fermentation routes through biotechnology will be limited in the near term. An additional complicating factor with vitamin C is that it is readily oxidized in solution.

Another approach to using biotechnology for improving the production of vitamin C with possibly more near-term potential has been proposed and is under development by Genentech. The approach uses rDNA technology to produce a key intermediate for the chemical synthesis of vitamin C. The following description of the process is adapted from a recent paper by Genentech staff (4).

At present most vitamin C is produced by a modification of the Reichstein-Grussner synthesis that involves a microbial fermentation and a series of chemical steps (as shown in Figure 1). The last intermediate in the Reichstein-Grussner synthesis is 2-keto-L-gulonic acid (2-KLG), a compound that can easily be converted into L-ascorbic acid through a simple acid- or base-catalyzed cyclization.

An alternative microbial route to 2-KLG is shown in Figure 2. Organisms from the coryneform group of bacteria (*Corynebacterium*, *Brevibacterium*, and *Arthrobacter*) as well as species of *Micrococus*, *Staphylococcus*, *Psuedomonas*, *Bacillus*, and *Citrobacter* are able to convert 2,5-di-keto-D-gluconic acid (2,5-DKG) into 2-KLG. Furthermore, species of *Acetobacter*, *Gluconobacter*, and *Erwinia* can efficiently oxidize D-glucose to 2,5-DKG. Thus, 2-KLG can be produced from D-glucose via 2,5-DKG by a cofermentation of appropriate microorganisms from the above two groups or a tandem fermentation process.

Although the tandem fermentation represents a considerable simplification in the route from D-glucose to L-ascorbic acid, the Genentech goal was to simplify this process further by combining the relevant traits of both the *Erwinia* sp. and the *Corynebacterium* sp. in a single microorganism. To accomplish this goal, Genentech

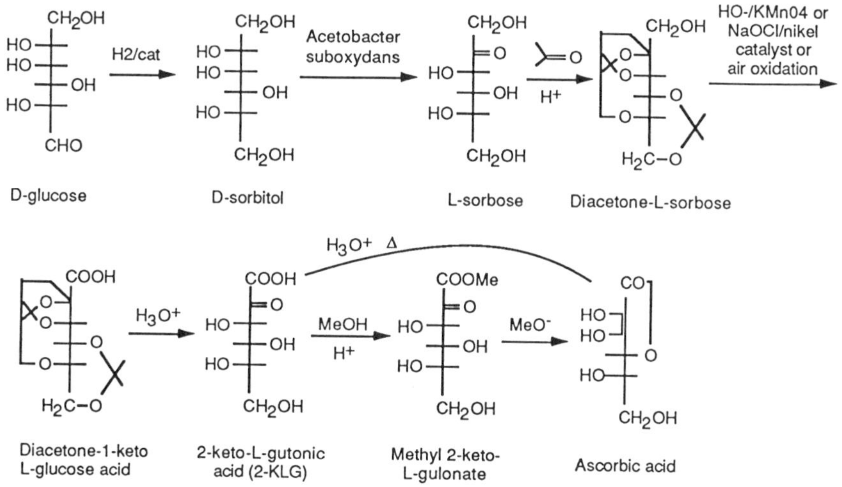

Source: Adapted from Ref. 4

Figure 1. Steps in the Production of Vitamin C.

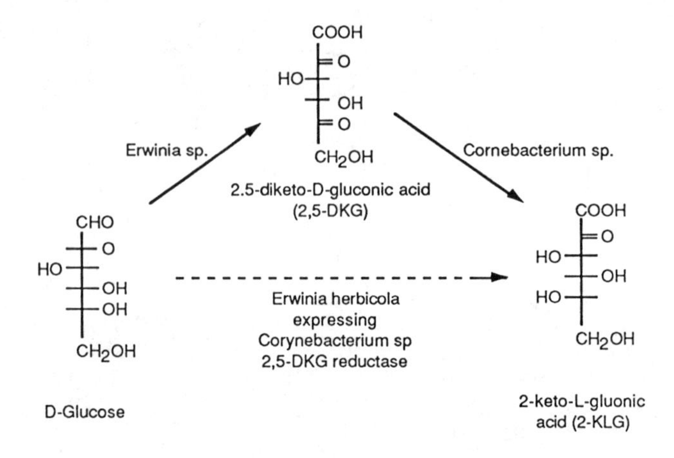

Source: Adapted from Ref. 4

Figure 2. Tandem Fermentation Route and Direct Route from Glucose to 2-KLG.

identified the 2,5-DKG reductase in the Corynebacterium sp. that was responsible for converting 2,5-DKG into 2-KLG. The gene for this reductase was then cloned and expressed in Erwinia herbiocola, a bacterium of the family Enterobacteriaceae that is able to convert D-glucose into 2,5-DKG. The resultant organism is able to convert D-glucose into 2-KLG in a single fermentation (Figure 2).

In this example, separate biochemical pathways from the two organisms were joined by cloning and expressing a gene from one in the other. However, even this simple case of metabolic pathway engineering is extremely complex. According to Genentech scientists, the dehydrogenases that serially oxidize D-glucose to 2,5-DKG acid are membrane-bound periplasmic enzymes linked to the electron transport chain, whereas the 2,5-DKG acid reductase is soluble, NADPH-linked, and cytoplasmic. Therefore, an effective interface between these two topologically separate pathways requires transport of 2,5-DKG acid across the inner cell membrane, a step that may be mediated by a specific permease.

In addition, efficient bioconversion of the glucose to 2-KLG acid requires an active NADPH regeneration system and some means of 2-keto-L-gulonic acid efflux from the cell. Exchange of metabolites between the periplasmic space and the medium, via diffusion through porin channels, is also necessary. The use of this or any similar scheme for the cost-effective production of L-ascorbic acid will require that all these elements, not just the immediate enzymatic activities, be present and appropriately regulated.

Genentech formed a joint venture to pursue this invention with Lubrizol Corporation. The joint venture is known as GLC Associates, and GLC is reported to be working with Pfizer to continue process development.

Phenylalanine

Recombinant DNA technology offers the potential for improving the yield and reducing the cost of producting key enzymes for use in specific steps in a synthesis pathway. Such advances can lead to new combined synthesis/enzymatic transformation processes that may be competitive with or superior to direct synthesis or direct fermentation processes. One excellent example of such a process is the production of phenylalanine.

The worldwide demand for phenylalanine has grown dramatically from less than 50 mt/yr in 1980 to more than 3000 mt in 1985. By 1995, demand should exceed 5000 mt with a total value of $50 to $100 million. This rapid growth occurred because L-phenylalanine (L-PHE) along with aspartic acid is one of the building blocks for the synthetic sweetener, aspartame. Prices for L-PHE have dropped over the last five-year period from well above $50/kg to ≤ $20/kg by mid-1985. More than 30 companies worldwide have developed various process routes to produce L-PHE, including

- Chemical synthesis starting with glycine.
- Fermentation.
- Chemical synthesis/enzymatic transformation, "the cinnamic acid route."

- Chemical synthesis/enzymatic transformation, "the phenylpyruvate route" starting with hydantoin and benzaldehyde.

Recently, the most interest has focused on the last three routes (see Figure 3). Developers of each process route claim to have "the lowest cost route." SRI recently prepared very preliminary process designs and cost estimates to determine how the processes compare from an economic viewpoint. The estimated plant gate selling prices were calculated using a discounted cash flow procedure and a 30% rate of return and were done in mid-1985.

The comparison was prepared in sufficient detail to include estimates of major operating costs (primarily feedstocks and labor) and preliminary estimates of capital investment (with an accuracy of no better than +50%, -25%). The process designs for the two bioconversion routes with immobilized enzymes are based on patent descriptions and literature descriptions and probably represent what might be achieved for second-generation plants or optimized existing plants (after several years of operation, perhaps with process modification). See references (5-14) for background on L-PHE production.

The results of the analysis are summarized in Table I. The design for the fermentation process is based on achieving a concentration of 30 g L-PHE/L in the fermentation broth (over a 72-hr batch cycle), a value that we believe is now being achieved or exceeded. The fermentation facility is designed with nine 120,000-L fermentors so that one fermentor can be unloaded every shift. The fermentation section of the plant is comparable in size to existing large antibiotics fermentation facilities. The yield from glucose was assumed to be 0.15 kg L-PHE/kg glucose.

In all base case designs, we assumed an 80% recovery of L-PHE in the process train downstream of the fermentor or bioreactor. The recovery train after fermentation includes cell harvesting by centrifugation, ion exchange, and crystallization (at 0° to 5°C). A higher yield of L-PHE from glucose, an improved level of recovery, and a lower glucose price could easily lead to a drop in the plant gate selling price of several dollars per kilogram. However, also note that, for a preliminary study of this nature, our base case capital cost estimate could be low by 50%. A 50% increase in capital cost to $35-$40 million would raise the estimated plant gate selling price to $19-$20/kg.

The plant using the cinnamic acid (CA) route was designed using the following assumptions:

- 95% utilization of cinnamic acid.
- 20:1 molar ratio of ammonia to CA.
- 75% conversion per pass of CA.
- Two-step crystallization for CA recovery and PHE recovery.
- Steam stripping for 99% recovery of ammonia and CO_2.
- 12 packed bioreactors operating in 6 parallel trains.
- Refilling of each bioreactor once every 14 days on a rotating basis.

  Enzyme loading in bioreactor of 1500 PAL units/L for a conversion of 15 g PHE/L/hr (at peak). (Design of 50% activity basis.)

Production of L-PHE by Aerobic Fermentation Process

Glucose + Cells + Nutrients + O_2 → L-PHE + More Cells + By-Products

Production of L-PHE from trans-Cinnamic Acid

t-Cinnamic Acid + Excess Ammonia $\xrightarrow[\text{phenylalanine ammonia-lyase}]{}$ L-PHE

Production of L-PHE by Transamination of Phenylpyruvic Acid with Hydantoin and Benzaldehyde as Starting Materials

Hydantoin + Benzaldehyde → Phenylhydantoin + Water

Phenylhydantoin + Water → Phenylpyruvic Acid + Urea

Phenylpyruvic Acid + Aspartic Acid $\xrightarrow[\text{aspartate phenylalanine transaminase}]{}$ L-PHE + Oxaloacetic Acid

Oxaloacetic Acid ↓ Pyruvic Acid

Figure 3. Three Routes Considered for Producing L-Phenylalanine.

Table I Comparison of SRI Base Cases for Production of 2000 mt/yr L-PHE

	Fermentation Route		Cinnamic Acid Route		Hydantoin/Phenylpyruvic Acid Route	
	$/kg	%	$/kg	%	$/kg	%
Materials and supplies						
Major feedstock(s)	3.35[a]		6.20[b]		7.45[c]	
Other	1.57		0.40		0.20	
Total materials and supplies	4.92	31.9	6.60	47.7	7.65	55.0
Labor	1.77	11.5	1.30	9.4	1.24	9.0
Utilities	0.51	3.3	0.35	2.5	0.45	3.2
Fixed costs	0.48	3.1	0.31	2.2	0.25	1.8
Total operating costs	7.68	49.8	8.56	61.8	9.59	69.0
Capital related charges and income tax	7.75	50.2	5.28	38.2	4.32	31.0
Total revenue requirement	15.43	100	13.84	100	13.91	100
Sources of revenue						
L-PHE	14.94	96.8	13.84	100	13.91	100
By-products	0.46	3.2	--		d	
Estimated total capital investment ($$10^6$)	~ 24		~17		~14	

[a]Glucose at $0.40/kg (or $0.18/lb).
[b]Cinnamic acid at $5.25/kg (or $2.40/lb).
[c]Hydantoin at $4.40/kg (or $2.00/lb); benzaldehyde at $1.65/kg (or $0.75/lb); acetic anhydride at $0.96/kg (or $0.44/lb); aspartic acid at $2.75/kg (or $1.25/lb).
[d]Sale value of pyruvic acid assumed equal to recovery cost.

Cinnamic acid loss is assumed to be kept to a low level by using a preliminary controlled pH (approximately 2) and temperature (3°C) crystallization (after steam stripping). The effluent from the first crystallization step is concentrated and then crystallized at a pH of approximately 5 and a temperture of 0° to 5°C. This route as well as the hydantoin/phenylpyruvic acid route are sensitive to feedstock costs. A decrease from \$5.25/kg to \$4/kg for cinnamic acid would decrease the PHE price from about \$14/kg to \$12.25/kg. A price of <\$4/kg for cinnamic acid seems unlikely because the raw material costs to produce cinnamic acid (benzaldehyde at \$1.65/kg and acetic anhydride at \$0.96/kg) are equivalent to about \$2/kg.

A 50% increase in capital cost would increase the base case plant gate price from about \$14/kg to \$16.5/kg. A 10% increase in cinnamic acid consumption would increase the PHE price by about \$0.5/kg.

The major assumptions used in the design of the plant for production of L-PHE from hydantoin, benzaldehyde, and aspartic acid are as follows: For synthesis of phenylpyruvic acid, we assumed

- Batch synthesis process for precursors
  - Overall yield of 95+% of theoretical to phenylhydantoin.
  - Overall yield of 90+% of theoretical from phenylhydantoin to phenylpyruvic acid.
- Recovery and recycle of acetic acid.
- No by-product credit taken for acetic acid formed from acetic anhydride addition.

For conversion of phenylpyruvic acid and aspartic acid, we assumed

- Bioreactor productivity of approximately 18 g PHE/L/hr (4 columns in parallel).
- 98% overall conversion.
- No by-product credit taken for pyruvic acid. Recovery cost assumed to be offset by revenue from sale.

The impacts of changes in hydantoin prices, changes in the capital investment cost, or recovery are as follows. A 50% increase in the investment would increase the plant gate selling price from \$14/kg to about \$16/kg. A drop of \$1/kg in hydantoin cost would reduce the PHE price by slightly less than \$1/kg. A 10% change in recovery from L-PHE would change the plant gate selling price by \$1.5/kg.

The results of our analysis show that, by any of the three routes considered, it should be possible to produce L-PHE at an actual selling price of \$15 to \$20/kg and receive an excellent return on investment. A price of less than \$15/kg also appears possible, but the probability of achieving such a level cannot be predicted based on the level of detail of the current analysis. At the level of design detail possible in a preliminary evaluation this nature, the differences in process economics between the routes do not appear to be great enough to allow ranking of the processes.

Choice of a process route for L-PHE production will be influenced by technical as well as economic/market issues, including

(1) Raw materials availability and price (numbers of existing and/or potential suppliers, or captive feedstock position).

(2) Potential market impacts of a completely "natural" source of feedstocks (glucose from starch) versus synthetic chemicals.
(3) Royalty levels and other licensing conditions.
(4) Extent of involvement in the L-PHE/aspartame business and other amino acid businesses (e.g., a producer of aspartame could favor the hydantoin/phenylpyruvic route if planning to captively produce both PHE and aspartic acid).

Improvements in process technology and operating economics are also possible for all three base case designs.

Finding Applications of Biotechnology for Chemicals

The three previous examples show that biotechnology offers real commercial opportunities, but predicting the extent and timing of impacts is difficult. In all three cases, considerable prior knowledge existed about the bioconversion steps that eventually were identified and selected as the basis for the new processes. These processes were not developed based on short-term research programs. Without the extensive prior knowledge gained from research, developments to date would not have occurred as rapidly.

In collecting the information for preparing these examples and from numerous other studies related to chemicals, we have reached the following conclusions regarding bioprocesses for chemicals:

- Most knowledgeable technical people in the chemical industry now view biotechnology just as one of several techniques, albeit very powerful, for improving chemical processes and products or for finding new processes and products. It is certainly not going to change the chemical industry overnight.
- Identifying and evaluting new processing options that include bioconversion steps add considerable complexity, time, and cost to R&D. Staff technical capabilities must go well beyond traditional disciplines in chemistry and chemical engineering.
- Successful R&D teams to date have been multidisciplinary and include bioscientists and biochemical engineers as well as more traditional chemists and chemical engineers. The larger multidisciplinary teams and associated facilities usually significantly increase the costs for chemical R&D.
- Successful R&D teams are not only multidisciplinary, they are often composed of researchers from different organizations. Single companies cannot always find or afford to keep on staff all the necessary technical expertise and facilities, especially with the realities of chemical industry restructuring.
- Finding applications for biotechnology in chemicals production requires the same basic inputs as other chemical R&D (time, money, good people, and facilties); no shortcuts are available.

Literature Cited

1. Murray, J. R. Bio/Technology 1986, 4, 293.
2. Fong, W. S. "Acrylamide by Enzymatic Hydration of Acrylonitrile," SRI International Process Economics Program View No. 84-2-4, 1985.
3. Wasselle, L. A. "Ascorbic Acid," SRI International Process Economics Report No. 144, 1980.

4. Anderson, S.; et al. Science 1985, 230, 144.
5. Backman, K. C. European Patent Application 0 140 606, 1985.
6. Calton, C. J. Bio/Technology 1985, 3, 761.
7. Calton, C. J. Biotechnology 1986, 4, 317.
8. Finkelman, M.A.J.; Yang, H-H. European Patent Application 0 140 714, 1985.
9. Hamilton, B. K; et al. Trends in Biotechnology 1985, 3, 64.
10. Klausner, A. Bio/Technology 1985, 3, 301.
11. Rozzell, J. P. U.S. Patent 4 518 692, 1985.
12. Schruber, J. J.; Vollmen, P. J.; Montgomery, J. P. European Patent Application 0 140 713, 1984.
13. Tsuchida, T.; et al. European Patent Application 0 138 526, 1985.
14. Yamada, S.; et al. Applied and Environmental Microbiology 1981, 42, 773.

RECEIVED September 14, 1987

Chapter 29

New Diagnostic Tools from Biotechnology

Martin Nash

Corporate Development, Synbiotics Corporation, San Diego, CA 92127

Recent years have seen the development and evolution of new tools in the diagnostic field, tools which derive from biotechnology. Generally, these are concentrated in three technology/product clusters. There are others - but these three are particularly notable. In part this is because they are quite different from each other, and in part because they are at quite different stages in their evolution as useful technologies. Specifically, they are monoclonal antibody development, nucleic acid probe development, and biosensor technology.

Monoclonal antibodies and nucleic acid (DNA or RNA) probes are emerging as diagnostically approved, commercial products in diagnostic testing markets. Biosensors have not yet reached that stage of development, but their general form can be discerned on the horizon and they are included in this analysis. I shall discuss these technologies and products in the light of industrial expertise rather than scientific expertise; that is I shall compare and contrast the products of these technologies according to the characteristics that are commonly accepted as determinants of success in the diagnostic industry. No matter how marvelous a technology or how intriguing its scientific possibilities, to become a widely utilized and successful diagnostic product it must meet certain tests of utility, and meet them in a way which provides differential advantage over competing products. This chapter explores how these tools of biotechnology fare against each other in a variety of product characteristics. It will look broadly at the areas in which the products of the new technologies will be employed in diagnostics, will assess their advantages (both absolute and relative) and will examine some of the criteria for commercial success which inescapably apply to any diagnostic product.

First, a consideration of monoclonal antibodies. The method for obtaining monoclonal antibodies is well known. The intent of this outline is to serve as a general reminder of many of the important

0097-6156/88/0362-0350$06.00/0

milestones in producing a monoclonal antibody-based product, and also to remind of the point that many of the steps are not well automated and so remain labor intensive.

PRODUCTION OF MAbs
Isolate Antigenic Determinant
Immunize Mice
Harvest Antibody Producing Lymphocytes
Harvest Cultured Myeloma Cells
Mix
Fuse
Screen Immunoglobulin Producing Hybridomas
Inject In Host Mice or Grow In Tissue Culture
Harvest
Purify From Ascites
Attach Reporter Groups

The business of harvesting, for most laboratories the business of screening, and certainly the feeding, housing, injecting, and finally sacrificing of the mice, contain a substantial labor component. Many would be quite surprised to contrast the salary industry must pay to a person who spends the work day sacrificing mice with the salary paid to post-doctoral workers in laboratories. Suffice it to say that the work of sacrificing mice and harvesting ascites fluid from their bodies is not sufficiently stimulating nor exciting to draw people to perform it for its own sake. At current production quantities, the body cavity of the mouse remains the production vessel of choice for monoclonal antibodies, however, when new products involving vaccines and therapeutic products come to market, other production schemes will have to evolve with them. The recognition of the fact that it is inefficient to produce 10, 20, and ultimately hundreds of kilograms of monoclonal antibody in mice is what has stimulated the great interest in hollow fiber, fermentation and other large scale mammalian cell culture production methods, and in fact what provided much of the excitement behind the activities of companies offering such products. In any case, issues of practicality of various production methods are the substance of another discussion, and the rest of this paper will assess conditions impacting diagnostic products currently being produced and expected to be produced in the reasonably near future.

We find in turning to the utilization of monoclonal antibodies in the diagnostic testing market there is relative concensus on their current, and for the next few years at least, projected financial activities. Sales actually began in the early 1980's and were more or less a million dollars in 1980. Depending on which survey one reads, sales in 1985 were in the vicinity of $50 million. Projections for 1990 are of course more challenging, but among the various market research and forecasting organizations there is rather remarkable agreement that sales of monoclonal antibody-based diagnostics will be in the $200 million area, with estimates ranging from about $175 million to about $210 million. These figures are worldwide with a bit over half the total concentrated in the United States. It is interesting that the same forecasters project that growth of revenues overall in the diagnostic reagent marketplace will be about 8% per year between now and 1990. The projections work out to 15% to 20% per year for monoclonal antibody-based diagnostics. If there was any question why companies are aggressive in MAb product development, these growth data supply the answer.

Monoclonal antibodies are primarily employed in two broad product classes; replacement products and novel products. The early monoclonal antibodies, as would be expected with any new technology, were largely replacements for the polyclonal antisera component in such products as pregnancy tests. Such substitutions carried the least technological and market risk yet provided more "sizzle with the steak" and often lower cost than their predecessor products. The market has progressed, and the science underlying it has progressed, to permit the production of antibody specificities which make it possible to both replace and even expand product lines. This is important to industry because it makes it possible to avoid competing on price and begin to release products based on improved performance characteristics, which are commonly translated into higher prices than the products they replace. An example of such a development is the replacement of T-cell tests which simply enumerate the number of T-cells in a patient sample with monoclonal antibodies which make it possible, and in fact even easier, to enumerate total T-cells, total B-cells, and in fact the ratios of certain significant subsets of T-cells. Finally, companies can produce monoclonal antibody products which perform tests which are not replacements for older assays but rather are products uniquely possible as a result of the specificity of monoclonal antibodies. It is in such products that the highest sales value can be derived from the technology; and so these are the most attractive to companies. Examples of such tests are the tumor marker tests currently emerging in the cancer market and the tests for natural killer cells emerging in clinical immunology, and such tests as the Synbiotics' test for the heartworm parasite. Parasitology has been a troublesome area for antibody companies as parasites commonly pass through different life stages where different antigens are expressed. Further, parasites are typically difficult to distinguish with polyclonal antibodies due to cross-reactivity. The heartworm is deposited in the bloodstream of several mammals by mosquitos; it then colonizes the heart muscle in numbers from 30 to a few hundred. The Synbiotics' assay is a 20 minute test for a shed antigen.

There are, however, areas of testing which provide medically useful information, but which are not amenable to any form of antibody analysis. These include analysis of genetic abnormalities or disease susceptibility at the gene level. Such assays, and others which I will discuss presently, are thought to be the province of DNA and RNA probe technology.

In turning our attention to probes, first we should note that not all probes are created equal, nor in fact in the same way. "Long" DNA probes are the more traditional entity, while "short" DNA probes are the more recent development produced synthetically, often on DNA synthesizers currently manufactured by some eight companies.

In "long" probe production, DNA from a target of interest (for example, a virus or pathogenic bacterium) is isolated. The broth is purified to obtain DNA free of debris, which is then inserted into a suitable vector (eg plasmid, bacteriophage). These molecules are then cloned thereby replicating the inserted DNA and this DNA of interest is then removed by restriction enzymes. The resulting material is purified (for example on a gel) and then labeled with signal-generating reporter groups (or their precursors) using polymerases or other appropriate enzymes. Briefly, note that there are a fair number of steps, and that many of the steps involve purification of a solution, steps which are not well liked by manufacturing concerns since they are both expensive and exceptionally troublesome.

PRODUCTION OF "LONG" PROBES
Isolate DNA Fragment from Target Molecule
Selectively Restrict, Isolate and Test Target Fragment
Insert DNA into a Vector Molecule, e.g. Plasmid
Propagate Host Molecule
Enzymatically Restrict DNA
Purify
Identify and Isolate Desired Probe Fragment
Enzymatically Incorporate Reporter Group
Isolate Labeled Probe

The production cycle for "short" DNA probes is considerably simpler, and does not involve the complex purification steps. To produce "short" probes using chemical DNA synthesis, one begins by identifying a desired sequence. The chemist must know the sequence desired before he can program the microprocessor in the synthesizer. It is certainly worth remembering that this requirement for sequencing the target and identifying the particular variable region of interest does not exist for "long" probes. A single "short" probe synthesis can produce enough material for millions of tests, typically in a single day.

PRODUCTION OF "SHORT" PROBES
Obtain Target Sequence
Chemically Synthesize Modified Probe
Purify
Attach Reporter Groups

Using nucleic acid probes to detect RNA rather than DNA in a patient sample is an interesting and in some ways advantageous approach. Under some conditions, RNA probe duplexes are more stable than DNA probe duplexes, so the assay can be simpler.

Further, the RNA probes have advantage in that the user does not have to pre-treat the specimen to separate strands of double stranded native DNA. This advantage saves time and reduces the technical complexity of the assay. It also means that in the final assay the probe need not compete with complimentary strand to anneal to the target strand. RNA probes are commonly intermediate between long and short DNA probes in length. RNA probes tend to generate somewhat less signal than either antibodies or "long" probes, but more than "short" probes. It is not necessary to determine a target sequence prior to producing an RNA probe.

There are currently two RNA probes approved for diagnostic use, both from GenProbe in San Diego. The assay employs solution hybridization, thus has substantially shorter performance times than DNA probe assays, and provides more sensitivity than DNA probes because there is substantially more target RNA than native DNA in the sample. There are technical problems to be overcome, and RNA probes are not used to detect the presence of viruses since their current methodology detects ribosomal RNA and viruses lack ribosomal RNA. Nevertheless, their ease of use, speed, and sensitivity do impart some inherent advantages to RNA probes over DNA probes.

Technical and production aspects do not in themselves determine the desirability of a technology to industry. They must be viewed through the filter of characteristics which apply to all in vitro diagnostic tests and determine their suitability for development into product. These are not the only elements which enter into a product decision, but they do enter into all decisions:

BASIS OF COMPETITION
Cost of Production
Ease of Production
Procedural Ease
Sensitivity
Specificity
Technical Familiarity
Time of Product Development

COST OF PRODUCTION

o As used in this context, cost of production will apply to those costs directly attributable to production of the antibody. It does not encompass acquiring capital equipment since tax/depreciation impacts vary widely among companies for the same plant and equipment. Additionally, the product volume against which equipment can be written off is conjecture at present. In the case of the MAbs, this term presumes murine production yielding some 5 ml of ascites fluid per mouse and 5 to 10 mg per ml of antibody.

EASE OF PRODUCTION

o Ease of production skews the mix of desirable products toward the better known of the technologies in question (MAbs) and the more automated (short probe). As diagnostic companies are not primarily pursuing new information of a scientific nature, but rather the most efficient and failure-proof production technology, anything which involves a great deal of "art" or extreme complexity will be relatively disadvantaged. Art particularly is anathema to rational production decisions and to the regulatory and supervisory mechanisms in the healthcare industry of most countries. This acts to the detriment of such products as "long" probes.

PROCEDURAL DIFFICULTY

o This aspect eases this consideration into the area which companies call the marketplace and consumers call the laboratory. The term could also be called laboratory procedural ease. It is common for tests which are complex to be run less successfully by technicians in the diagnostic laboratory, who are usually not interested, nor should they be, in blazing new research trails or in creating unique diagnostic events. Perhaps the most overwhelming advantage of MAbs over any type of probe assay is that they are used in familiar assay formats, formats compatible with existing instrumentation.

SENSITIVITY

o This is commonly, if incompletely, defined as positivity in the disease state. Monoclonals have an intrinsic advantage over the competing probe methodologies discussed herein simply because of the number of copies made of DNA, RNA and proteins respectively. One copy of DNA makes many copies of RNA, which then makes many copies of protein. Monoclonal antibodies able to recognize each manifestation of the protein thus have an inherent advantage in signal generation. In this respect, MAbs have the advantage of thousands of antibody binding sites per target entity. RNA probes similarly have more RNA to "look at" than DNA probes have DNA.

SPECIFICITY

o Defined as negativity in non-disease, this is a critical but complex issue. Probes are inherently specific for the target of interest since they look directly at the DNA, hopefully at the region which codes for the identity of the organism in question or the stretch which defines the target of interest. In the "long" probe embodiment, and in the RNA probe embodiment, often the target is not sequenced and thus other regions are included in the area with which the probe is reactive. This causes most investigators to report that it is difficult to detect one or few base pair differences between different organisms with a "long" probe.

o Monoclonals are able to recognize a target protein site consisting of as few as five amino acids. This is useful when a site consisting of five amino acids is unique to the target, but can in fact be something of a problem if the identical sequence is present on an unrelated protein. The example is that one of the commonly used monoclonal antibodies to human T-cells was found after some two years of wide-spread use to be present on normal brain tissue. That is to say that the high specificity imparted by recognizing an antigenic site of as few as five amino acids can lead to cross-reactivity with the same site in unrelated molecules or tissues. This problem is minimized in diagnostic assays, particularly of fluids, by restricting the application to that for which the test is shown to be specific. It is, however, not infrequently a complication in research applications. Nature has generally been kind to immunologists in producing markers which permit monoclonals to detect all relevant strains but no others. Companies producing monoclonal antibodies must presume that this situation will hold in future.

TECHNICAL FAMILIARITY

o This refers to the familiarity of the customer with a given test protocol. Many existing diagnostic techniques (for example ELISA) are somewhat complex procedures and yet are routinely practiced. Procedural difficulty is not, therefore, in itself a barrier to the entrance of a product into the diagnostic marketplace. It is more procedural familiarity that can retard new entrants. Clearly, monoclonal antibodies, and to a lesser extent RNA probes when used in solution hybridization testing, hold an overwhelming advantage over DNA probes which must be baked onto filters or used on difficult-to-handle solid state surfaces.

TIME OF DEVELOPMENT

o For MAbs this is defined as the period from identification of the immunogen and its purification to the release of a product; for "long" probes it is the period from availability of a DNA fragment to the product; for "short" probes it is the time from which the target sequence is known to the product. Regulatory issues are explicity omitted from this timetable.

More generally, DNA probe tests will involve challenging extraction, sample purification, hybridization, and some substantial form of)preferrably non-isotopic) amplification of signal. As a result, basic instrumentation work needs to be done. This will retard test development.

So the question is, Where does the advantage lie in the "current state of the art"? As we have defined the common bases of competition, the advantages are described as follows:

BASIS OF COMPETITION	ADVANTAGE
Cost of Production	Short Probes, RNA Probes
East of Production	Short Probes
Procedural Ease	Antibodies
Sensitivity	Antibodies, Long Probes
Specificity	Roughly Equivalent
Technical Familiarity	Antibodies
Time of Product Development	Short Probes, RNA Probes

As is common with data when presented in this format, the particular characteristics are not weighted. When a company weights these various criteria, particular significance is given to the customer's ability to utilize the product successfully (procedural ease, technical familiarity), to ability to make the product successfully, and to cost of production. In none of the cases described above is the cost of production prohibitive for tests carrying the sorts of profit margins typical in the diagnostic industry, so for the purposes of this evaluation cost of production is roughly the same between "short" probes and RNA probes and is not really significantly more for antibody production in the quantities required for a diagnostic assays, so this element is not controlling. Clearly then from the weighted analysis of these bases of competition, the current advantage is with monoclonal antibodies. A factor which explains every bit as much as their experience advantage why they are the biotechnology tool most commonly employed by companies today.

Considering inherent technology - the two tools differ as well.

On the far left are products with clear MAb inherent advantage - far right, probe advantage resides. This table illustrates product areas where one or another leads.

APPLICATIONS OF MAbs AND NUCLEIC ACID PROBES

Tumor Associated Antigens   Cancer Diagnostics   Cancer Susceptability   Latent Viral Infection

TDM   Microbiology   Disease Susceptability

Cell Typing

Virology

Immunoassay Replacement   STD   Antibiotic Resistance   Genetic Screening

HLA Typing

MAb Advantage ⟶   ⟵ Probe Advantage

It is not uncommon to hear it said that probes are today where monoclonals were five years ago, a comment meant to indicate that probes are coming along roughly the same development track. It can be argued that biosensors are coming along the same track but are a couple of years farther back still.

Biosensors are an ingenious combination of high technology elements. Physically, biosensors normally appear as a probe coated with an immobilized biochemical. The solid state probe makes possible signal detection and amplification while the coating biochemical substance imparts specificity and makes the entire device sensitive to biochemical parameters. The concept of biosensors, as with monoclonal antibodies and nucleic acid probes, is sufficiently familiar to this audience that I shall not dwell on it at great length. The fundamental structure, by way of illustration, can be pictured by imagining a pH probe, over the surface of the probe stretch a membrane and on the membrane immobilize an enzyme specific for some substrate in the serum, plasma, whole blood or whatever is being tested. In such model systems as urease and penicillic acid, the event which takes place is that the reaction of the enzyme results in a change of pH in the solution, which change is detected by the pH probe and reported as a positive test.

This technology has a number of features which endow it with a special allure. One is that it is easy to make such devices the size of a needle due to the miniaturization of solid state electronics. Another is sensitivity, in the system illustrated just previously the enzyme could be expected to turn over many times, thereby generating a great deal of signal. Biosensors are commonly described with sensitivity that reaches to the single molecule level. Another attractive feature is that this single molecule of target substrate can be located and specifically recognized in a solution which has been only minimally prepped.

However, reality intrudes and there are few biosensor products currently approved for diagnostic use. In part, this is because the typical immunological interaction of antibody and antigen gives rise to such a slight change in charge that it cannot be sensed for one antigen antibody event. In the drug testing area, it is difficult to get an enzyme to break down drugs specifically. One approach to solving this problem is to attach biological cells, cellular components, antigens or antibodies onto the sensor's surface. Then the reaction between the immobilized species and the molecule to be sensed in the fluid under test could set up a detectable electrical or optical signal.

Like nucleic acid probes, however, it appears that the promise of the technology is so great that the problems will continue to be addressed until such time as they yield. The potential for biosensors is that they can virtually eliminiate wet chemistries and many other forms of wet assay. One imagines a testing event where an appropriate biosensor is simply exposed to a drop of blood from a fingerstick. Use of such a technology for drug monitoring, for example, would permit the drug level in blood to be assessed directly from the fingerstick. This would permit the drug to be used closer to the real therapeutic range, rather than having it be administered on a "guess work" or "average" style.

It is certain, however, that our timeline analogy does apply and that biosensor technology is not for 1988 or even 1989.

This review concludes with a glimpse into the future, specifically into the future for MAbs. Perhaps the most exciting development on the reasonably near-term horizon for this technology is in the area of anti-idiotypic antibodies, for example in vaccines.

In the past, vaccines consisted of antigenic preparations composed of killed infectious disease organisms, attenuated infectious disease organisms (live), or extracts obtained from such organisms. Such vaccines have been highly successful in suppressing or eliminating such diseases as small pox, diphtheria, rubella, polio, and others. Although successful, such vaccines suffer from a number of serious shortcomings.

One approach to overcome some of these problems was to develop an antigenic preparation composed of chemically synthesized portions of the immunogenic protein molecules. Antigenic portions of immunogenic protein molecules were determined, sequenced, and chemically synthesized. Antigenic preparations composed of chemically synthesized subunits were then used to immunize subject animals. These immunized animals are then screened, firstly to determine if they produce antiserum to the infectious disease organism, and secondly to determine if this antiserum can neutralize the infectious disease organism. Antigenic preparations composed of chemically synthesized peptide subunits can sometimes be used effectively, but not without other drawbacks.

A new alternative being pursued by Synbiotics and other firms to the approaches described above for producing a vaccine is the use of antigenic preparations composed of anti-idiotypic monoclonal antibodies.

The process of immunogenesis causes each antibody to have a unique antigen binding site structure, which may be termed its idiotype. Polyclonal antibodies frequently are unsuitable for use in vaccine applications as described above due to, among other things, bleed-to-bleed variation in the host animals' response, and due to the low proportion of specifically useful antibody in the polyclonal mixture.

In the case of an idiotypic monoclonal antibody and its corresponding anti-idiotypic monoclonal antibody, each has a unique antigen binding site which can serve as a template for the immunogenic production of the other. The two antigen binding sites are complementary to one another. On the other hand, the antigen binding site of the anti-idiotypic monoclonal antibody will exhibit a structural congruence with at least one epitope of an antigenic site found on the infectious disease organism or pathogenic agent which stimulated the immunogenic production of the corresponding idiotypic antibody. This structural congruence is the basis for the utility of anti-idiotypic monoclonal antibodies and antibody fragments as vaccines.

In development, the process begins with selection of an antigen. The first step is to produce a monoclonal antibody to the antigen. The investigator then produces monoclonal antibodies to the monoclonal idiotypic antibody. Exhaustive screening will select out those anti-idiotypic antibodies complementary to the hypersensitive region of the idiotypic antibody. When introduced into the host, the anti-idiotypic antibodies stimulate an immune response - so act as a vaccine.

The development of anti-idiotypic antigens as vaccines will allow production of highly purified, specific antigens which will be totally non-infective and non-immunosuppressive in the animals to be treated.

RECEIVED October 30, 1987

Chapter 30

Creating Value with Agricultural Biotechnology

Developing World Applications

Peter S. Carlson

Crop Genetics International Corporation, 7170 Standard Drive, Hanover, MD 21076

Agricultural Biotechnology in the Developing World

In the absence of real world examples or concrete results, the potential utilization of plant biotechnology in agriculture, particularly in the developing world, is like a Rorschach ink blot test; the viewer experiences subjective rather than objective reality. The problem is particularly acute in attempts to divine the utility of new technologies in altering the productivity of small holding, primarily marginal farmers in the developing world. Plant genetics in actual fact, will probably not play a major role in changing the condition of such indivviduals when compared to political and economic forces. However, the role of plant genetics appears real and not inconsequential: improve seeds of planting materials are an often cited need for these farmers. Improved seed probably holds as much potential for helping small farmers as any other single agronomic input. What types of genetic alterations are appropriate for the needs of developing world agriculture? Certainly, the agronomic situation of the small, marginal farmer is different from that of the large commercial farmer common in the developed world. It is impossible to transfer standard developed world farming technologies if there is not the ready availability of chemical and physical means to manipulate the environment. Additionally, the economic (e.g., market price, access to credit) and politiccal (e.g.,, land tenure arrangements) constraints are beyond immediate control.

Small farmers have responded with a diversity of solutions to their individual situations. It appears as if there is one central theme in all of the various viable solutions: to develop a defensive stance. In agronomic terms this may mean a diversity of crops for home consumption and for the market. In biological terms this implies genetic variability. Some risk is

0097-6156/88/0362-0361$06.00/0

acceptable, but not too much. Some harvest is preferable to no harvest, and when there is little control over the environment, or limited agronomic inputs, the genetic composition of the crop becomes an important consideration. This defensive stance may alter the plant breeder's strategy: increasing yield, while continuing to be an essential breeding objective, becomes of secondary importance when compared with survival and adequate levels of production under the entire range of possible environmental (physical, chemical, and biological) conditions. Hence, a major concern for crop improvement must be to maintain and increase the range of genetic variability which permits a crop species to survive and be productive under widely different conditons. The production and analysis of genetic variability and its incorporation into adapted varieties is an appropriate goal to aid these farmers. The ingerent strength of conventional plant breeding techniques and new biotechnical manipulations for the small farmer is that they are potentially scale neutral, politically neutral, relatively inexpensive for the farmer, andd can involve few ties with commerical networks. These qualities make plant improvement programs and their results accessible to small farmers and acceptable to governmental agencies and international developmental efforts. While the crop improvements provided by the "green revolution" have not always been of this nature, there is no reason why the technologies of breeding and genetics cannot accomplish these goals.

Commerical incentive is the driving force behind the development and application of biotechnology in the developed world. I doubt this engine will work in less-developed countries primarily because the financial opportunities do not balance the science and market risks. The burden of application must be shouldered by the public sector. This task demands uncommon wisdom and forethought in formulating policy. I base my conclusion on the following considerations:

1. Science does not exist in a vacuum; economic, social and political forces determine which technologies are utilized. Inexpensive, proven technologies are, under ordinary conditions, preferred by the market. Biotechnology is in many instances neither an inexpensive nor proven methodology for crop improvement.

2. Currently utilized agricultural technologies work. In the developed world, the food machine isn't broken! In fact, the machine is running too fast already. The world suffers from massive overproduction of foodstuffs and declining prices for such products. What needs fixing is the production/delivery/market infrastructure in nations experiencing famine, and

more rational ecological policies to reverse the expanding desert and shrinking tropical jungles. Indeed, a policy using biotechnology to increase developed country food output might make a bad situation worse.

3. The use of biotechnology to create novel crop varieties is generally not a business with attractive return on investment. No matter how carefully crafted the patent protection may be, as a practical matter for many crops the farmer's field is a genetic Xerox machine.. The onetime sale by the biotechnology breeder of a new variety can not support a price structure adequate to cover research costs. The extra effort to hybridize a new variety to create a "repeat sales" opportunity is substantial. Seen in this context, agricultural biotechnology may be an "orphan technology."

4. The farmer is a price and risk sensitive consumer, and is extremely reluctant to increase the portion of his costs devoted to planting materials at the beginning of the growing season. Biotechnology is an expensive undertaking and an additional cost in varietal production. The added value created by biotechnology manipulations will not be reimbursed by the market.

5. Technology transfer is not a generic exercise. Manipulations which ahve been defined for use with tomatoes will not be directly applicable to bananas. Goals for the improvement of corn are inappropriate for cassava. Agronomic goals for rice production in Japan are wholly different from those in India. In this light, the question of who "owns" or "controls" new technology is absurd. The underlying science of biotechnology can not and will not be effectively protected. Additionally, the developing world has access to the world patent literature and is under little or no constraint to abide by intellectual property laws in force in the developed nations. Private industry in the developed nations will not undertake work aimed at modifying tropical crop species in the absence of protection and a clear return on investment.

6. The information base for tropical crops is much smaller than that for the annual temperate species making technical applications more diffficult. Crop productivity or crop yield is a quantity often expressed in bushels per acre, tons per acre, kilograms per hectare, or jin per mu. This notation is a physical description of a biological endpoint, for productivity is the sum of the biological

> components which result in a population of mature plants. Although this expression is often a satisfactory one for farmers, ecconomists, or statisticians it hides the extent of our biological ignorance. Not enough is known about the underlying biology which directly or indirectly determines crop yield. Furthermore, what is known is inadequate for our needs. Contemporary research efforts have revealed and will continue to reveal new methods for genetic manipulation with higher plants. These new methods, should be of direct utility to plant breeders and agonomists in the developing world. To adequately utilize the techniques, it is important to have a clear biochemical description of what needs to be altered. What are the biological limitations of plant growth and productivity, and how does this biology interact with the environment? Defined genetically, encoded processes can be genetically manipulated while kilograms per hectar cannot.

International agricultural biotechnology laboratories located in the developing world should focus their effort on the extension and application of existing technologies to tropical crop species and the particular needs of tropical agriculture. The extension andd application of existing technologies is, at the current time, a more important and clearer goal than the invention of new science. It is the strategy that the CGIAR Institute have successfully utilized.

An Example of Biotechnology Extension and Application

The utilization of plant tissue and cell culture techniques in plant improvement may well have its greatest impact in those crop species whicch are clonally propagated, e.g., bananas, sugar cane, yams, cassava, potatoes, tree species, etc.[1] Prolific plantlet production from meristems and other tissue sources facilitates rapid propagation of plants with elite genotypes in crop species not easily propagated via seeds. Meristem culture techniques also provide the basis for elimination of systemic diseases from planting materials.

Plants regenerated from longer term callus or suspension cultures often show an increased genetic variation which has been reported to be of value to breeding programs in perennial crop species. Genetic variation produced during tissue culture of sugar cane has resulted of several new disease resistant clones[2,3,4]. Similar variation resulting from the culture of potato protoplasts is now being utilized by breeders as a source of disease resistance and greater adaptability to environmental extremes[5]. The increase in genetic variability due to passage of cells through tissue culture has also been described in other crops such as rice, oats, wheat, lettuce and tomatoes[6].

The primary application of increased genetic variability resulting from tissue culture has been the recovery of new genotypes with increased tolerance to plant diseases[6]. Tissue culture variation is superimposed directly upon existing qualities of a cultivar. This theoretically allows the introduction of alterations in single traits without the segregation of a vast array of alleles as in a sexual cross[2,6]. Passage of plant tissues throgu *in vitro* culture combined with selections of the resulting genotypes for their response to plant pathogens constitutes a method for improving a well adapted or successful variety by the addition of single disease response characteristics.

Bananas and Black Sigatoka

Bananas are an unusual agronomic crop since only clones selected from nature are currently cultivated. Commercial banana clones are vegetatively propagated, parthenocarpic triploids and are sexually sterile. Consequently, the standard techniques of plant breeding based upon sexual crosses can not be utilized for banana improvement[7]. Only two major clones (Gros Michel and Cavendish) have been acceptable for the export trade[7], and are important sources of export income for a number of developing nations. Diploid bananas, which are sexually fertile,, are not grown for export since their fruit size and plant vigor are unsuitable.

No crop species better illustrates the biological dangers of a genetic monoculture than bananas. Panama disease, or fusarial wilt (caused by *Fusarium oxysporum Cubense*), became a serious problem in Gros Michel in the 1940's and by 1955 had made cultivation of this cultivar unprofitable. Approximately 100,000 acres of bananas were destroyed or abandoned in central America because of this disease[8].

Cavendish clones resistant to Panama disease were substituted for Gros Micchel during the 1950's and 1960's creating a new monoculture of the variety. There are several different Cavendish subclones which vary in height but are all essentially genetically identical[9]. No known natural or selected clones exist as possible commercial alternatives to the cavendish cultivars[10].

The Cavendish cultivars are susceptible to a destructive and costly disease. Since 1933, Sigatoka leaf spot or Yellow Sigatoka (cause by *Mycosphaerella musicola*), has been a potentially serious disease in the Central and South American regions. However, routine and inexpensive control measures were able to contain the spread and moderate the destructiveness of this pathogen[11]. Black leaf streak, a more severe strain of M. musicola, was present in the Pacific region for many years, but was not found in Western Hemisphere[12]. In 1972, a more virulent Sigatoka leaf spot pathogen, apparently caused by a mutation or sexual recombination of M. musicola, was identified in Honduras[13]. This new fungal leaf spot pathogen, termed Black Sigatoka (Mycosphaerella fijiensis var. difformis) has greatly increased disease control costs andd currently threatens the export banana industry in the Americas. Since the original epidemic in Honduras in 1973-74, this disease has spread throughout central and South America and the Caribbean Basin region.

New genetic variation is needed in Cavendish cultivars to protect this industry against another devastating epidemic. Genetic resistance to Black Sigatoka in Cavendish cultivars would also reduce production costs. Sources of genetic resistance to Black Sigatoka have been identified in several selected clones of the diploid banana species Musa acuminata[14]. However, this genetic resistance cannot be incorporated into Cavendish varieties because of sexual sterility[15]. An alternative method for genetic manipulation of Cavendish clones wouuld be to induce somatic mutations for Black Sigatoka resistance by using ionizing radiation and other mutagens on plantlets, followed by selection of increased disease tolerance. However, several attempts at using mutation breeding methods to recover Black Sigatoka tolerant subclones of Cavendish bananas have not been successful[15].

Genetic variation occuring in tissue culture holds promise as a useful technique for recovering Black Sigatoka resistant Cavendish clones. The use of tissue culture is a realistic alternative because:

1. Genetic variation occuring in cell and tissue culture is a proven technique for the production of disease resistant or tolerant genotypes in a range of crop species[2,3,4,5,6].

2. Even though our knowledge of banana tissue culture is not complete, there has been some progress over the last several years[18]. It is now possible to regenerate large numbers of plants from callus cultures of Cavendish bananas and to identiffy somaclonal variants within this population of regenerated plants[20]. We can expect the remaining major problem, that of plantlet regeneration, from suspension and protoplast cultures, to be defined within several years if an intensive effort is initiated[19].

3. The Black Sigatoka fungus appears to produce a toxin. Related fungi in the genus Cercospora produce a phytotoxin, Cercosporin, which can readily be isolated and potentially used as a direct selective agent for increased disease resistance or tolerance[20,21,22,23].

4. Resistance to Black Sigatoka in wild diploid banana species is controlled by one or only several genes. This relatively simple genetic system is well suited for manipulation by tissue culture techniques[10].

Several different selection systems could be employed to produce Black Sigatoka resistant banana clones:

1. A large number of plants could be regenerated from callus cultures and examined for their reaction to Mycosphaerella fijiensis var difformis by direct inoculations. Techniques for greenhouse inoculations have been established[25]. If insufficient variability is produced by passage through culture, chemical and physical mutagens could be employed. This approach is currently available[19].

2. The phytotoxic chemical or chemicals produced by M. fijiensis var difformis could be used as a selective agent in tissue culture to recover resistant cells and calli. These resistant calli could be regenerated into plants and examined for tolerance by direct inoculation with the disease fungus.

3. Cercosporin, a phytotoxin produced by a related fungus, is known to act by producing singlet oxygen and superoxide radicals when exposed to light[25,26]. The Black Sigatoka toxin appears to act in a similar fashion[21]. Tobacco cells resistant to a superoxide radical producer, the herbicide paraquat, show a slight increase in tolerance to Cercosporin. Selection of banana cells which are resistant to paraquat followed by regeneration of entire plants and examination of the reaction of these selections to Cercosporin and to M. fijiensis var difformis by direct inoculation could be accomplished. A similar protocol ccould be established with agents that produce singlet oxygen radicals.

4. Conditional lethal mutants (temperature sensitive) of M. fijiensis var difformis could be produced for use in co-culture selection systems. Banana suspension and callus cultures could be co-cultured with conditional lethal mutants of the pathogen can attack the plant cells. Growth of the pathogen

could then be halted by shifting the co-cultures to non-permissive conditions. Banana cells and calli able to resist infection by the pathogen and survive in tissue culture could be regenerated into plants and examined for plantlet resistance by direct inoculation.

5. Black Sigatoka resistant diploid banana species could be established in callus and suspension culture and the response of these tissue cultures to M. fijiensis var difformis and its toxin could be analyzed. The characterization will be utilized to develop selective systems not obvious from our current knowledge of the host-pathogen interactions.

6. Protoplast fusion could be attempted between the Cavendish variety and resistant diploid banana species. This approach would constitute a last resort since protoplast technologies appear more remote with this genus than do suspension culture methodologies.

Once Black Sigatoka tolerant plantlets have been recovered, evaluation of yield and other agronomic characteristics of the selected subclones would be undertaken. Subclones which maintain the characteristics of the parental Cavendish cultivar and demonstrate resistance or tolerance to Black Sigotoka would then be chosen for mass production via rapid shoot tip propagation.

This type of technology extension is unlikely to be undertaken by a private company for the following reasons:

1. The project requires tremendous up front research costs with no guaranteed success.
2. There are no ready customers willing to pay the price for an improved CCavendish clone, or for the research costs.
3. An improved Cavendish clone represents a one-time sale since it is not legally protectable.

Consequently, this project appears ideal for execution in a publicc sector laboratory in the developing world.

It is important to realize that bananas have enormous importance in the developing world beyond the export value of the crop. Of the approximately 12 billion bananas produced annually, 80% are consumed locally in the domestic market of the producing country. Domestic consumption is primarily of a large number of varieties different from the Cavendish produced for export. The majority of these local cooking banana and plantain varieties are also susceptible to Black Sigatoka, threatening an important developing world food source. Once proven methodologies and manipulations have been defined for recovery of Musa sp tolerant to Black Sigatoka, they could be rapidly transfered to regional or national laboratories for use with locally adapted and accepted cultivars.

Although this banana project is presented only as an example, it embodies criteria appropriate for emphasis in developing world institution. These criteria include:

1. A clear developing world need that cannot, or will not, be filled by existing institutions.
2. Minimal resulting social disruption or impact.
3. Technical feasibility by extension of existing technologies.
4. A relatively short horizon time.
5. Definition of routine manipulations transferable to regional or national efforts throughout the developing world.

REFERENCES

1. Genetic Engineering for Crop Improvement, 1981. The Rockefeller Foundation Working Papers, New York City.
2. Heinz, D.J., M. Krishnamurthi, L.G. Nickell & A Maretzki, 1977. Cell, Tissue and Organ Culture in Sugarcane Improvement. In: "Applied and Fundamental Aspects of Plant cell Tissue and Organ Culture" (eds, J. Reinert & Y.P.S. Bajaj), Springer-Verlag Berlin, pp. 3-18.
3. Krishnamurthi, M., 1974. Notes on disease resistance of tissue culture subclones and fusion of sugarcane protoplasts. Sugar Cane Breeder's Newsletter 35:24-26.
4. Krishnamurthi, M., 1980. Annual Report of the Agricultural Experiment Station of the Fiji Sugar Corporation, Forward & pp. 18-20.
5. Shepard, J.F., D. Bidney & E. Shahin, 1980. Potato protoplasts in crop improvement. Science 28:17-24.
6. Larkin, P.J. & W.R. Scowcroft, 1981. Somaclonal variation - a novel source of variability from cell cultures for plant improvement. Theor. Appl. Genet. 60:197-214.
7. Simmonds, N.W. and K. Shepherd. 1955. The taxonomy and origins of the cultivated bananas. J. Linn. Soc. Lond. Bot. 55:302-312.
8. Stover, R.H. 1962. Fusarial wilt (Panama disease) of bananas and other Musa species. Phytopath. Paper 4, Commonwealth Mycological Institute, Ken, Surrey, England pp. 117.

9. Simmonds, N.W. 1954. Varietal identification in the Cavendish group of bananas. J. Hortic. Sci. 29:81-88.
10. Rowe, P. 1981. Breeding an "intractable crop": Bananas. The Rockefeller Foundation Working Papers, New York City.
11. Stover, R.H. 1962. Intercontinental spread of banana leaf spot (Mycosphaerella musicola. Leach), Trop. Agric. Trin. 39:327-338.
12. Stover, R.H. 1971. Banana leaf spots caused by Mycosphaerella musicola: contrasting features of Sigatoka and black leaf streak control. Plant Dis. Rep. 55:437-439.
13. Stover, R.H. and J.D. Dickson 1976. Banana leaf spot caused by Mycosphaerella musicola and M. fijiensis var difformis: a comparison of the first Central American Epidemics. FAO Plant Prot. Bull. 24(2):36-42
14. Vakili, N.G. 1968. Responses of Musa Acuminata species and edible cultivars to infection by Mycosphaerella sp. Trop. Agric. Tain. 45:13-22.
15. Menendez, T. and K. Shepherd, 1975. Breeding new bananas. World Crops, May-June, pp. 104-112.
16. Berg, L.A. and M. Bustamante, 1974. Heat treatment and meristem culture for the production of virus-free bananas. Phytopath. 64:320-322.
17. Ma, S.S. and C.T. Shie, 1974. Meristem Cultures of bananas. J. Chinese Soc. Hor. Sci. 20:6-12.
18. Vessey, J.C. and Rivera, 1980. Meristem Culture of Bananas. Turrialba 31:162-163.
19. Carlson, P.S. 1985. unpublished results.
20. Calpouzos, L. and A.T.I.C. Corke, 1963. Variable resistance to Sigatoka leaf spot bananas. Univ. Bristol Annual Report Agr. Res. Sta. 1962:106-110.
21. Calpouzos, L. 1966. Action of oil in the control of plant disease. Ann. Rev. Phytopath. 4:369-390.
22. Calpouzos, L. 1981. personal communication.
23. Daub, M.E. 1982. Cercosporin, a photosensitizing toxin from Cercospora. Phytopath. 72:370-374.
24. Goos, R.J. and M. Tschirch 1962. Establishment of Sigatoka disease of bananas in the greenhouse. Nature 194:887-888.
25. Daub, M.E. 1982. Peroxidation of tobacco membrane lepids by the photosensitizing toxin, Cercosporin. Plant Phys. 69:1361-1364.
26. Duab, M.E. and P.S. Carlson, 1981. Technologies and Strategies in Plant Cell Culture: New Approaches to old Problem. Envr. Exptl. Botany 21:269-275.

RECEIVED September 3, 1987

CHEMICAL INFORMATION

Chapter 31

Biotechnology Information: An Introduction

Ronald A. Rader

OMEC International, Inc., 727 15th Street, N.W., Washington, DC 20005

Biotechnology, no matter how broadly or narrowly defined, is a major activity in the U.S. involving industry, government, and non-profit institutions (1). Many different definitions of biotechnology exist, and like the term information resource, it is more readily recognized when encountered than defined. Biotechnology may be defined as the controlled application of intact biological organisms or isolated cellular components to solve problems or obtain desirable benefits (2). This usually involves use of microorganisms, plant or animal cells, organelles, enzymes, or other constituents to obtain a transformation or product. Many make distinctions between "old biotechnology," such as baking, brewing, cheesemaking, composting, and other processes which have been used for millenia, and "new biotechnology" involving more precise genetic and cellular manipulation and control, based on current knowledge of biology and chemistry. This papers, like most discussions of biotechnology, concentrates on newer biotechnology.

An information resource may be broadly defined as an organized information source which may be used systematically to obtain information meeting a specific need. Generally, this definition includes the full variety of information systems and services available, but excludes most primary sources of information, such as particular journals, technical reports, and proceedings. Many types and forms of information resources are available, including: bibliographic, numeric, and factual databases; information centers and libraries; index and abstract services and publications; bibliographies; culture, germplasm, and specimen collections; referral centers and clearinghouses; directories and guides; and management information systems. Note that use of an information resource may involve interaction with databases, hard copy documentation, and/or individuals. In all cases, a limiting factor in the value and utility of an information resource is the ability of the user to interact with the resource.

As demonstrated by two recent books, Information Sources in Biotechnology (3) including descriptions of most commercially available information resources, and the Federal Biotechnology Information Resources Directory (4) covering U.S. government sponsored

0097-6156/88/0362-0002$06.00/0

biotechnology-relevant information resources, there is a wealth of information available to support biotechnology. This consists primarily of the mature infrastructure of information resources in the chemical, biomedical and life sciences. However, problems and pitfalls arise when trying to obtain specific, desired information, especially in biotechnology. Here, information resource accessibility, organization, relevance, costs, ease of use, format, subject orientation, and a multitude of other factors affect their value and uses. These factors affect both use of internal and external information resources. Most successfull organizations eventually establish in-house dedicated information resources, such as an information center/library, and/or employ at least one partially dedicated employee to deal with their information retrieval and organization needs and problems.

This is the first collection of papers concerning biotechnology information published in the U.S. Papers discuss:

1) The overall status of the infrastructure of information resources supporting U.S. biotechnology efforts.
2) The critical role and potential contributions of information specialists to biotechnology, especially when integrated into the research and development process.
3) Information resources for the assessment of hazards from biotechnology products, and especially for genetically engineered, novel organisms to be released into the environment, from the perspective of the Environmental Protection Agency.

Considerable investments have propelled U.S. biotechnology and related sciences and industries to a current position of world preeminence. To a large extent, biotechnology and genetic engineering developed from federally funded research programs. Currently, federal agencies spend over $2 billion in biotechnology and related areas (5-6). Private industry probably spends on the order of twice or more this amount annually. Much innovative biotechnology research and development occurs in small start-up companies, usually founded by scientists with venture capital funding. Large established chemical, pharmaceutical and other companies have become involved in biotechnology through acquisitions of successfull startups, joint ventures, licencing of technology, and establishment of in-house research and development efforts.

The reader should note, that the information resources, services and expertise provided within the authors' organizations are not those typically found in most biotechnology-intensive organizations. The companies represented here are among the elite of U.S. biotechnology. They are well funded, have large staffs, have major products and technologies entering the market, and function competitively on a worldwide basis (7). Similarly, the Environmental Protection Agency has considerable resources.

Literature Cited

1. Office of Technology Assessment, Commercial Biotechnology: An International Analysis, OTA-BA-218, Government Printing Office, January 19, 1986.

2. Definition from BioInvention (formerly Biotechnology Patent Digest), OMEC International, Washington, DC (functionally the same as the many other definitions in use).
3. Crafts-Lighty, A., Information Sources in Biotechnology, 2nd edit., Nature Press, New York, 1986.
4. OMEC International, Federal Biotechnology Information Resources Directory, Washington, DC, 1987.
5. Office of Technology Assessment, Public Funding of Biotechnology Research and Training, workshop held Sept. 9, 1986, Washington, DC (in press).
6. Perpich, J. G., "A Federal Strategy for International Industrial Competiveness," Bio/Technology, vol. 4, pp. 522-525, June, 1986.
7. OMEC International, Inc. is a member of the Porton International Group, the largest private investor-held biotechnology company in the world.

RECEIVED July 30, 1987

Chapter 32

Status of the Infrastructure of Information Resources Supporting U.S. Biotechnology

Ronald A. Rader

OMEC International, Inc., 727 15th Street, N.W., Washington, DC 20005

The infrastructure of information resources supporting U.S. biotechnology needs improvements in a number of areas. Existing resources do not meet the needs of many users and uses. Information resources need to be developed for safety assessment, international competition, public information, and other areas. Biotechnology-involved organizations need to initiate and improve their information centers, systems, resources, expertise and services. Information resources are a limiting factor affecting organizations' and national innovation, competitiveness, decision making, protection of intellectual property and obtaining patents, regulatory affairs, research, development, and commercialization in biotechnology.

The U.S. biotechnology information infrastructure is the sum of all the nation's readily available information resources, services, and professional expertise, both within organizations and publicly available. Although not contributing directly to the bottom line, readily accessible, high quality information resources, services and professionals have a great impact on organizations' and nations' capabilities, productivity, actions and reactions, and competitiveness.

Chemistry and toxicology/pharmacology are areas where a strong infrastructure of information resources exists in the U.S. Chemical, toxicology, and pharmacology information resources are available in abundance and tailored to diverse needs, often with high degrees of specialization and sophistication, such as in-depth subject indexing, registry and nomenclature systems, substructure searching, and structure and activity predictive systems.

The state of the infrastructure of information resources in biotechnology may be compared with that of toxicology and related life and chemical sciences as of about ten or more years ago (1). Prior to the mid-1970's, there were very few toxicology information resources available. Federal information activities mandated by the

0097-6156/88/0362-0375$06.00/0

Toxic Substances Control Act, the National Cancer Act, and other societal efforts to regulate and define chemical-related public and environmental health threats from the mid-1970's to the present have resulted in a very healthy array of information resources. Similarly, activities and resources in other life sciences, especially the biomedical sciences, have grown and a healthy array of information resources developed, many with federal support.

Yet, few and inadequate biotechnology-oriented information resources exist. More and better information resources are required to support the development of biotechnology into the $40-100 billion industry it is predicted to be in the U.S. in the year 2000. In general, those biotechnology information resources which are available lack the sophistication required for many uses and users. There are a number of factors which have contributed to this situation, and a number of factors which should lead to development and availability of more and better biotechnology information resources.

Sources and Flow of Information in Biotechnology

A excellent overview of information sources and information flow in biotechnology has recently been published in second edition (2). Brief descriptions of most private sector information sources may be found here, along with much introductory and explanatory text. The biotechnology information marketplace is easily seen to be characterized by a very large number of primary sources of information, such as journals, meetings, conferences and proceedings, books, technical reports, and trade publications. Other sources of information for biotechnology include a number of often costly newsletters, consultants and experts. A large proportion of biotechnology-oriented information resources are devoted to commercial and competitive news and information reporting, as demonstrated by the large number of company directories available, and the commercial orientation of a large proportion of available specialized biotechnology abstracting and indexing services, as shown in Table I.

Table I. Some Major Biotechnology-specific Secondary Sources

Name	Orientation	Source
Abstracts in BioCommerce	Commercial	Britain
BioBusiness	Commercial	U.S.
Biotechnology Research Abstracts	Scienfific	U.S.
Current Biotechnology Abstracts	Scientific	Britain
Derwent Biotechnology Abstracts	Scientific	Britain
Pascal Biotechnologies	Scientific	France
Telegen	Commercial	U.S.

Note: All but BioBusiness, a database, are available in publication and database form.

Biotechnology is a multidisciplinary, fragmented activity involving the interface of many scientific and commercial activities. The diverse sciences upon which biotechnology builds and to which the biotechnologist needs information access includes many chemical and biological sciences and technologies, including microbiology, biochemistry, genetics, molecular biology, toxicology, pharmacology, and bioengineering. Also, biotechnologists require information about regulations, safety assessment, environmental effects, commerce, funding sources, patents, and other types of information. This fragmentation becomes even more apparent, when one realizes the diverse application areas of biotechnology, including industrial activities in pharmaceuticals and diagnostics, agriculture, food, energy, waste processing, and commodity and specialty chemicals. The fragmented nature of biotechnology, combined with its relative newness as a distinct activity, is a contributing factor to the lack of biotechnology information resources.

This often fragmentary interplay of disciplines and specialists is reflected in the numerous professional and trade organizations representing scientists and institutions involved in U.S. biotechnology. To date, neither professional nor trade associations in biotechnology have taken significant active roles in developing specialized information resources, a common situation for such organizations in many other fields.

Secondary sources, notably abstracting and indexing publications and bibliographic databases, are information resources routinely organizing and summarizing information about documents and their contents. Besides scanning of journals, these are usually the main means for keeping up with developments and for retrospective searching of the literature. Some major secondary publications and databases specifically oriented to biotechnology are shown in Table I. Examination of these and others reveals that they universally are spin-off or derivative (subset) products from major broader coverage scientific information services, and/or are primarily oriented to covering commercial news and activities in biotechnology.

Indexing and subject access in most secondary and other biotechnology and related information resources is rather primitive. Subject indexes either involve very simple and general classification schemes, employ keywords (no controlled indexing), or employ the classification and indexing schemes of their parent broader coverage resources. There has been little development of classification and indexing systems specifically for biotechnology information and access to the literature. The lack of classification and nomenclature schemes adversely affects the whole information infrastructure by making information retrieval and resources coordination more difficult and haphazard, and keeping information organization and exchange on very basic, simplistic level.

In the area of protein and nucleotide sequence databases, current resources are struggling to keep up with the published data, have reduced the amount of other information (annotations) recorded, capturing only the minimum data, and do not cover patents and commercial products. In fact, there are very few biotechnology information resources, whether bibliographic or other types, which seriously deal with biological technologies and products, rather than broad basic science or commercial activities.

Biotechnology Information Resources Marketplace

Biotechnology information resources have yet to successfully establish their niche in the U.S. marketplace for a number of reasons. Many biotechnology information resources find greater interest, market penetration, and their major markets in foreign countries, primarily Japan and Western Europe. Perhaps, the most important reasons for a sluggish U.S. demand and market for biotechnology information resources are: a general lack of recognition of the value of information resources as a strategic long- and short-term asset; lack of knowledge and exposure to specialized information resources among those most involved in biotechnology; and lack of highly visible national programs and activities.

U.S. organizations involved in biotechnology, including most biotechnology companies, are relatively weak in information resources, capabilities, and expertise. Biotechnology companies with a library/information center or an even partially dedicated information professional are a distinct minority. This situation occurs even in well-funded biotechnology companies with considerable research and development activity. In some cases, a local university library performs online searching and fulfills document requests on demand. For the most part, information handling is a haphazard and unorganized activity. Exceptions to this situation may be found in the established pharmaceutical and chemical firms becoming involved in biotechnology, most of which have information centers/libraries and information specialists thoroughly integrated into their research, development, marketing, and regulatory affairs efforts (3).

Biotechnology executives and researchers, when questioned about the need for information services and resources within their organization, very often reply that they are on the cutting-edge or forefront of their particular areas of research and development, go to all the right meetings, and keep in touch with the right people. Many fail to recognize the value and provide support for building and providing information resources, services, and expertise within their organization. Executives and researchers complain they have more information than they can assimilate, and mistake this for the information they may really need and that others in their organization should have long-term ready access to.

Biotechnology companies, on the whole, do not budget for information handling and organization, as do more established companies involved in pharmaceutical and chemical research and development. This is probably due to: their relatively recent entry into the commercialization and regulatory phases of product and process development; their not making profits, yet; and the history of most start-up companies' researchers and executives coming from academia or other biotechnology companies. Established chemical and pharmaceutical industries spend on the order of 2% of their research and development budget for library/information center and related resources and staff, but this is not observable in biotechnology companies.

A situation of information rich vs. information poor may arise or presently exist in biotechnology. An elite of larger biotechnology and other companies may be better able to conduct cost-effective research, commercialization, regulatory affairs, obtain and defend patents, and survive in the world marketplace. With the history of most significant biotechnology innovations and developments arising from small companies, universities, and research institutions, this may have broad strategic implications for these organizations and U.S biotechnology.

Some biotechnology companies are finding that they need to develop their own information resources. Some major biotechnology companies have become information vendors through commercialization of initially in-house information resources. Examples include Abstracts in BioCommerce, originally developed by Celltech in Britain, the AGRIBUSINESS database developed by Pioneer Hybrid, and the BioScan corporate activities directory developed by Cetus Corporation. Many companies are finding that organized information is a marketing asset. Often, companies distribute extensive bibliographies, and some operate online electronic mail networks relating to their products. These trends will likely continue.

The very nature of biotechnology complicates information handling and the protection of inventions through patents. For example, there are many ways to define and characterize biotechnology-related organisms, their products and components, and processes. One can identify and describe organisms and their products based on sequences and structures of DNA/RNA and proteins, uses and applications, observable characteristics and appearance, metabolic activities, and other parameters. Terminology used in biotechnology is far from standardized, and may be purposefully ambiguous or unclear to broaden and obscure boundaries of patent coverage (4). Only a few of many patent and intellectual property issues in biotechnology have been resolved in the U.S. and foreign countries' courts.

International Competition in Biotechnology Information

The U.S. presently is the leader in most aspects of biotechnology, due primarily to the considerable basic biomedical and life sciences research efforts of the federal government and a strong entrepeneural industrial sector of biotechnology start-up companies which have built upon this research (5). Similarly, many large and established chemical, pharmaceutical, biomedical, agricultural and other U.S. firms have become very involved in biotechnology research, development, and commercialization.

However, a number of foreign governments have targeted biotechnology as an important area where they are developing coordinated national efforts to challenge U.S. research and market preeminence. Development of information resources is formally recognized as an important component in these efforts.

The European Communities (Common Market) has sponsored the European Biotechnology Information Program (EBIP), recently accorded permanent funding status and renamed the Biotechnology Information Service, within the the British Library for several years (6). EBIP sponsors an annual meeting concerning biotechnology information, provides information services on demand, assists inquirors with information acquisition, and is actively analyzing and reviewing the information requirements of its member countries' research and commercial institutions. EBIP has sponsored studies, including assessments of the feasiblity of a computerized information system for European culture collections (collections of viable samples of microorganisms) and an information system on enzymes and enzyme engineering. The U.K. has recently implemented online access to its various culture collections' holdings.

The Japanese government has well established and coordinated industrial biotechnology research and development programs and research centers with a number of associated specialized information centers and activites. A branch of the Japanese government has recently outlined development plans for an integrated protein data network. These foreign government-supported efforts are too new to assess their impact on international competition, but are worthy of our attention. Also, most of the specialized secondary information resources shown in Table 1 and many others originate in Europe.

International competitiveness and encouragement of innovation are ever growing issues in the U.S. Information resources are not a solution to U.S. problems in these areas. However, information resources need to be recognized as a limiting factor for competitiveness and innovation at both the organizational and national level.

Federal Biotechnology Information Resources and Activities

The federal government is the single main organization responsible for and involved in biotechnology. Biotechnology originally developed from federally funded research, which remains the primary impetus for biotechnology research and development activity in the U.S. and the reason for acknowledged U.S. leadership in the field. Federal agencies spent over $2 billion dollars for biotechnology and related research in Fiscal Year 1986 and this level of spending will likely be maintained (7-8). Despite major U.S. interests and investments in biotechnology, generally, the federal government has not initiated development of biotechnology information resources to support national needs and federal mandates.

OMEC International, Inc. has recently completed its Federal Biotechnology Information Network (FBIN) project with partial federal funding. This has resulted in publication of the *Federal Biotechnology Information Resources Directory* (9), describing over 470 federal biotechnology-relevant information resources, and the *Federal Biotechnology Program Directory* (10), describing over 470 biotechnology-relevant research, regulatory, technology transfer and other federal programs and activites. Together, these provide the first comprehensive description of the infrastructure of federal resources and programs supporting and affecting biotechnology, exclusive of facilities.

From this project and other experience, a number of general conclusions may be reported regarding federal biotechnology information resources and activities:

1) There has been no significant development or discussion of new, needed information resources for biotechnology (with some exceptions noted below).
2) Most federal biotechnology-related information resources and programs are not specific for biotechnology. Rather, they support underlying or related basic research, or more generalized regulatory or other agency activities.

3) Existing biotechnology-related information resources, on the whole, are relatively stagnant, receiving little additional funding for qualitative or quantitative improvements.
4) There exist insufficient information resources to appropriately support biotechnology-related public health and environmental safety assessments. Information resources do not exist or are not readily available to assist persons in information gathering and assessment to evaluate the effects of releases of genetically engineered or other novel microorganisms and their products in the environment and marketplace.
5) Many agencies formerly active in chemical and biological information resources development and information dissemination are now significantly less active in these areas. This is most notable among the regulatory agencies. This general situation may be due to the political climate for deregulation. Many policy and programmatic decision-makers are not favorably disposed to information resources, recognizing that information resources are required and may be used to support development of regulations and spot potential and developing problems.

Major ongoing federal biotechnology-specific information resources and activities include: GENBANK and other nucleotide sequence database systems; the Protein Identification Resource (PIR) protein sequence database; the Microbial Strain Data Network (MSDN), a directory to culture collections' holdidngs; and the National Library of Medicine's biotechnology information research program and Biotechnology Information Resources Directory, to be an online database and published directory of worldwide information resources.

<u>Biotechnology Safety and Oversight Information Resources</u>

The lack of biotechnology information resources, accessible information, and infrastucture development is already having an adverse impact on U.S. biotechnology. This is most obvious in the related areas of regulation and oversight of research and premarket testing, safety and hazard assessment, information dissemination, and public (mis)perception and (mis)understanding of biotechnology-related hazards. New, innovative technologies, and especially biotechnology, require well-developed, comprehensive, coordinated, science-based regulations to establish public and industry confidence in regulatory and oversight actions and procedures. Currently, important regulatory and safety assessment are performed on a case-by-case basis by a handfull of persons with experience and/or credentials in this area. The Biotechnology Sciences Coordinating Committee (BSCC) has been formed and a coordinated framework for federal regulation is being put in place. However, there are few, if any, information resources available to assist in assessments of novel biotechnology products and organisms or make this information available to the biotechnology community and general public.

The lack of biotechnology product and process safety-related information resources is likey to make itself more evident as more legal, regulatory, and safety-related delays, uncertainties, and misjudgements. Even at this early stage in the development of U.S. biotechnology, a number of procedurally based, obstructive lawsuits have successfully diverted and delayed federal, academic, and industry testing and commercialization plans. Both small biotechnology companies and large, established chemical firms have made significant mistakes in the design of premarket testing strategy, protocols, and information provided (or not provided) to government agencies and the public.

Although the slowly advancing unresolved and uncoordinated nature of regulation and oversight within and among the federal and other government agencies is a major factor in regulatory and judicial delays and uncertainties, the general lack of organized and accessible information is surely a strong contributing factor. No fatal or other significant biotechnology-related accidents or adverse environmental modifications have occurred yet, but there are ample examples to be taken from the chemical industry of unidentified and misassessed hazards resulting in mishaps, public and environmental health hazards, and corporate liabilities. In partial response to this situation, OMEC International has recently published _Biotechnology Regulations: Environmental Release Compendium_, a compilation of U.S. federal, state, and local regulations, laws, and guidelines concerning releases in to the environment of genetically engineered microorganisms (11).

The NRC Committee on Biotechnology Nomenclature and Information Organization

A workshop sponsored by the National Library of Medicine (NLM) of the National Research Council Committee (NRC) on Biotechnology Nomenclature and Information Organization was held in May 1986 (12). Various subcommittees examined the state and relevance of chemical and biological nomenclature, the organization of biotechnology information, and developed a number of recommendations. Major recommendations included:

1) All federal agencies involved in biotechnology should continue current and initiate new programs and activities in biotechnology information. This could involve the establishment of information centers of excellence in biotechnology which might develop and provide information resources, conduct research related to biotechnology information, and provide referral services.
2) The NLM should catalyze national and international efforts to coordinate and develop standardized subject vocabularies (for terminology and subject indexing schemes) for biotechnology diciplines, and a uniform nomenclature in the form of registries for organisms, clones, genetic elements, and other biotechnology materials and products.
3) The NLM should establish a "database of databases" for biotechnology and expand its role as an information resource center. This would involve expansion of the DIRLINE database, NLM's online directory of biomedical and other information resources. Work in this area will be initiated this Fall.
4) NLM should develop a cross-referencing system and a thesaurus (subject classification scheme) for biotechnology information resources. A cross-referencing system would work in tandem with the "database of databases" to facilitate use of common data elements, compatibilities, and data sharing among databases, and

also aid searchers in identifying and locating sources of desired types and forms of data and information.

5) The NLM should facilitate networking among database systems and establish "transparent" interfaces among them.

The report emphasized that the federal government needs to recognize the importance of biotechnology information as a national resource vital to science, technology, commerce and other national interests. The Committee recognized the need for deficit and federal budget reduction, but reported that the economic advantages of developing, processing, and disseminating biotechnology information far outweigh the costs. Biotechnology deserves a high standard of information resources and federal involvement in these, much as other developing technologies have a federally-sponsored common denominator of information resources.

The Committee reported that vocabulary in biotechnology is suboptimal. This includes the terminology used by scientists, such as the fabricated terms used for transposable genetic elements, the undeveloped or nonexistant nomenclatures for biotechnology products and processes, and the biological and chemical nomenclatures now in use. Registries need to be developed for clones, genetic elements, and other materials used in biotechnology to provide unique and unambiguous identifiers and descriptions. Biological nomenclature currently provides taxonomic descriptions of whole organisms and does not extend to their components or below the species level, which is the level at which biotechnology functions. Similarly, chemical nomenclature is not oriented to complex macro- and multi-molecular biological materials. These nomenclatures break down when applied to recombinant organisms, cell lines, genetic elements, modified proteins, antibodies and other biotechnology materials.

Congressional Activities

New programs and significant reorientations of funding and priorities within federal agencies are difficult without Congressional mandates or other high-level directives. As discussed above, much of the U.S. infrastructure of information resources in the chemical and related life sciences may be traced to laws passed in the mid-1970's. Congressional actions are likely to be required to initiate similar activity in biotechnology information.

Rep. Pepper has introduced the National Biotechnology Information Act (H.R. 393) in Congress. The bill would establish a National Center for Biotechnology Information within the National Library of Medicine and provide additional funding of $10 million/year. The bill does not contain much detail about specified programs and activities. It is primarily oriented to the molecular biology and biomedical research communities and National Institutes of Health (NIH) activities. Activities mentioned in the bill and supporting materials include nucleotide and protein sequence databases, development of information resources for gene mapping, and the coordination and integration of computer-based information resources. Resources and programs for safety assessment, international competitiveness, public information, technology transfer, regulatory coordination, classification schemes and registries are not specifically addressed. It will be interesting to follow the evolution of this bill and the level of effort to be directed to the critical applied and technological information needs of biotechnology. Those concerned with biotechnology information should take note of this bill and participate in its formulation and debate.

Recommendations

The author endorses the NRC Committee's recommendations, especially the first calling for recognition of biotechnology information as a national asset and establishment of information centers of excellance. The NRC Committee had a distinct biomedical orientation, properly reflecting the interests of its sponsor and the predominance of biomedically-oriented biotechnology within the federal and private sectors. Many of the same findings also apply to critical biotechnology information needs for agriculture, commerce, energy, and defense.

Besides the Committees recommendations, and the general requirement that biotechnology organizations upgrade their information resources, the author suggests prompt federal and private sector attention to:

1) Establishment of series of information centers collecting, translating, organizing, and assessing foreign biotechnology scientific and commercial information and developments;
2) Extensions of indexing and classification schemes used by established information resources, especially abstracting and indexing services, to better cover biotechnology;
3) Execution of user needs surveys, market studies, and assessments of available options and priorities in U.S. biotechnology information resources development;
4) Assessment by the federal government of the cost-effectiveness and appropriate means to assist the development, improvement and public release of private and nonprofit sector information resources;
5) Support for development and implementation of biotechnology information resources within the National Agricultural Library (NAL), Department of Commerce, and Department of Energy to parallel and keep up with the development of biomedically-oriented information resources;
6) Establishment of at least one information center and bibliographic and factual databases concerning the safety, risk assessment, and regulatory affairs of biotechnology products, processes, and materials; and implementation of an emergency response-capable information center and online database for biotechnology and industrial microbiology.
7) Development of knowledge-based and expert systems, and other information resources to supplement U.S. manpower and educational deficiencies and needs in bioprocessing, fermentation, and other areas of relative foreign dominance in biotechnology (5, 13); and
8) Establishment of federal and private sector clearinghouses to facilitate public access to biotechnology and related information.

In summary, biotechnology is a relatively new, diverse, major scientific and commercial activity in the U.S. and throughout the world. The infrastructure of information resources supporting U.S. biotechnology needs improvements on a number of levels, requiring efforts by all involved organizations - the federal government, the biotechnology and information industries, and research institutions. Greatest needs are for establishment and expansion of information collection, organization, and services within biotechnology-intensive organizations, especially U.S. biotechnology companies and research institutions, and the recognition and coordinated action of federal agencies to promptly address biotechnology information resource needs and problems. Federal implementation of safety, regulatory, and international information resources is required to protect the considerable U.S investment in biotechnology. Although many author recommendations concentrate on the federal role and activities, the private sector needs to become involved in all aspects of these activities to assure understanding of biotechnology as a diverse technological and commercial, and not just as a biomedical research-oriented activity.

Literature Cited

1. Kissman, H. M. and Wexler, P., "Toxicology Information Systems: A Historical Perspective,"Journal of Chemical Information and Computer Sciences, 25(3), pp. 212-217, Aug. 1985.
2. Crafts-Lighty. A., Information Sources in Biotechnology, 2nd edit., Nature Press, New York, 1986.
3. Brown, H. D., "A Drug is Born: Its Information Facets in Pharmaceutical Research and Development," J. Chem. Info. and Comp. Sci., vol. 25, pp. 218-224, 1985.
4. Meyers, N., "Biotechnology Patents: Don't Say Just What You Mean," Nature, vol. 324, p. 504, Dec. 11, 1986.
5. Office of Technology Assessment, Commercial Biotechnology: An International Analysis, OTA-BA-218, Government Printing Office, January 19, 1986.
6. Cantley, M., "Bio-Informatics in Europe: Foundations and Visions," Swiss Biotech., vol. 2, no. 4, pp 7-10, 13-14, April, 1984.
7. Office of Technology Assessment, Public Funding of Biotechnology Research and Training, (in press), workshop held Sept. 9, 1986, Washington, DC.
8. Perpich, J. G., "A Federal Strategy for International Industrial Competiveness," Bio/Technology, vol. 4, pp. 522-525, June, 1986.
9. OMEC International, Federal Biotechnology Information Resources Directory, Washington, DC, 1987.
10. OMEC Int., Federal Biotechnology Programs Directory, Washington, DC, 1987.
11. Strauss, H. S., Biotechnology Regulations: Environmental Release Compendium, OMEC Int., Washington, DC, 1987.
12. Committee on Biotechnology Nomenclature and Information Organization, National Research Council, Biotechnology Nomenclature and Information Organization, National Academy Press, 1986.
13. Zaborsky, O. R., and Zubris, D. K., Biotechnology Engineers: Status Report 1985, OMEC Int., Washington, D.C., 1985.

RECEIVED July 30, 1987

Chapter 33

Information Resources and the Assessment of Risk for Modern Biotechnology

Mark C. Segal

Office of Toxic Substances (TS–796), U.S. Environmental Protection Agency, Washington, DC 20460

Developing information resources to assess deliberately released genetically engineered microorganisms is challenging due to a lack of relevant historical data. Predicting dissemination and effects will rely on accurate knowledge of a new organism's intrinsic characteristics and the use of data for existing related organisms. Development of microbial ecology will aid in evaluating establishment and spread of environmentally released microorganisms. The use of existing microorganisms as analogues requires the compilation and indexing of relevant features and selection of appropriate characteristics for comparison with the engineered microorganisms. The most directly relevant information will be manufacturer supplied data on their products. New world-wide data networks are being established for improved data access and user maintained data systems will allow manipulation of acquired data while protecting proprietary information.

Modern biotechnology has moved out of the research laboratory and into the commercial world, bringing with it potential for great societal benefits. However, as product development has become more ambitious, some biotechnology applications are beginning to push against the boundaries of knowledge needed to predict the actions of living materials released to the environment. The use of contemporary technology permits construction of new organisms not expected to be found in nature. Though many believe that the probabilities are generally remote, problems could arise from the release of such new organisms. For deliberately released microorganisms, the public, and the government, do not have the benefit of precedent since there is no established history for the modern biotechnology industry, with respect to these organisms. What experience exists is an insufficient predictor of the safety of many new biotechnology products. A company's or industry's track record must be supplemented by experience in the form of technical information needed for regulatory decision making, in order to assure the public that they are receiving safe, as well as useful new products.

The following discussion illustrates how various data sources can contribute to the evaluation of modern biotechnology products. Many of these sources exist but have not been considered useful for this purpose until recently. Most of the discussion will deal with the sources of data supporting the evaluation of hazards and risk assessment from the perspective of a regulatory agency like EPA. Many of the specifics apply primarily to the use, as a regulatory vehicle, of the Toxic Substances Control Act (TSCA), which applies only to products used for purposes other than foods, drugs, cosmetics, pesticides and others uses subject to other federal statutes. Though this paper will not focus on data used to assess the benefits of products, it should be noted that much information may have utility for both hazard evaluation and benefits assessment.

Statutory Concept of "New". TSCA has a statutory requirement that the "newness" of an organism be ascertained in order that the applicability of key parts of the law can be determined. One section of this law requires that a maker submit a notification to EPA prior to the production of a "new" substance. That notification, called a Premanufacturing Notice, or PMN for short, is then subject to a risk/benefit evaluation to be completed in 90 calendar days. Substances that are not new, i.e. were in production prior to the implementation of TSCA or were formerly subject to a PMN, may be regulated under other parts of TSCA, but only if there is an indication of potential or observed increased risk from continued or expanded use. No *a priori* suspicion of risk is associated with the PMN. Thus, newness for microorganisms, in the regulatory sense, is not strictly a scientific concept but a means of product classification for statutory purposes. Determining newness has been basically a straightforward concept for use with traditional chemical substances but has proved difficult for microorganisms. A taxonomic approach for microorganisms was used when proposals for other approaches were found to suffer greater problems than the taxonomy based one.

Thus, "new" for TSCA occurs if the derivation of an organism results from the mixing of distinctly different parental types. While recognizing that often this means the introduction of only a few genes into an otherwise well characterized recipient, an operational interpretation of distinctly different has centered on the transfer of genetic information between microbial genera. The actual risk, if any, from such a transfer is the subject of an evaluation under the PMN portion of TSCA only after the issue of newness has been resolved.

Data Needs for Risk Assessment

Data needs can be readily subdivided into two major categories: data for product identity and for hazard and exposure assessment. Some data can serve both purposes, since intrinsic characteristics related to hazard may contribute to the unique set of discriminating features needed for identification.

Identification. The first and most obvious need of any regulatory authority is the need to confirm the identity and composition of the subject material. With traditional chemical products, there are accepted, reproducible standard tests to reveal properties that establish a product's identity. Products of biotechnology are often living organisms that don't usually lend themselves to the same analytical testing approaches used for chemicals. For microorganisms, the descriptors needed for confirmation of identity are usually physiological and biochemical. Though the tests needed to reveal these organism specific properties may be viewed as analogous to chemical analytical tests, the results are often variable for identically named cultures.

The identity of microorganisms is usually tied to their taxonomy. This assignment of a taxonomic position may be accomplished in a variety of ways. Traditionally, a battery of tests is usually required to separate similar species and strains, with key results leading to a classification in a determinative fashion. More recently, these same test results may be evaluated by statistical means giving a numerical taxonomy. A variety of molecular analyses, including nucleic acid hybridization and ribosomal RNA sequence analyses provide relationships at several taxonomic levels. Use of several of these techniques for the same organism may yield differing taxonomic assignments due to biases of the classification systems. With the recent tendency for constant revision, most acknowledge that bacterial taxonomy is complex and thus developing a fixed standard set of data needed to confirm the identity of new bacteria is currently frustrating.

This is further compounded by the commercial desire to mix genetic material from taxonomically distinct organisms. Using traditional determinative taxonomic approaches, such manipulations involving a few key traits can produce an organism that appears taxonomically unique compared with its parents.

Because of the difficulty in assigning a taxonomic position to modified bacteria, especially those derived from recent environmental isolates, any such organism needs to be associated with the sum of its properties. It is essential that any description of a new microorganism contain full phenotypic information on the construct or a means to relate such information from the source organisms, especially the major parent, and the new construct. In addition information on the genetic manipulations done to create the new product are needed to complete the picture.

Data used by a submitter to determine the identity of its new product, and discussion of the basis for such an identification will need to be supplied to the Agency. The manufacturer will propose a taxonomic assignment based on its own assessment. Use of more than one taxonomic method may yield conflicting identifications. In order to assign an appropriate identity for statutory purposes, EPA may be placed in a position of choosing from among the proposed names or independently making an appropriate assignment.

The Agency must be able to know, unequivocally, that the organism, which is the subject of a notification is uniquely identifiable and is the same as that for which testing was done. Some quality control information needs to be supplied. The stability of the cultures and the conditions of maintenance are important considerations for verifying that the product will exhibit the traits described for it. Finally, there is a need for methods to detect and monitor the new organism once it is released.

Such identification data are the responsibility of the manufacturer. They may be generated by the manufacturer's lab or done under contract. Published information generated independently by a third party rarely is sufficiently definitive to adequately identify the specific product.

Risk. Since risk is a function of both hazard, which involves intrinsic features of the organism, and exposure, which includes environmental and organism specific features, many different data elements are essential to a complete assessment. Data specifications for hazard and risk evaluation may vary in style, format and relative standardization among statutes, but the overall needs are basically the same. In the recent OSTP FEDERAL REGISTER notice (1) dealing with government wide biotechnology activities, details of data needs were described and can be summarized as follows for organisms intended for direct release.

Some hazard information will ultimately have to be provided to the responsible agency ensure that the product organism will cause no unreasonable adverse effects to man or the environment. For microorganisms, this requires determination of the following: the likelihood that the organism produces an effect that is deemed undesirable or inappropriate; determination of the existence of a potential target for that effect; determination that there is a mechanism for bringing the organism and target together in location and time; and determination that the expressed result ,if it occurs,is sufficiently serious to warrant action.

For many microorganisms, the existence of traits that are potentially detrimental is evident, as with overt pathogens or where characteristics responsible for potential effects are part of the usual description of the microbe. Some have unusual roles in the biogeochemical cycle which, if misplaced, could have significant effects. A manufacturer would be expected to be aware of these obvious characteristics.

For most microorganisms, however, the potential for adverse effects is not so apparent. In many cases, the need for data on undesirable effects will best be met by direct examination of the new organism. When such examination of important traits is found to be impossible or impractical , data from closely related organisms can provide a useful fall back position. The selection of these comparison organisms and the features that may be appropriate indicators of the performance of the new product organism is likely not to be as straightforward as it might appear. Taxonomically related microbes don't necessarily have to function similarly under similar environmental conditions.

The exposure potential includes factors such as the ability of the
released organism to establish itself, increase in number and move from its release site. Data relevant for exposure evaluation may be intrinsic, such as temperature sensitivity, production of survival structures like spores, measurable growth rates in various media, etc. The movement of susceptible populations, rate of release, climatic variables all are environmental factors that influence this potential as well. Many of these considerations will vary with a specific use of a product.

Sources of Data

For the near future, reliance on data from anything but tests on the new organism will be suspect, unless the organism is derived from well known parental genotypes and the modifications are so accurately described that the result of the modification is very predictable. The manufacturer of a new organism, especially one derived from a recent environmental isolate, will likely be the primary data source for evaluation of newly derived microorganisms. Data generated for the parents should be useful in evaluating results obtained for the derivative. In some cases, these data may be found in the open literature. Often, however, published results are not complete data, but summaries of work for which additional supportive information may be needed to make the data useful for purposes other than those intended by the author. Makers of new organisms may desire to refer to such published data for convenience, but may wish to acquire unpublished archived data for their parental strains as well.

The basic utility of data obtained from organisms other than the new construct or derivative should be to permit the initial formulation of intelligent and relevant questions that narrow a subsequent field of inquiry. The current state of the art of microbial ecology doesn't generally permit the accurate prediction of survival and effects of released microorganisms without direct experimentation using the organism in question. However, the design of experiments and selection of appropriate test methods can be guided by information from taxonomically or ecologically related organisms.

Where does one go for data relevant to a new organism if testing of that organism is not an acceptable option and appropriate parental data are not available from, or to, the maker? If the parental organisms were derived from stock cultures, useful information may have been collected at the time of isolation of the progenitor strains of the parents. These data might be stored in the private files of the original isolator, or archived at the culture collection receiving the progenitors for deposit. Sometimes the original isolate was obtained from a remote location along with other cultures, and characterized some time after isolation by a second or third party. It may be that traits present in the parent organisms were first and only characterized in the progenitors at the time of isolation. Certainly, most relevant environmental information would likely have been recorded at that time. This information need not have been carried forward, however, especially if the isolate evolved into a laboratory curiosity which only later was shown to have commercial potential. How, then to locate the data important to environmental assessment, but not necessarily to commercial development?

One mechanism being developed is a worldwide network of microbial data sources (2). This network, called the Microbial Strain Data Network (MSDN) , will make location and access to microbial data easier by electronically linking the cooperating data sources. Some of the data sources will contain information in an electronic form, so that transfer of data can be facilitated by the MSDN communications system. Other sources will have data in paper form. Like all members of the network, their data will be catalogued and the significant information about the data source, the metadata, will be retrievable using the Central Directory of the MSDN. A secretariat will be maintained to assist in specific inquiries. The network structure will be such that a secretariat maintaining the Central Directory will be located at Cambridge in the U.K., and will be electronically linked via a DIALCOM system to regional "nodes" on all continents. Written inquiries will be answered at the closest node or referred to the secretariat. Participants will have the option of using the secretariat, making their own central directory searches, or going to the nearest node for assistance.

The current status of the MSDN is that it is expected to begin service in the fall of 1987 after a year of demonstration. The communications system is now in place, and participating nodes are being recruited. Personnel for the secretariat have been selected and trained. The design for the Central Directory has been worked out and is being implemented. A coding system for catalogued traits has been adopted. The communications system is called the CODATA network, system 42 of the DIALCOM complex of data systems. The CODATA system now allows for electronic transfer of data among participants.

A coding system has been developed enabling all participants to assign standardized numeric values to traits or conditions recorded in their files (Rogosa, M.; Krichevsky, M.I.; Colwell, R.R., Coding Microbiological Data for Computers, Springer-Verlag, in press.) Using the test data base today, a participant can determine a code for a specific trait,or traits, query the system to see which of the data sources maintain information on those traits - and for which group of organisms - and can then send an electronic message to the data source contact person asking for more information, or access to the data. The source, in return, can electronically mail a response to the inquirer. As envisioned, the completed network will have numerous sources identified, each with the ability to "mail" responses almost instantaneously to an inquiring scientist, a manager of a regional node, or the manager of the whole network. Computer conferences are also possible, to permit online give-and-take.

Another part of the system already operational is the Hybridoma Data Bank (3), a catalogue of characterized hybridomas, with information on reactivity, source, special characteristics, etc. This special network has its own nodes in the U.S., Japan and France. In the U.S. it is administered through the American Type Culture Collection (ATCC). Data for a sampling of several hundred hybridomas from four major sources are now accessible on-line, with several thousand available for loading into the system as time and funds permit.

Using the CODATA network components both the maker and the government will have easier access to unpublished data on organisms of interest. Through the MSDN,the maker may locate potential parental organisms possessing desirable traits while the government may be able to identify existing strains for which studies have been done that are relevant to questions being asked about related new organisms which have not been as thoroughly studied.

NIH has developed, over the course of a decade, a data system specifically designed to accommodate microbial strain data. This system,called MICRO-IS (4), was primarily supported by FDA and other government agencies, but has also been used by academic investigators. It is now being adapted to run on microcomputers for use by EPA. This stand-alone version will allow us to analyze data obtained through the MSDN or other non-commercial data sources and compare these to data supplied by the manufacturer as confidential business information. By retaining confidential data on securable removable media, we can protect the confidentiality of the information, yet take advantage of the comparison data downloaded from sources participating in the MSDN. To make the last easier, the MICROIS stand alone software is expected to be made available to MSDN users, thus reducing the difficulty of conversion of data to a usable format.

Published Data Sources. The MSDN is primarily designed to assist in the acquisition of unpublished data from sources like culture collections, government laboratories and academic institutions. Sometimes, useful data will have been published, though rarely in the context sought by the data user.

Sophisticated use of commercial and governmental on-line search services may be able to locate such data. At EPA, we routinely use such services for obtaining data for toxic chemicals. Search strategies based on CAS Registry number and keywords are prepared by information specialists working directly with scientists. Some scientists have become efficient on-line searchers themselves. A mechanism to assist scientists in searching for relevant organism oriented data is also needed. It appears that, with very little modification, these bibliographic search methods are readily adaptable to organism, versus chemical, oriented data retrieval.

To assist in recognizing the taxonomic and nomenclatural relationships among bacteria listed in the literature, BIOSIS has been developing a Taxonomy Reference File that could make on-line selection of appropriate references easier than now. This system attaches an alphanumeric identifier to all species, and subspecies, of bacteria, based on currently accepted names. Past names are cross referenced and a hierarchy of taxonomic relationships is maintained on-line. EPA is participating in the initial test phase of the TRF. In addition, EPA working with the National Library of Medicine (NLM) has developed a software device for constructing search strategies and efficiently retrieving abstracts, called CSIN (Conry, T.J. and J.M. Hushon 1986 Critical issues in microcomputer gateway design: the Micro-CSIN experience. Presented at American Chemical Society 192nd National Meeting, Anaheim, CA). This software helps choose the appropriate commercial data bases and vendors, allows for construction of queries using Boolean logic, logs on, searches and downloads from each data base in succession, organizes the results and allows for editing, all in a single session. It has been adapted to run on microcomputers. This bibliographic search aid places control over the search in the hands of the evaluator of the data. It shortens the waiting time to minutes, if necessary, and brings much of the library to the desk top. Due to statutory time constraints on analysis, we at EPA have come to depend on the on-line searches for our chemical evaluations and will no doubt do the same with those for biotechnology.

Strategy for Information Use

Because of the newness of the industry and the state-of-the-art of microbial ecology, initially we will desire to perform case-by-case analyses relying, primarily, on data supplied by the submitter and on the opinions of recognized experts in selected disciplines pertinent to the product and use under evaluation. Our information strategy for biotechnology risk evaluation cannot, however, be geared simply for the momentary support of current application review. Instead ,we must aim for the construction and use of comprehensive support data bases to improve and facilitate analysis for similar subsequent products and uses. We are therefore establishing the widest number of information options possible to support this endeavor.

As indicated above, data generated for the submitter will be the key for the first few case-by-case analyses.

Often, we expect, the data will come from studies proposed to EPA in advance of their submittal for review by the Agency. In some cases, EPA personnel may be asked to critique the study design by the products maker., although such advice cannot have the effect of a priori approval of results. The maker may request and receive approval for limited field release to acquire data needed to support a subsequent application. EPA will eventually receive these data and may desire to use them for purposes beyond specific application evaluation. It may be that the research initiated by the maker provides insight into basic ecological processes as well as the performance of a specific new organism. Working within the statutory limits on use of such submitter generated data, the agency will want to use such basic data as part of its overall data base.

The NIH designed data system, MICRO-IS, mentioned above, should allow ready conversion of these data into computer usable formats. Companies may now claim a wide variety of data "confidential business information" (CBI), although EPA is encouraging companies to limit the extent to which this designation is applied to their data. The stand-alone version will allow tight security control over the data, making working within statutory CBI data limitations a reasonable prospect. By and large 'health and safety' data cannot be claimed proprietary. By tagging the proprietary data, inadvertent release of the information, or improper use in support of a competitors application can be avoided. This will free the remaining data to be used for baseline information as we build our survival and dissemination models, for example. Thus, as we gain experience with each submittal, we have a feed-back loop that can be activated to ensure consistency of analysis and an ability to learn from previous analyses.

The feedback mechanism may also serve to permit future modeling activities that could free us from case specific analysis for each new submittal. Data bases can be constructed which will merge information from all sources, open literature, non-proprietary manufacturer data and MSDN identified files. Exposure models are the most likely prospects for the foreseeable future. We have experience from the fields of plant pathology, epidemiology and environmental chemistry that may show us how to begin to predict the distribution of released non-pathogens. Modeling ecosystem effects will depend on improved microbial ecological knowledge just now beginning to be developed. The experience of macro-ecosystem ecologists may assist in that area, but the current belief is that such experience will have limited applications to most microbiological effects questions.

Summary. We are planning for a time when biotechnology is routine and the evaluation of risk from deliberate release of modified microorganisms can be simplified by the use of predictive tools based on data from previous experience. Since that experience has not yet come, we must be cautious with our first examples. Therefore our current priorities in data management for biotechnology risk analysis include developing mechanisms to acquire needed data, synthesizing a knowledge base, ensuring a feedback mechanism for use of the acquired data and ultimately developing a predictive capability supported by the enlarging knowledge base.

Literature Cited

1. Coordinated Framework for Regulation of Biotechnology, Office of Science and Technology Policy, Federal Register 1986, 51:123, pp 23302-23393

2. Hill, L.R. ; Krichevsky, M.I. Needs and Specifications for an International Microbial Strain Data Network; United Nations Environmental Program: Nairobi, 1985

3. Blaine, L., Biotechnology 1984, 3, 338-341

4. Krichevsky, M.I. Anal. Chim. Acta. 1981, 133, 747-751

RECEIVED August 6, 1987

INDEXES

Author Index

Affiliation Index

Subject Index

G

N

T

U

V

Production by Barbara J. Libengood
Indexing by Keith B. Belton
Jacket design by Carla L. Clemens

Elements typeset by Hot Type Ltd., Washington, DC
Printed and bound by Maple Press, York, PA